An Introduction to Broadband Networks

LANs, MANs, ATM, B-ISDN, and Optical Networks for Integrated Multimedia Telecommunications

Applications of Communications Theory
Series Editor: R. W. Lucky, *Bellcore*

Recent volumes in the series:

BASIC CONCEPTS IN INFORMATION THEORY AND CODING: The Adventures of Secret Agent 00111
Solomon W. Golomb, Robert E. Peile, and Robert A Scholtz

COMPUTER COMMUNICATIONS AND NETWORKS
John R. Freer

COMPUTER NETWORK ARCHITECTURES AND PROTOCOLS
Second Edition • Edited by Carl A. Sunshine

DATA COMMUNICATIONS PRINCIPLES
Richard D. Gitlin, Jeremiah F. Hayes, and Stephen B. Weinstein

DATA TRANSPORTATION AND PROTECTION
John E. Hershey and R. K. Rao Yarlagadda

DIGITAL PHASE MODULATION
John B. Anderson, Tor Aulin, and Carl-Erik Sundberg

DIGITAL PICTURES: Representation and Compression
Arun N. Netravali and Barry G. Haskell

FIBER OPTICS: Technology and Applications
Stewart D. Personick

FUNDAMENTALS OF DIGITAL SWITCHING
Second Edition • Edited by John C. McDonald

AN INTRODUCTION TO BROADBAND NETWORKS: LANs, MANs, ATM, B-ISDN, and Optical Networks for Integrated Multimedia Telecommunications
Anthony S. Acampora

AN INTRODUCTION TO PHOTONIC SWITCHING FABRICS
H. Scott Hinton

OPTICAL CHANNELS: Fibers, Clouds, Water, and the Atmosphere
Sherman Karp, Robert M. Gagliardi, Steven E. Moran, and Larry B. Stotts

PRACTICAL COMPUTER DATA COMMUNICATIONS
William J. Barksdale

SIMULATION OF COMMUNICATION SYSTEMS
Michel C. Jeruchim, Philip Balaban, and K. Sam Shanmugan

An Introduction to Broadband Networks

LANs, MANs, ATM, B-ISDN, and Optical Networks for Integrated Multimedia Telecommunications

Anthony S. Acampora

Columbia University
New York, New York

PLENUM PRESS • NEW YORK AND LONDON

Library of Congress Cataloging-in-Publication Data

Acampora, Anthony S.
An introduction to broadband networks : LANs, MANs, ATM, B-ISDN, and optical networks for integrated multimedia telecommunications / Anthony S. Acampora.
p. cm. -- (Applications of communications theory)
Includes bibliographical references and index.
ISBN 0-306-44558-1
1. Broadband communication systems. I. Title. II. Series.
TK5102.5.A285 1994
621.382--dc20 93-41272
CIP

ISBN 0-306-44558-1

A Division of Plenum Publishing Corporation
233 Spring Street, New York, N.Y. 10013

Printed in the United States of America

To my wife Margaret and my children Anthony and Rose, for the patience and understanding which they extended to me during the many hours which I consumed in preparing the original manuscript. Further, to my many friends and colleagues who have provided me with so much intellectual stimulation throughout my entire career.

Preface

This is an elementary textbook on an advanced topic: broadband telecommunication networks. I must declare at the outset that this book is not primarily intended for an audience of telecommunication specialists who are well versed in the concepts, system architectures, and underlying technologies of high-speed, multimedia, bandwidth-on-demand, packet-switching networks, although the technically sophisticated telecommunication practitioner may wish to use it as a reference. Nor is this book intended to be an advanced textbook on the subject of broadband networks. Rather, this book is primarily intended for those eager to learn more about this exciting frontier in the field of telecommunications, an audience that includes systems designers, hardware and software engineers, engineering students, R&D managers, and market planners who seek an understanding of local-, metropolitan-, and wide-area broadband networks for integrating voice, data, image, and video. Its primary audience also includes researchers and engineers from other disciplines or other branches of telecommunications who anticipate a future involvement in, or who would simply like to learn more about, the field of broadband networks, along with scientific researchers and corporate telecommunication and data communication managers whose increasingly sophisticated applications would benefit from (and drive the need for) broadband networks. Advanced topics are certainly not ignored (in fact, a plausible argument could be mounted that all of the material is advanced, given the infancy of the topic). However, the objective is to provide a gentle introduction for those new to the field. Throughout, concepts are developed mostly on an intuitive, physical basis, with further insight provided through a combination of performance curves and applications. Problem sets are provided for those seeking additional training, and the starred sections containing some basic mathematical development may be safely skipped with no loss of continuity by those seeking only a qualitative understanding.

Telecommunication networks have emerged as a strategic component of the worldwide infrastructure needed to support economic development, scientific discovery, educational opportunities, and social advancement. Driven by the proliferation of high-speed local data networking and the potential of new multimedia service offerings, broadband telecommunication networks have become the focus of research, development, and standards activities worldwide. Underlying the field of broadband telecommunications are the concepts of universal interface and bandwidth-on-demand: all types of traffic are to be presented to the network in a common self-routing packet format, and high-bandwidth applications are distinguished from low-bandwidth applications only on the basis of the frequency with which the packets are generated by the users.

This book is focused on the principles of operation, architectures, performance management and traffic control strategies, protocols, emerging standards, and future directions for integrated broadband networks capable of supporting voice, data, image, and video telecommunication transport and services. The range of coverage spans from high-speed networks in use today (such as Local Area Networks) through those that will emerge over the next five years (Wide-Area ATM networks), up to and including those that might form the telecommunication infrastructure of the next decade (optical networks). The impact of rapidly advancing technologies, including VLSI, lightwave devices, and software, is addressed, and the genesis of broadband networks is explored. The relationship between LANs, MANs (FDDI, IEEE 802.6 DQDB), WANs, Asynchronous Transfer Mode (ATM), Broadband ISDN, and all-optical networks are developed, and several ATM switch architectures are presented and their performances compared. Traffic-related management strategies are described, including call setup procedures, signaling, congestion control, admission control, flow control, performance management, and network resource allocation. Traffic descriptors, important for developing call control procedures that guarantee variable/continuous bit rate service qualities, are discussed, along with possible rate-enforcement procedures. In an extension of conventional broadband network architectures, the concept of the "optical ether" is introduced, and possible second- or third-generation architectures for ultrabroadband all-optical or lightwave networks are presented, discussed, and studied. Finally, several categories of broadband multimedia applications are described.

Readers will gain an understanding of fundamental principles, underlying technologies, architectural alternatives, standards, and future directions for broadband networks. They will develop familiarity with LANs, MANs, WANs, FDDI, DQDB, ATM, and B-ISDN, as well as with traffic control and performance management issues and strategies. They will be exposed to concepts and approaches for self-routing ATM switches, and they will also be exposed to fiber optics principles, lightwave device technologies, and the potential of all-optical networks. Finally, they will be able to assess the opportunities for broadband multimedia service offerings.

The book contains eight chapters, starting with an introductory chapter that describes concepts and principles generic to all broadband networks. The origins of congestion are explored, and basic congestion control procedures are established. Various types of telecommunication traffic are characterized, and the concepts of universal packet format and bandwidth-on-demand are introduced. The driving forces behind broadband networks are discussed, along with a description of the types of services and user applications enabled by broadband networks. The need for asynchronously multiplexed transmission is described, along with the basic concepts behind segmentation and reassembly of information-bearing telecommunication signals. The distinction behind connectionless and connection-oriented service is described, and communication protocol layering is briefly discussed.

Chapter 2 contains a review of LANs, including performance criteria and evaluation methodologies. Although LANs are well treated elsewhere and are, in many regards, distinct from broadband networks, their treatment has been included here to illustrate several relevant and fundamental concepts, to assist the student or practitioner having no prior exposure to this field, to introduce some of the mathematical techniques used to study packet-oriented transport networks of any type (LAN, MAN, broadband), and to present the principles behind the use of ATM (an important technology for broadband networks) for LAN service. This chapter may be skimmed or skipped by those already familiar with LANs, or only its starred sections may be omitted by those uninterested in the mathematical details.

Chapter 3 contains descriptions of Metropolitan Area subnetworks: Fiber-Distributed-Data-Interface and the IEEE 802.6 Distributed Queue Dual Bus, including operating principles, channel access protocols, performance, protocol data units, offered services, and integration into the broadband network environment.

Chapter 4 is devoted to principles and architectures of the ATM switch. Statistical smoothing at the input and output are distinguished, and several switch architectural alternatives are presented and compared, including the Banyan switch, the fully connected Knockout switch, the multistage Batcher–Banyan switch, the Tandem Banyan switch, and the Shared Memory switch.

Chapter 5 is devoted to ATM, the underlying technology for B-ISDN, including cell format, virtual paths and channels, the adaptation layer, signaling, and control.

Chapter 6 covers the principles of traffic control and performance management for broadband multimedia telecommunications. Congestion control that involves traffic descriptors, admission control at call setup time, and rate-enforcement procedures are contrasted with those that also involve flow control at the ATM cell level, buffer allocation and management, and selective cell loss. Quality-of-service metrics are developed, and the effects of time delay in a high-speed wide-area network are assessed.

Chapter 7 presents advanced concepts for all-optical or "passive" lightwave networks having an enormous capacity potential that is measured in units of terabits per second. Capabilities of state-of-the-art lightwave devices such as tunable semiconductor lasers, tunable optical filters, wavelength-division multiplexers, and fiber-optic amplifiers are described, and principles of optical communications are presented. Constraints imposed on lightwave network architectures by lightwave device limitations are discussed, and candidate architectures that employ wavelength-selective routing with multihop access stations are presented. Techniques that exploit the relative independence between the physical topology of the optical medium and the logical connectivity among the access stations are presented, and logical reconfiguration via wavelength reassignment is presented as a technique to optimize overall capacity, adapt to network failures, and modularly expand the network.

Chapter 8 explores a range of possible applications enabled by broadband networks: a more natural mode of people-to-people communications; the access to (and manipulation of) multimedia electronic libraries by people and machines; wide-area distributed computing; delivery of entertainment HDTV signals; simplification of network management and control algorithms; execution of a rich set of diagnostic routines for improved network reliability; new educational delivery mechanisms; scientific computation and three-dimensional image rendering; and medical diagnostics and treatment programs.

Contents

Chapter 3. Packet Switch Interconnection Fabrics

Chapter 4. Metropolitan Area Networks

Chapter 5. Broadband ISDN and Asynchronous Transfer Mode

1

Introduction

1.1. Megatrends in Technology

The field of modern telecommunications is being rapidly transformed by megatrends in three underlying core technologies: microelectronics, photonics, and software. In the field of microelectronics, advances in materials, design methodologies, high-resolution photolithography, and fabrication processes have produced high-gate-count very large scale integration (VLSI) circuitry characterized by ultrahigh reliability and capable of performing highly sophisticated functions at fast, real-time clock speeds. Submicron complementary metal oxide semiconductor (CMOS) technology permits single-chip custom integration approaching an equivalent count of one million gates if the circuit pattern is regular (e.g., memories, certain types of packet switches) and can be operated at a clock speed of 100–200 MHz. Emitter coupled logic (ECL), while not permitting as high a degree of integration as CMOS, can operate at clock speeds approaching 1 GHz with moderate functional complexity and even higher speeds with further-reduced functional complexity. Gallium arsenide (GaAS) technology, while not as mature as CMOS and ECL silicon technology, is capable of operating at clock speeds in excess of 10 Gbit/sec. These capabilities, in turn, have had profound impact on the fields of consumer and military electronics, computers, and telecommunications. For example, in the field of telecommunications, circuit boards populated by VLSI chips perform the routing, protocol processing, storage and media access control needed to enable low-cost Local Networking among desktop computers and workstations distributed throughout large office buildings and college campuses. Until quite recently, this type of functionality was feasible only in software running on a minicomputer at execution speeds that were but a tiny fraction of the speed at which these operations are performed today in VLSI-dominated hardware designed specifically for the intended application. Moreover, knowledge-based computer-aided design tools (or expert systems) permit workers to custom-design sophisticated circuitry, fabricate VLSI chips, and rapidly prototype, debug, and evaluate ideas for new telecommunication products and applications. Standard programmable logic arrays provide even greater design flexibility, but do not offer the degree of functionality enabled by custom designs.

Similarly, advances in the field of photonics have had significant impact on telecommunications and, here, we may thus far have witnessed only the tip of the iceberg. The great appeal of lightwave technology as it applies to the field of telecommunications is the low loss, low dispersion, and extremely high bandwidth afforded by single-mode silica-based optical fiber and associated passive components (e.g., couplers, power splitters and combiners). When combined with narrow-linewidth single-frequency semiconductor lasers and low-noise optical receivers, it becomes possible for optical fiber links to support very high point-to-point digital data rates over very long unrepeatered distances. The speed–distance product, long regarded to be the essential figure of merit of a point-to-point fiber-optic transmission system, has steadily increased over the past 15 years from the 2 Gbit-km/sec produced by multimode fiber systems using light-emitting diodes to about 200 Gbit-km/sec as produced by a single-mode fiber system using single-wavelength lasers. Substantially higher figures of merit are expected through deployment of optical amplifiers. As impressive as these gains have been, however, applications of lightwave technology to telecommunications have been largely limited to transmission systems, with optical fiber systems affording distinct technical, performance, and economic advantages relative to copper and radio-based systems. The architecture and very nature of the telecommunication network have not yet been fundamentally altered by photonics, but advances in optical amplifiers, wavelength-agile photonic components (lasers and optical filters), and passive optical components promise new applications to all-optical or lightwave *networks*, as opposed to lightwave transmission systems. A lightwave network allows the signals of many users geographically dispersed over some large service region to simultaneously share the bandwidth of a common optical medium (fiber, splitters, couplers, etc., which may be configured into a tree, bus, or some other arbitrary physical topology) in such a way that the identity of each signal is maintained and each signal is distributed by the tributaries of the medium to its correct receiver or set of receivers. Wavelength multiplexing is one technique that may accomplish these objectives. Such networks have the potential to unleash the vast bandwidth of the all-optical medium, three to four orders of magnitude greater than that occupied by a lightwave transmission system, thereby enabling the creation of an all-optical telecommunication infrastructure characterized by low-cost bandwidth and stimulating bandwidth-intensive end-user applications and other services needed to better support network operations (such as fault diagnostic routines).

In the software domain, powerful desktop workstations, file servers, object-oriented programming, distributed operating systems, distributed data bases, and distributed computing are creating the demand for telecommunication networks with features and capabilities far surpassing those intended for a voice and voice-grade data-dominated traffic environment. New applications requiring the assembling and processing of vast amounts of digitally encoded information from geographically dispersed sources and data bases, the cooperation among computing resources separated by large distances, and the involvement of human workers networked by their workstations and collectively contributing to the resolution of scientific, business, political, and educational issues are transforming the functionality, architecture, and components comprising both citywide and worldwide telecommunication infrastructures. These same distributed software capabilities also afford a more thorough integration of computers and computing technologies into the control fabric of the telecommunication infrastructure.

1.2. The Evolving Architecture of Telecommunication Networks

This book is, most emphatically, *not* about these core technologies but, rather, is focused exclusively on telecommunication *networks*: the new system-level concepts, operating principles, architectures, performances, applications, and opportunities made possible by rapid advances in the aforementioned underlying core technologies. Networks intended to provide service over a local area (single building or small campus), metropolitan area (medium- to large-size city), and wide area (spanning thousands of kilometers) will all be treated. In all cases, the networks to be considered can be classified by the descriptor *broadband* and exhibit some common characteristics:

1. The network supports bandwidth-on-demand at each access port or access station to which a generic "user" (e.g., telephone, workstation, personal computer, host computer, network feature processor, network control computer, network data base, gateway to another network) attaches. As shown in the example of Figure 1.1, each "user" is attached to the network via a high-speed access link operating at some specified data rate. All information presented by the user is sent to the network via these access links in some common packet format. In the example of Figure 1.1, we assume that this common format consists of a sequence of short, fixed-length "cells" of information [as we shall see, this is the basis of asynchronous transfer mode (ATM)], and the user is responsible for transforming the signals generated by the actual terminating equipment (e.g., workstations) into this common format. In such an approach, large data blocks or continuous bit rate information streams are segmented and delivered to the network as a sequence of fixed-length cells; the larger data blocks are then reassembled from the cells arriving at the receiver. The terminating equipment controls its *effective bandwidth* by means of the *frequency* with which cells are generated, accessing more or fewer of the

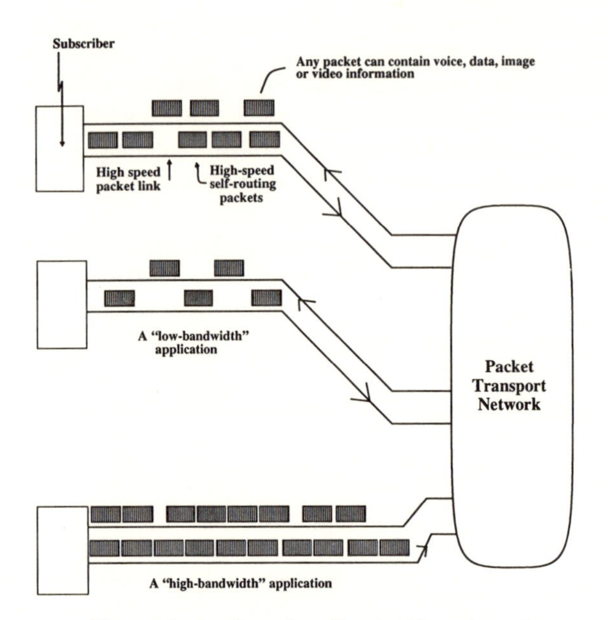

Figure 1.1. An illustration of bandwidth-on-demand.

fixed-length time slots appearing on the access links. The only feature that distinguishes a low-bandwidth application from a high-bandwidth application is the frequency of cell generation, and users can access bandwidth-on-demand at any effective data rate up to the maximum speed permitted by the access link. Within the network, all cells are handled autonomously. The network operates only on the cell *header* which contains the information needed by the network to deliver the cell to its intended receiver. The network disregards the nature or contents of the cell payload, which may contain voice, data, image, or video information (except, possibly, to satisfy the different quality-of-service requirements of different traffic classes, as will be discussed in Chapter 6).

2. In general, a given user attaches through one of a limited number of types of access links, each characterized by a unique format. Each type of access link therefore appears as a *universal port* to the class of users connected through that type of link. The link format may be a low-level specification including, for example, bit rate, packet or cell size, and location and contents of the payload and header information fields. Alternatively, a link format may also include higher-level functionality such as flow control. For a given format, bandwidth can be accessed on demand up to the

maximum bit rate specified for that format. Several universal formats or interfaces are provided to avoid encumbering a simple termination with the sophistication demanded by a complex termination. (For ATM, at least two types of universal ports are permitted, both of low-level functionality: one operates at 155 Mbit/sec, the other at 622 Mbit/sec).

3. A generic "user" can simultaneously generate independent multimedia information signals, each signal intended for a distinct receiver. The packets or cells created from these signals arrive at the network time-multiplexed through that user's access port, and the network uses the information contained in the cell header fields to rapidly direct each cell from the time-multiplexed sequence to its intended destination. Thus, a user chooses the receiver for each of its signals merely by placing the correct information into the header fields of the corresponding cells. This allows a user to rapidly direct each message of a message string to that message's intended receiver, or to concurrently maintain a communication link or connection with each of several other users, rapidly accessing a given connection and its respective receiver by again placing the correct information into the corresponding cell headers. Similarly, a given user may concurrently receive information signals from a multiplicity of senders. Each user may generate or terminate varying combinations of voice, data, image, and full-motion video traffic, and the network access stations must accept such fully integrated traffic, support the quality of service guaranteed to each traffic type, maintain the distinguishability among the messages multiplexed through a common access port, and maintain synchronization among the diverse traffic types that may be destined for (or arrive from) any one receiver (or sender). The integration of bursty traffic of variable bit rate with various types of continuous bit rate traffic, and the ability to rapidly forward each message of a sequence to a distinct receiver (essential for distributed processing, data communications, and other types of highly multiplexed applications including multiplexed voice and video transport), are hallmarks of emerging broadband networks.
4. Some universal ports may support circuit-switched connections for which the required network transport resources are reserved, or scheduled, at call setup time. Other universal ports support a packet-oriented transport in which all traffic types are segmented at the source into a stream of cells, with each cell being autonomously relayed through the network prior to reassembly at the destination (this type of port was demonstrated in Figure 1.1). In the latter case, either datagram (connectionless) or virtual circuit (connection-oriented) services may be provided to the user, as will be explained in Section 1.6.
5. The telecommunication network can be separated into two components: (a) a transport network responsible for moving information among geo-

graphically remote user and equipment locations, relaying packets through switching nodes, and reliably delivering applied information to the intended receivers; and (b) a service network containing the feature processors, distributed data bases, and applications that are accessed by subscribers through the transport network, along with the distributed network control and management environment. The transport network is hardware-based in that the real-time access speeds are too fast to permit header processing, packet relaying, and protocol-related decisions to be made by software-driven processors. The overall supervision, management, and control of the network are, however, enabled by a system of software-driven processors that are interconnected via the transport network but do not directly process the high-speed, real-time, user-to-user information.

Today's telecommunication infrastructure is characterized by two distinct elements: transmission systems and switching systems. Transmission systems are responsible for the movement of properly formatted information on a point-to-point basis. Typically, as shown in Figure 1.2a, information arising from a multitude of sources are multiplexed onto a common transmission medium, which carries the information to a geographically remote destination where it is detected and demultiplexed into its constituent components. Here, multiplexing may be done in the time-division mode, in which the identity of each source of information is maintained by assigning that information to a particular, periodically recurring time slot of a larger transmission frame (more will be said of time division multiplexing later). In a variant of traditional synchronous time division multiplexing, the identity of the individual sources might be maintained by segmenting the informational content of each source into fixed or variable length packets, each of which carries a unique identifier within its header field. The packets associated with different sources are asynchronously time multiplexed via a device known as a statistical multiplexer (again, more later; this is the approach taken by ATM). Frequency division multiplexing is an alternative to time division multiplexing in which the identity of each source is maintained by modulating that source's information onto a unique carrier frequency, or channel, of a multichannel carrier transmission system (typically done for radio and satellite transmission links).

The end points of the transmission system might be either remote terminals which gather and distribute information to end-users, or switching nodes which terminate some multitude of transmission links, as shown in Figure 1.2b. It is the responsibility of the switching system to route the information corresponding to each end-user source to the correct outbound transmission system by forming the correct interconnection patterns among its input and output ports. For example, in a wide-area network, the connections $s_{1,1}, s_{2,1}, \ldots, s_{N,1}$ in some given locality might be carried to the switch by a common transmission system and then de-

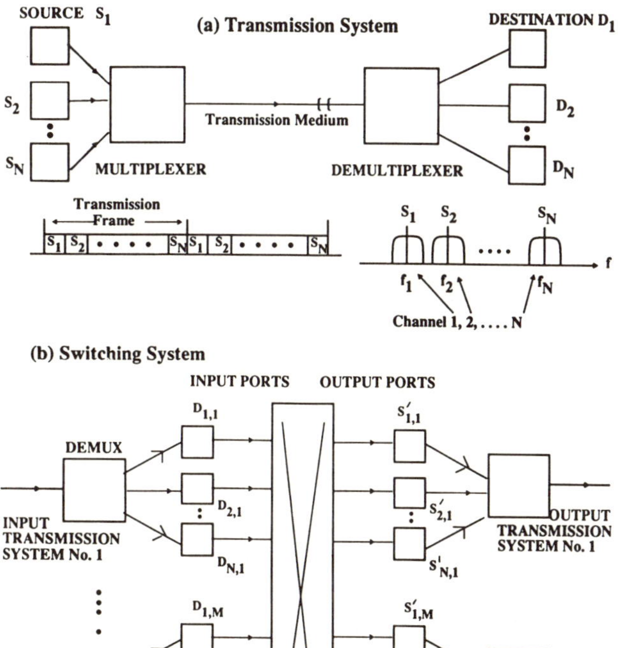

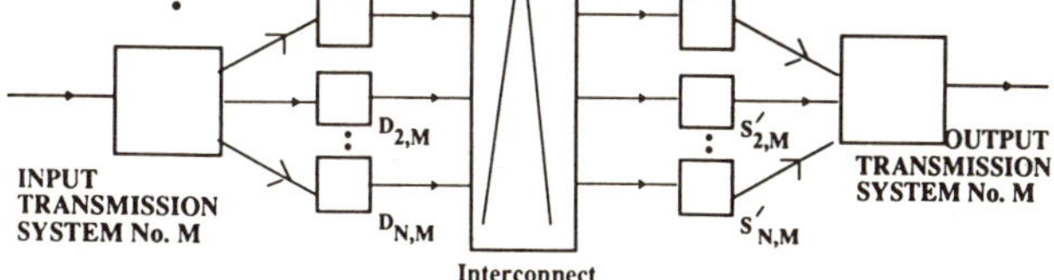

Figure 1.2. Transmission and switching system block diagrams.

multiplexed, switched, and remultiplexed for delivery, each to a destination in a different locality. The switch is reconfigurable; that is, the interconnection patterns among its input and output ports can be changed in response to changing patterns among source–destination pairs.

In today's telecommunication infrastructure, large, centralized switching systems of this type are interconnected by point-to-point transmission systems such that a connection originating in one locality might be multiplexed, transported, demultiplexed, switched, remultiplexed with other connections, transported, and so forth, many times before arriving at its ultimate destination in some geographically remote locality. Network services (such as administration and billing, diagnostics, and connection setup) and customer services (such as voice store-and-forward) are provided via dedicated feature processors, typically co-located with the switches within large switching offices. Control over these processors is typically exercised via a separate signaling network, an overlay to the communication network as embodied in the switching and transmission systems of Figure 1.2.

By contrast, the integrated broadband telecommunication infrastructure, drawing heavily on VLSI, lightwave, and software technologies, will be characterized by the following functional elements: transport, interfaces, and distributed services. The traditional boundary between switching and transmission systems is already becoming blurred, and, in the broadband era, the functionality of these systems will be replaced by the transport network which has exclusive responsibility for the movement of information from geographically dispersed input ports to the correct output ports. The transport network may contain traditional point-to-point transmission systems and centralized switching systems, supplemented by (or entirely displaced by) multipoint transmission systems. Such multipoint systems allow information-bearing signals to enter and leave the transmission medium from multiple, geographically distributed access stations, and hardware-based intelligence located within each access station is capable of making local routing decisions (accept a segment of information, ignore that information, regenerate and relay the information, etc.), effectively creating a distributed switch. Users (which may, themselves, be gateways to other networks) will be interconnected by the transport network and attach via bandwith-on-demand universal ports. Services provided by the network attach via universal ports of the same design as those by means of which users attach. Control information is transported among the users and feature processors over the transport network, i.e., the transport network does not distinguish between control and communication signals. Similarly, the transport network does not distinguish a feature processor from any other user; it is responsible exclusively for delivering the information entering at the universal ports to the correct outputs, but is not concerned with the informational *content* of the signals. In this sense, as shown in Figure 1.3, the distributed transport network views all information as having originated in some generic distributed *application*, the components of which are attached through the universal access ports. Different applications might be distinguished by the amount of bandwidth drawn from each port, but even this is ignored by the transport network as long as the total demand per port (in either direction) does not exceed the capacity of the port. End-users derive enhanced services from the network by interacting with the applications provided by the network. Alternatively, large distributed end-user organizations may install their own shared applications, each of which may itself involve cooperation among geographically separated software routines. In this manner, the transport network might provide a "virtual private network" for large, geographically dispersed end-user organizations. The fact that multiple end-user organizations are, in fact, sharing a common transport network is totally transparent to the end-users. Administration and billing, connection setup, diagnostic routing, and other network services are sup-

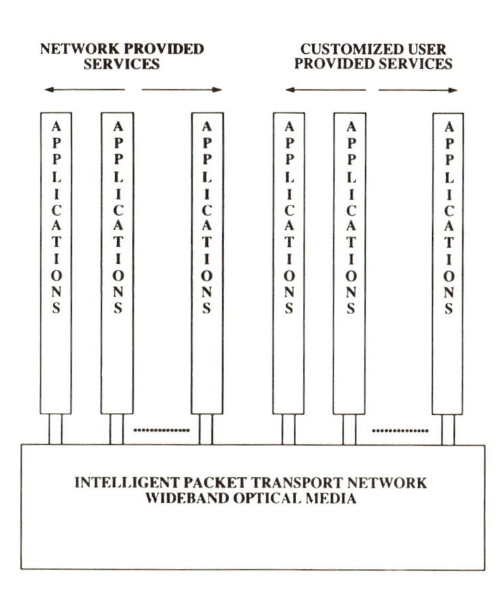

Figure 1.3. A conceptual representation of a distributed communication network.

ported as yet additional distributed applications running over the transport network.

The transport network may assume one of several forms. As shown in Figure 1.4, the transport network may contain some number of geographically dispersed self-routing packet switches. These switches, using custom-designed VLSI circuitry, read the header fields of each arriving packet, making the appropriate decision with regard to output link. Contention for the output links is handled via smoothing buffers contained within the switches (more will be said of this later). The switches are interconnected by point-to-point transmission facilities which serve exclusively to carry packets to the next relaying point, and, as shown in Figure 1.4, distributed access to transmission facilities is not permitted. The network is "transparent"; it does not process the user-applied data produced by the local area networks (LANs), private branch exchanges (PBXs), or other user devices attached to the network, but simply transports the information to the correct output. Higher-level protocol functions (retransmission requests, selection of flow control parameters, etc.) are performed externally to the network, using any protocol model agreed on by the two or more participants involved in any given connection (there is always at least one transmitting and one receiving participant). The transport network is totally devoid of software, insofar as the actual transport of applied data is concerned; software processing of packet headers as required for switching is simply too slow at the real-time data rates of the interconnecting transmission links. Software for non-real-time supervisory functions (e.g., maintenance, diagnostics, call setup) resides within feature processors

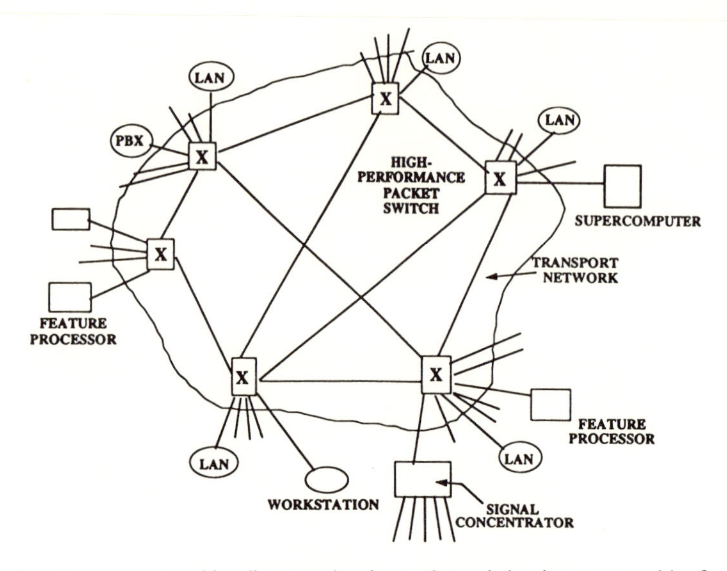

Figure 1.4. A transparent multimedia network using packet switches interconnected by fixed-point transmission links, surrounded by its supervisory feature processors and signal-supplying users.

which are external to the transport network and which connect to the transport network through the universal ports, appearing to the transport network like any other group of users.

An alternative architecture for the transport network appears in Figure 1.5. Here, the transport network consists of a passive optical medium containing fiber, optical couplers, splitters, power combiners, etc., and a system of geographically distributed access stations. All logical functions (packet switching, buffering, etc.) are performed exclusively within the access stations which contain all of the real-time electronics; the "optical ether" serves simply to move information among the access stations. For example, many high-speed channels might be wavelength multiplexed onto the "optical ether," each providing a dedicated path between two (one transmitting and one receiving) access stations. The user pair connected by a given optical channel may be changed with time by tuning the optical transmitters and receivers contained within each access station to different wavelengths. Traffic generated by a given user is transparently transported from that user's access station to the access station of the receiver. The access stations provide real-time transport, again devoid of software data-handling (software is too slow). Non-real-time supervisory functions are provided by software and processing elements distributed among the access stations to create a distributed management environment which does not directly process the real-time high-speed traffic. Enhanced services (data base access, shared computing facilities,

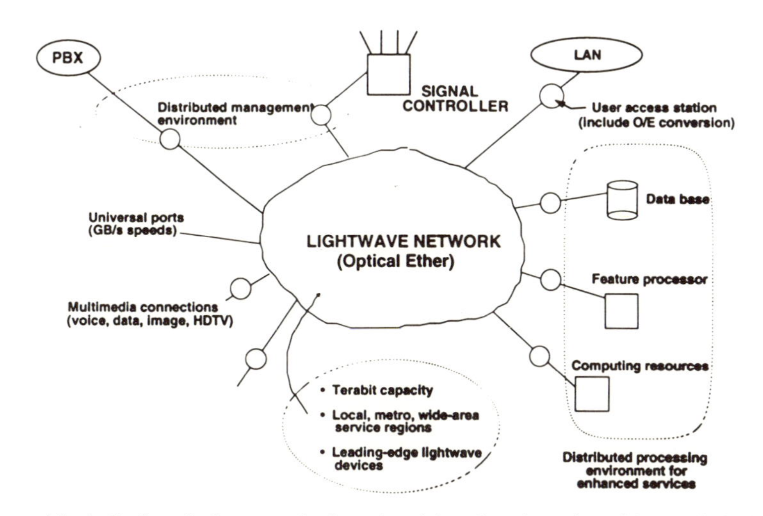

Figure 1.5. A distributed telecommunication network based on the notion of the "optical ether."

etc.) are provided via equipment, external to the transport network, each component of which connects to the transport network via one of the distributed access stations.

1.3. Broadband Networks: Driving Forces

Interest in broadband networks first began to develop in the early 1980s. At that time, LANs were finding their way into offices, universities, and small industrial parks, and the need developed to extend the intrapremises, data-oriented services provided by LANs to communities of users who might be geographically located throughout cities (metropolitan area networks) or larger regions (wide-area networks). The intrapremises data services provided by LANs include, for example, terminal-to-host access, host-to-host file transfers, distributed processing, shared data base access, and electronic mail. The LAN equipment used to support these services is well matched to the bursty data nature of the telecommunication traffic: a single, high-speed channel, time-shared by user access stations, which implements the rules for accessing the channel, with information presented in a format suitable for packet switching. Packets generated by the data device (e.g., terminal, host, PC, file server) attached to a particular access station are sequentially addressed to different destinations by placing the appropriate information in the packet header, thereby avoiding the unacceptable time delay associated with reserving channel time for each packet by means of a centralized setup procedure. Each access station reads the header of every packet broadcasted onto the shared channel and accepts those packets determined to be intended for local reception. The ability of each access port to support multiple simultaneous sessions and the high peak data rate afforded by the common shared channel are, perhaps, the primary distinctions between data services offered by a LAN and other types of data services such as might be offered, for example, by a digital PBX.

In this context, the concept of the metropolitan area network (MAN) began to develop as a means to enable interpremises extension of LAN services. As shown in Figure 1.6, the MAN consists of a larger, interpremises version of the LAN shared channel, with one access station of each LAN serving as the concentrator for interpremises traffic. In this way, hosts and data bases connected by a LAN in one building may be accessed by terminals and workstations connected to a LAN in another building. The MAN would provide the interpremises inter-

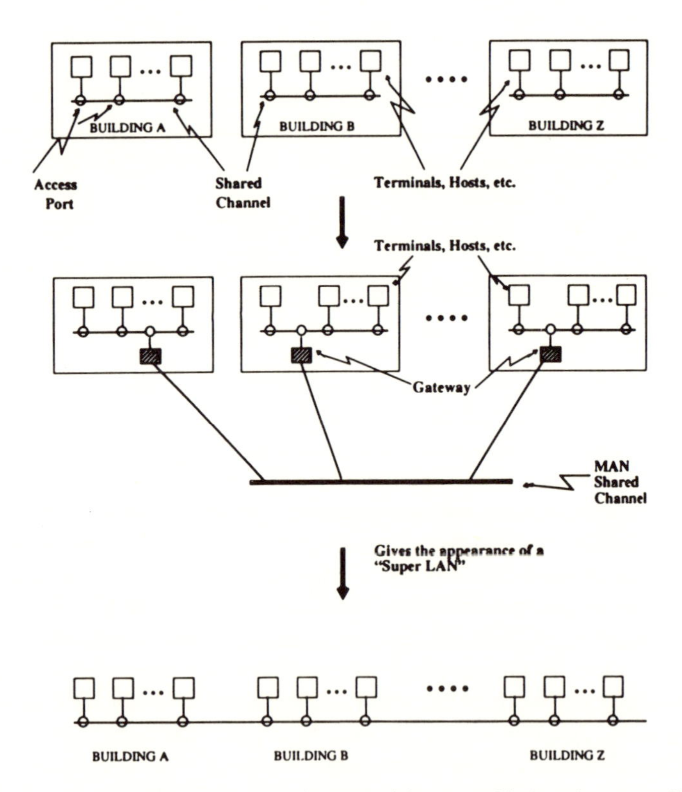

Figure 1.6. Use of a metropolitan area network to extend the geographical service range of local area networks.

connection among LANs, thereby creating the appearance of a "super-LAN" to extend LAN data services over metropolitan area geographies. This "super-LAN" would be a private network: owned, installed, operated, and maintained by a single entity and providing service to users affiliated with that entity.

Toward the mid-1980s, the common carriers began to develop an interest, first in MANs and soon thereafter in broadband networks to enable switched, high-speed public offerings. The interest of the carriers was motivated by (1) the increasing number of LANs installed within their service regions and the concurrent demand for LAN interconnects; (2) the concern that large corporate customers, demanding MAN-like service to interconnect their rapidly proliferating LANs, would install their own private MANs, thereby depriving the carriers of a potentially important source of new revenue; and (3) the opportunity to offer new broadband services, based on newly emerging technologies, to both business and residential customers. This latter consideration has since caused a marked departure from the earlier broadband driving force: the broadband network would no longer be limited to LAN interconnects and data-oriented services, but instead would be repositioned on a more pervasive basis as the enabler of integrated voice, data, image, and full-motion video services to large and small customers over wide service areas. In this context, networks intended for use either exclusively as metropolitan area LAN interconnects or as the point of entry to a pervasive, wide-area broadband infrastructure are referred to as metropolitan area subnetworks, MAN subnetworks, or simply as MANs.

Central to this change of emphasis is the notion of bandwidth-on-demand. As previously described, each subscriber would be provided with an access link capable of transporting digital information at a peak data rate commensurate with the bandwidth of that link. Multimedia information (voice, data, image, video) would be presented to the link in some specified self-routing packet format, and the frequency with which packets are generated and loaded onto the link would determine the effective "bandwidth" derived from the link.

The maximum rate of packet generation could be no greater than that which would cause the link to be loaded 100% of the time. For example, the packet stream generated by a 64 kbit/sec digital voice connection would load a 155 Mbit/sec link (one of the standard rates for ATM) only for a very small fraction of the time; a digital video connection, however, might load the same link for a significant fraction of the time. Voice, data, image, and video packets would all be multiplexed onto the link serving a particular subscriber and the packets corresponding to different types of traffic, or different packets of the same traffic class, could be forwarded to different network destinations. The only constraint is that the average packet loading does not exceed 100% on the link. (In practice, the average packet loading might typically be limited to approximately 80% of the link capacity in order to avoid excessively large queue buildup for bursty traffic.) The bandwidth-on-demand port to the network would accept and deliver properly formatted packets, independent of the packet content or traffic type, thereby assuming the characteristics of a "universal" information outlet, much analogous to an electric power outlet: the power outlet delivers electricity to any type of appliance (e.g., toaster, dishwasher), as long as the current delivered does not exceed the rating of the circuit; the universal information outlet can support the packets of any digital device (e.g., host, digital telephone) as long as the average rate of packet generation does not cause depletion of the bandwidth of the link.

The notions of bandwidth-on-demand and broadband multimedia networks are critically dependent on the provisioning of fiber-optic facilities to the home and office. For reasons to be described later, the bandwidth potential of a fiber-optic link is enormous, and the actual capacity of the link is limited by speed constraints of the terminating electronics and electro-optic components (e.g., semiconductor laser, photodetector, optical receiver). Unlike the twisted copper pairs which are wired to most homes and offices today and which are inherently limited to operating speeds of several megabits per second, the terminating equipment of a fiber-optic link could readily support speeds as high as several gigabits per second, more than adequate for broadband networks; and even this is but a small fraction of the inherent bandwidth of the medium.

Another important aspect of public switched broadband networks is the provisioning of enhanced features and services for multimedia traffic to complement the transport network for such traffic. For example, as an enhanced feature, a high-resolution facsimile service node might be provided which could accept a high-resolution document transmitted by a subscriber, continually retry delivery to busy facsimile machines, and duplicate and deliver the document to all recipients on some specified mailing list. More will be said later of other possible (and considerably more exciting) broadband applications.

In this book, we will develop the principles of broadband networks both for switched public services and, as a subset, for private LAN interconnects. In so doing, we shall develop the concepts of, and demonstrate the interrelationships among LANs, MANs, self-routing packet switching, ATM, Broadband Integrated Services Digital Network (B-ISDN), Distributed Queue Dual Bus (DQDB, the emerging IEEE 802.6 standard), Fiber-Distributed-Data-Interface (FDDI), and later-generation very-high-speed all-optical networks. Performance analysis will be summarized in graphical form to permit comparison among alternative approaches and architectures. Protocol details for emerging B-ISDN, DQDB, and FDDI will be presented as appropriate.

1.4. The Distinction between LANs and MANs

The distinction between LANs and metropolitan area subnetworks is primarily quantitative. In general, a metropolitan area subnetwork is quite similar to a local area network in that both are based on distributed switching and distributed transmission systems which, collectively, comprise the distributed transport network. For both, there is the notion of a shared transmission medium which combines and carries signals among a multitude of user pairs. For both, information is carried in a self-routing packet format, and the identity of the information which flows between each source–destination pair is maintained by appending to each packet some appropriate header identification field.

However, unlike the LAN, which is intended to provide data-oriented services among users who are spread out within a building or over a small campus, a MAN typically involves much larger distances: the coverage region of a LAN is typically under several kilometers, while the MAN typically covers a region spanning several tens of kilometers or even higher. Also, the transmission speeds exhibit order-of-magnitude differences, with the LAN operating, typically, at a rate of under 10 Mbit/sec and the MAN, typically, at a rate of greater than 100 Mbit/sec. These distinctions, longer distance and higher speed, collectively require that different strategies be used to enable the transmission medium to be shared among a plurality of users.

The aggregate capacity offered by a network is, by definition, the largest permissible value for the total volume of information carried by the network such that some specified quality-of-service objectives (e.g., delivery delay, fraction of lost or misrouted information) can be maintained, that is, the aggregate capacity is the largest possible value for the total average traffic presented by the individual sources. By virtue of the higher transmission speeds employed and the fact that multiple shared-media MAN subnetworks might be combined with remote statistical concentrators and packet switching nodes to form a wide-area broadband network, the aggregate capacity provided by the broadband network is enormous (tens or hundreds of gigabits per second). Although the aggregate capacity of a broadband network may be quite high, no one user can present a traffic load exceeding the capacity of the universal port. This capacity is, in general, slightly lower than the transmission speed of a single LAN or MAN subnetwork's shared channel, to accommodate inefficiencies associated with media access and packet headers. Thus, for a MAN, the link access speed is typically greater than 100 Mbit/sec but rarely greater than several gigabits per second. (The electronics and electro-optics which would be needed to drive the medium at a rate greater than 2–3 Gbit/sec would be regarded, as of this writing, as available only in research laboratories. This is the so-called fundamental electro-optic bottleneck; even as this speed is gradually increased with advancing technology, there will still remain an enormous mismatch between electro-optic data rates and the bandwidth of optical media, which constrains the permissible architectures of lightwave networks; more will be said of this in Chapter 7.)

Another quantitative difference between LANs, MAN subnetworks, and wide-area broadband networks concerns the expected number of subscribers. A premises-based LAN typically interconnects fewer than several hundred access stations, although this number may grow to be as high as several thousand for a particularly large installation. By contrast, a MAN subnetwork carries the inter-premises traffic of several or many LANs, and the total number of subscribers whose traffic must be managed is correspondingly higher.

Because larger geographical distances are involved, the user of the MAN subnetwork or the broadband network suffers a larger end-to-end propagation delay as compared with the LAN user. For the MAN subnetwork, this longer propagation delay requires that different media access schemes be used. For wide-area broadband networks, the longer propagation delay requires traffic management strategies which can tolerate appreciable signaling delay and which recognize that, because of the high capacities and long distances, much information is resident on the transport network at any given time. Longer propagation delay might also translate into longer latency insofar as cooperation among distributed computers is concerned, but this effect might be partially offset by the higher link access speeds involved.

Finally, unlike the LAN, which is always operated as a single-owner private network, the MAN subnetwork and wide-area broadband network might be either publicly or privately owned and operated. A public network, as previously noted, might provide virtual private network services for large users.

1.5. Some Possible Applications

As mentioned, the original application intended for broadband networks was to extend the on-premises data services enjoyed by LAN users across large geographies. Later, the notion of a universal information outlet began to emerge. Although these remain important, a sampling of other possible applications, spanning the end-user and network operations domains, might include:

1. *Provisioning of fiber-to-the-home.* In this variant of the classic chicken-and-egg dilemma, the advent of broadband networks is dependent on penetration of fiber-optic facilities to large offices, homes, and small offices. At the same time, economical feasibility of fiber-to-the-home, based on conventional star-on-star wiring topologies (one link per home/office, hubbed at a remote terminal, with local loops further hubbed at the central office) has yet to be demonstrated, especially if voice and voiceband data are the only types of traffic to be carried. Broadband networks may help to break this cycle by stimulating the market for new revenue-generating multimedia services while at the same time providing high-speed transport over a physical plant characterized by lower-cost distributed topologies. Physically, the plant might look much like the distribution network for cable TV signals but would borrow heavily from network notions of packet access, distributed ports, and shared media to provide two-way bandwidth-on-demand connections between many pairs of access ports rather than merely the delivery of

a limited number of signals broadcasted from some head-end to downstream receive-only stations.

2. *Enhanced network reliability.* Broadband is often viewed as a technology with the potential to offer multimedia applications to large and small users alike, by virtue of the economical provisioning of link access speeds over two to three orders of magnitude faster than narrowband, voice-oriented networks. Often overlooked is the opportunity to reserve a significant amount of capacity, taken from the large aggregate pool, to enhance network reliability. By exploiting the unprecedented availability of bandwidth, not only might shared backup facilities be reserved to spare failed resources, but a rich set of capacity-consuming diagnostic routines might be developed and installed to continually test and monitor the status of network hardware and software elements. The greater the capacity devoted to running test routines, the more diagnostic data that can be generated, the better the ability to detect even subtle bugs and failures, and the faster the network can be reconfigured to route information around the point of failure.

3. *Lightweight communication protocols.* Historically, as communication applications grew more sophisticated, a set of protocols or rules governing the flow of information and various quality checks were developed to enable reliable delivery of information in a relatively hostile network environment characterized by noisy, bit error-prone channels and congested network links and switching nodes. Protocols intended to ensure the integrity of delivered data, control the flow of traffic over congested elements, and monitor for data lost by buffer overlow and other causes, consume processing resources and constrain the sustainable rate of information flow among terminating devices to the fastest that the protocol processor will permit. For example, even over a high-speed high-quality link, execution of a protocol intended to protect information over a lower-quality link might limit the effective rate of information flow between two machines (the throughput) to a rate of perhaps a few megabits per second. Furthermore, increasing the link speed alone, but leaving the protocol unchanged, cannot provide for a higher throughput. While the development of faster, more customized processors will certainly relieve this "protocol bottleneck," a major improvement will also result from the development of so-called "lightweight" protocols which exploit the relatively benign characteristics of broadband networks to eliminate the need for extensive layers of protection. Since the broadband network will offer enormous aggregate capacity, it is possible to operate the network with a comfortable margin of safety, allowing it to carry a volume of traffic very large by today's standards, while at the same time maintaining a comfortable distance from the capacity limits. In this relatively uncongested environment, the protocols needed to protect against traffic overload (flow control, delivery prioritization, buffer overflow, etc.) might be scaled back or eliminated entirely. Furthermore, the fiber-optic medium associated with the broadband network is designed to produce a bit error rate orders of magnitude lower than that provided by radio or copper alternatives (10^{-12}

versus 10^{-6}, for example), and the requirements for error detection/error correction protocols might be relaxed considerably. Merely by permitting the streamlining or elimination of multiple protocol layers, broadband networks may permit an order of magnitude or more increase in throughput among communicating processors and workstations (megabits or tens of megabits to hundreds of megabits per second). Such an increase is essential to produce the tight computer-to-computer coupling required for efficient distributed processing.

4. *Broadband multimedia connections.* Voice and voiceband data are the dominant forms of end-user traffic carried by today's telecommunication network. End-users are offered (a) circuit-switched 3-kHz analog channels, (b) circuit-switched 64 kbit/sec channels, and (c) for the Integrated Services Digital Network Basic Rate Service, two independently switchable 64 kbit/sec channels plus a 16 kbit/sec channel that can carry packet or circuit-switched data. By contrast, the broadband network will offer bandwidth-on-demand, multimedia connections, which integrate voice, data, image, and full-motion video traffic through universal ports operating at link access speeds in excess of 100 Mbit/sec. Traffic, in each of its forms, will be reduced to a sequence of addressed data packets, with a full-motion video connection distinguished from a voice connection, as far as the network is concerned, only by its significantly higher rate of packet generation. Again, the use of VLSI renders practical the formatting of information in all of its multimedia forms into self-routing packets which can be carried over a common transport network. All forms of traffic can be carried on a common connection, and a multitude of connections, destined for the same or different network access ports, can be multiplexed through a common universal port.

5. *Access to distributed data bases, pictorial archives, and electronically published information.* Delivery of vast, electronically published information libraries to a large number of subscribers is currently limited by (a) insufficient end-user bandwidth provided by a narrowband telecommunication infrastructure; and (b) lack of suitable data base software. A high-quality black-and-white image (400 dots/inch, $8\frac{1}{2}$ by 11-inch size) contains about 15 million bits. Such an image could be delivered, in uncompressed form, to a human reader in about one-tenth of a second over the 155 Mbit/sec user port of a broadband network, but would require over 50 min via a 4800 bit/sec voiceband data link. Time of delivery over the broadband network is consistent with the patience of most people and the human responsiveness for archival browsing: the voiceband connection does not come close. Furthermore, only images containing fewer than 480 bits of information could possibly be compressed and delivered within one-tenth of a second by the aforementioned voiceband data link (such an image would be very uninteresting indeed, consisting for the most part only of large black blotches over a white background, or vice versa). The insufficiencies of narrowband networks are even more dramatic if color images are involved. With a broadband infrastructure, one can envision a system of distributed multimedia data bases and a software en-

vironment which enables each of a large number of users to synthesize needed information by accessing and assembling segments from several sources. For example, a user might electronically browse through a favorite new magazine as naturally and conveniently (and with the same graphical quality) as if hard copy were at hand; for a particularly interesting article, video segments from the evening news might be added. Not only does the broadband network provide the required link access speed for natural interaction but, also, it enables high-speed cooperation among data bases and processors to collect, assemble, and display the requested information.

6. *Entertainment video on demand.* Over-the-air and cable broadcast systems offer viewers the ability to select stations from a limited menu, with the content of the programming and hour of viewing entirely controlled by the network and/or cable operators. Broadband networks will provide adequate aggregate capacity and link access speed, along with two-way interaction, to permit viewers to select individual programs rather than broadcasted channels. Furthermore, each viewer can request a particular program for viewing at a time of day convenient to that viewer. In addition to handling an unprecedented volume of end-user traffic, the broadband network must, once again, allow programs resident in geographically dispersed data bases to be rapidly accessed and locally cached since, undoubtedly, multiple viewers will want to view the same program but with overlapping start times. The degree of difficulty encountered is strongly dependent on the start-time granularity.

7. *Natural, lifelike video conferencing.* Video conferencing has been heralded as a much-needed productivity enhancer because, in principle, it tends to avoid the need for time-consuming business travel. Today's bandwidth-constrained video conferencing services are characterized by large studios at each of several locations, with the physical presence of each participant required at one of the studios. One or more video monitors are present in each studio, enabling participants to see and hear participants at other locations. The quality of the image is often lacking because of the compression algorithms currently used to remove redundancy from the video signals and allow the signals to be carried over standard transmission facilities, typically at 1.5 Mbit/sec. To this is added a highly artificial environment in which the video camera at each remote location is steered toward the conference participant who is currently speaking or is speaking the loudest. The ability to judge the reaction of other participants and to engage in side discussion, a natural part of dynamic interaction, is totally lacking. Broadband networks will overcome many of these video conferencing shortcomings. Within a broadband environment, conference participants might not need to leave their individual offices. Each participant might select the video signal from any other participant (or, concurrently, from a group of geographically separated participants) for local display on a high-resolution monitor. Images of all participants might appear within a bar at the top or bottom of the display. The current speaker

might be highlighted by the display. Windows could be added so that a participant could watch the current speaker, while simultaneously conducting a side conversation with another participant. Other windows might contain electronic blackboards, and documents might be distributed by high-speed scanners and printers. In short, the abundant capacity of the broadband environment might be applied to enhance the naturalness of a multiparticipant video conference.

1.6. A Word about Transmission Formats and Traffic Types

For traditional byte-interleaved time division multiplexing, as illustrated in Figure 1.7, time on a transmission link is divided up into repetitive, fixed-length intervals known as transmission or switching frames, each of which is further subdivided into time slots capable of carrying one byte of information. The time slots are sequentially numbered throughout the frame. A time-multiplexed connection over such a transmission link consists of an allocation of one particular time slot from the switching frame, repetitively recurring on all successive frames. Information from the connection is broken down into byte-size units which are then carried in the assigned time slot on repetitive frames. Different time slots within the frame carry information corresponding to different connections. The identity of a particular connection is uniquely determined by the position of the assigned time slot within the switching frame. With such an approach, a network controller, or scheduler, is always needed to assign each newly requested connection to a particular time slot, and to return time slots to the unused pool, for possible reassignment, upon termination of a particular connection. Since each connection is uniquely determined by its assigned time slot, a switch or demultiplexer which might terminate the transmission line can be instructed to route the connection on the basis of position within the switching frame. The switching

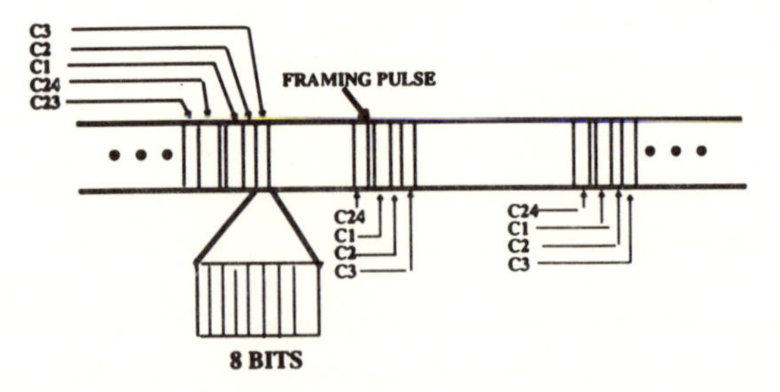

Figure 1.7. A typical byte-interleaved transmission format. Here, a given connection is uniquely determined by the fixed time slot assigned for the connection in each frame. A scheduler is always required to manage the time slot allocation.

sequence repeats frame after frame until a new connection is added or an existing one is dropped. Connections are assigned by the network controller to the time slots of two or more transmission links terminating on the same switch such that at no time are the time slots on more than one inbound link destined for the same outbound link. This may require the use of time slot interleaving within the switch to permit conversion of an inbound time slot assignment to a different outbound assignment. In this way, the network controller manages the competition among connections by means of reservation and time slot scheduling.

The number of time slots per frame, and the period of the switching frame, collectively determine both the data rate of the transmission line and the data rate of each connection. For example, in the DS1 transmission system, each 125 μsec switching frame contains 24 time slots, each one byte in duration, plus one additional bit, not associated with any connection, used for signaling and to delineate the frame boundaries. Each connection therefore consists of one byte every 125 μsec for a data rate per connection of 64 kbit/sec. The frame-mark bit, regularly occurring every 125 μsec, itself comprises an 8 kbit/sec signaling channel. Since there are 24 time slots per frame, each capable of supporting one 64 kbit/sec connection, the data rate of the transmission link is equal to 24 × 64 kbit/sec (= 1.536 Mbit/sec), plus 8 kbit/sec (the signaling channel), for a total of 1.544 Mbit/sec.

An alternative, hypothetical transmission system format, such as shown in Figure 1.8, might again use repetitive switching frames, each subdivided into time slots containing multiple bytes. Each assigned time slot might contain, for example, 53 bytes. Again, using a network scheduler, each requested connection would be assigned to a particular time slot of successive frames. In such a scheme, the 53-byte segments or cells corresponding to the various existing connections are said to be synchronously time multiplexed since the time slot assigned to a particular connection occurs regularly (synchronously), in the same position of

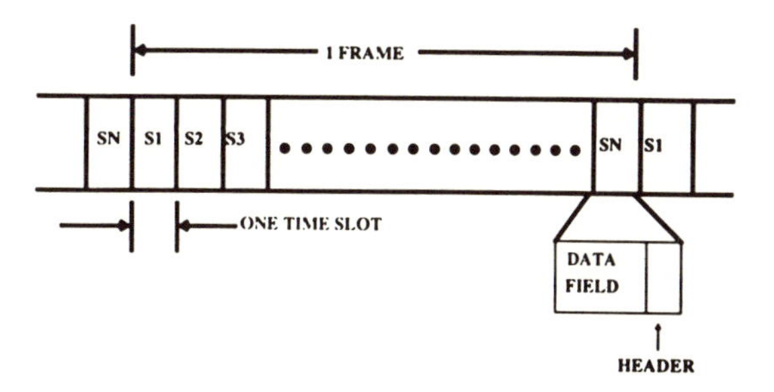

Figure 1.8. A hypothetical packet-interleaved transmission format. Here, a connection is identified either by a specifically numbered time slot reserved in each frame or by a connection identification number contained in the packet header.

successive frames. However, since each time slot can support a 53-byte segment, rather than a single individual byte, an alternative strategy becomes possible. Suppose that each 53-byte cell is divided into two fields, a data field containing user-supplied information and a header field containing network-supplied control information. Included in the header field is a number unique to each connection carried over the transmission link. For example, the data field might contain 48 bytes of user information, and the header might contain 5 bytes of control information, 3 of which are used to identify the connection. Then, it is no longer necessary to assign a given connection to a particular time slot within a switching frame; the time slot has become large enough to efficiently carry the information used to identify a given connection, and the need for both a synchronous time-multiplexed frame and a network time-slot scheduler disappears. The switch or demultiplexer terminating the transmission line would then "read" the connection identifier contained within the header, and make appropriate internal routing decisions in real time on each cell. The cells associated with a particular connection need no longer occur synchronously, but, rather, asynchronous patterns may develop in which the cells corresponding to a given connection are randomly interdispersed with the cells corresponding to other connections. In such a scheme, the transmission time slots occur regularly, or synchronously, but the payloads or cells associated with a particular connection occur asynchronously. In the above example, we have used parameters consistent with ATM, the emerging transport technology for broadband ISDN: 53-byte cells, each containing 48 bytes of "payload" plus a 5-byte cell header. Because cells corresponding to different connections occur asynchronously in time, and since no time slot scheduler is present, a terminating switch *must* contain smoothing buffers to temporarily store cells that simultaneously arrive on two or more inbound transmission links which are destined for the same outbound links, since only one cell can be present on any outbound link at any given time. Such competing cells are sequentially read onto the time slots of the outbound links.

Interest in asynchronous time multiplexing arises by virtue of the great variety of traffic types anticipated in the integrated broadband environment. With synchronous time multiplexing, each time slot of the switching frame supports a fixed data rate connection between two users, one transmitting and one receiving. This approach served well in an era when most telecommunication traffic was telephone-generated voice or modem-produced voice-grade data. Telephone traffic exhibits three unique characteristics. First, each telephone is involved in one and only one conversation at a time. Most telephone conversations involve two participants, and even when multiple participants are involved in a conference call, there is only one "conversation" in progress; a single telephone will not permit an individual to "time share" himself or herself among two or more conversations involving different groups of participants (such "time sharing" would, if possible, be quite unnatural for must of us). In somewhat more technical terms, the telephone does not support multiple simultaneous connections, a limitation completely compatible with natural human dialogue. Second, telephone traffic is *persistent*, that is, the typical conversation does not involve long idle periods punctuated by an occasional outburst of speech. Rather, one or more parties to a telephone conversation are usually speaking. Third, from a cold, technical perspective, all voice conversations look the same: continuous 64 kbit/sec information streams for digital voice (the telephone does not distinguish a beautiful, inspirational poem from the sound of a jackhammer breaking through the pavement!). Today's telecommunication infrastructure has evolved to reflect these unique properties of voice traffic: synchronous time multiplexing provides a persistent standard data rate connection between two end points. Where possible, data traffic (which has distinctly different characteristics from voice) is made to look like voice by means of a conditioning data modem. The volume of data handled in this fashion is sufficiently low that it can readily be carried over a network initially intended to support voice, although the resulting cost to the data user is unnecessarily high for the relatively low amount of bursty traffic carried over the network (data bursts are nonpersistent, but the network connection is nonetheless maintained even during long idle periods).

By contrast, the variety of traffic types to be supported in the broadband telecommunication environment mandates a different format. Some types of traffic are persistent, others are not. Among the persistent types (voice, full-motion video), the required connection data rates are vastly different (64 kbit/sec for uncompressed voice; greater than 100 Mbit/sec for uncompressed video, although codecs might compress this to 45 Mbit/sec, 1.5 Mbit/sec, or even lower, sacrificing quality and escalating cost along the way). The peak data rates required by nonpersistent types of traffic (typically called bursty traffic with connections characterized by occasional, randomly occurring short periods of activity) span several orders of magnitude. Telemetry traffic (temperature sensors, security monitors, etc.) may require a peak data rate during their active periods ranging from a few bits per second up to several hundred bits per second. Data rates for connections from remote terminals to a time-shared host range from a few hundred bits per second up to several hundred kilobits per second. Facsimile transmission may require anywhere from a few kilobits per second up to the megabit per second range, and interactions between mainframe computers may require peak data rates ranging from several tens of kilobits per second up to as high as 1 Gbit/sec. High-resolution image storage, retrieval, and display systems may require peak rates exceeding one Gbit/sec.

Moreover, in addition to spanning a peak rate range greater than nine orders of magnitude, the sources on nonpersistent traffic are characterized by the need to maintain multiple simultaneous connections from each terminating device. For example, unlike the telephone, a mainframe computer may need to concurrently maintain time-shared dialogues with many remote terminals and several other mainframes with which it is sharing tasks, memory, and data files. The frequency with which such a mainframe may generate time-multiplexed bursts of data (data packet), each intended for a different receiver with which it is in active dialogue, may be quite high (hundreds of kilopackets per second). To accommodate the needs of bursty data sources, two alternatives may be considered. First, a persistent, synchronous connection of the appropriate data rate can be established between each data device and each other device to which it is in active dialogue. Considering the range of data rates involved, and the infrequency of activity for each separate dialogue, this is not a practical solution. The second approach requires network equipment which can read packet headers in real time to effect the correct packet relaying decisions at each of the possibly several intermediate packet switching nodes along the various connection routes.

Carrying the packet switching approach one step further, the actual format of each packet, at the network transport level, can be standardized; this is the basis of ATM. A standard cell size is defined. Its format consists of a fixed-length data field and fixed-length header. The header contains, at a minimum, sufficient information to allow the network to deliver the packet to the correct receiver and for the receiver to identify the source. Typically, an activity bit is also included (or something equivalent) such that the network knows if the time slot contains an active or empty cell. Large data packets are divided or segmented into multiple fixed-size cells, each with the appropriate routing header. As shown in Figure 1.9a, the destination address (or its equivalent in the form of a connection identifier) must be carried in each segment produced from a given packet. Since the large packet undergoing segmentation may not contain an integral number of cells, the last cell produced from the packet may be only partially filled. Similarly, packets smaller than the cell size will only partially fill one cell. All cells generated by one data device will be discharged onto the network through a network access station at a peak data rate sufficiently high to satisfy peak rate needed by the highest speed packet generated anywhere within the network (alternatively, the access speed for all cells will be selected and standardized at a rate sufficiently high to satisfy most packets; those packets requesting a higher peak rate must either "throttle back" or use specifically provisioned facilities, or else not be supportable over the network). Once inside the network, cells from multiple sources are statistically multiplexed onto network-internal transmission facilities and relayed along their correct routes by packet switching nodes.

At the receiving end, a data device accepts the time-multiplexed cells produced by those sources with which it is in active dialogue, and reassembles the packets from which the cells were created. The cell header field is used to deliver each cell to its intended receiver. Since cells corresponding to different sources (and, therefore, different packets) may arrive intermingled in time, the receiver must sort the cells on the basis of the originating sender, as shown in Figure 1.9b. Thus, in general, each receiver must be capable of reassembling several packets

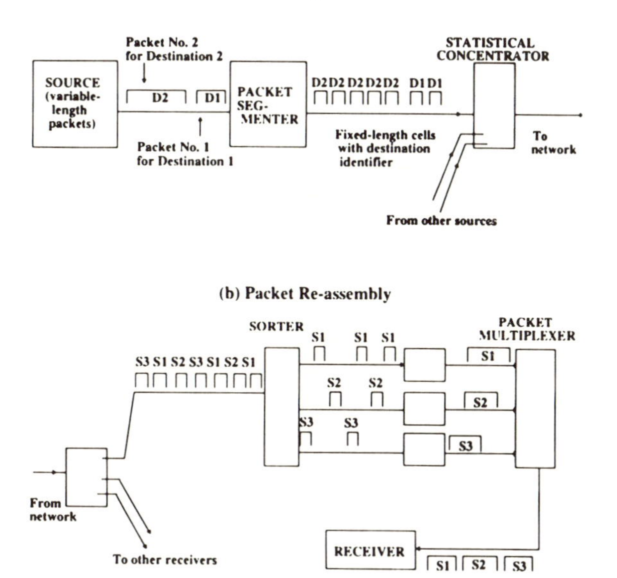

Figure 1.9. An example of packet segmentation and reassembly for Asynchronous Time Multiplexing.

in parallel, adding to one of the packets being reassembled each time a cell corresponding to that packet is received. Clearly, this sorting operation will require that each cell header must contain not only destination information but also source information or other information from which the identity of the cell sender can be deduced by the receiver. Reassembled packets are then time-sequenced for delivery to the the terminating device.

A possible format for the data cell is shown in Figure 1.10. We see that to enable packet segmentation, each cell corresponding to a given packet must contain a destination field, and to permit packet reassembly at the receiver, a source field must also be provided. In addition, the activity bit is useful to the network switches for identifying those time slots containing active cells. The activity bit and source and destination identifiers can be explicitly carried as shown in Figure 1.10 or, as an alternative, a connection identifier can be assigned at call setup time which uniquely identifies the sender and receiver. Each cell created from packets corresponding to a given connection will bear the same connection identifier, which can then be used by the network switching, multiplexing, and demultiplexing equipment for relaying purposes (remember that the connection identifier uniquely identifies the receiver) and by the receiver for reassembly purposes (remember that the connection indentifier uniquely identifies the source). It is this latter strategy of assigning a connection identifier to each active con-

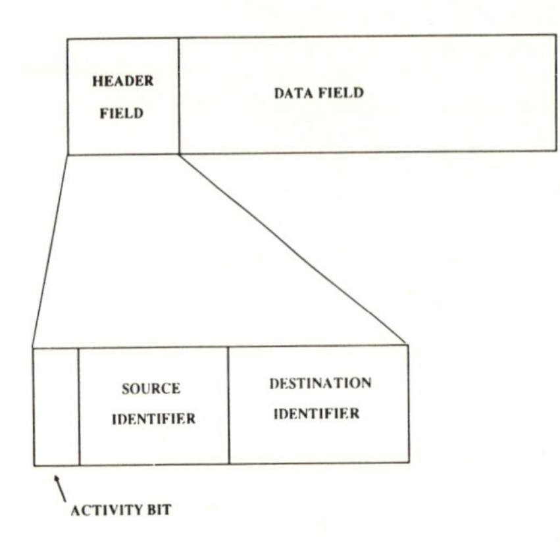

Figure 1.10. A possible format for the information cell or packet. (Note: The header field may not explicitly express source, destination, and activity, but contain other information from which these can be derived in real time using custom VLSI.)

nection which has been adopted for ATM; by using connection identifiers rather than explicit source/destination fields in the cell headers, it becomes possible for the network to support an arbitrarily large number of connections, even though the connection identifier field is of constant length (we shall see later that connection identifiers can be reused in geographically disjoint portions of the network). Also, as we shall see, use of a connection identifier greatly facilitates congestion control in a broadband network through a procedure known as admission control.

Finally, persistent types of traffic can be carried over the asynchronously time-multiplexed network by segmenting the traffic streams produced by the persistent sources into fixed-length cells, each with an appropriate header to permit relaying and reassembly; the segmentation and reassembly process are virtually identical to those needed for variable-length data packets.

Thus, we see that asynchronous time multiplexing of small, self-routing cells provides a "universal" mechanism for integrating and transporting both persistent and nonpersistent traffic, and both low and high peak rates, over a common telecommunication network. The network does not know, or care, about the "payload" of each cell, nor does the network know or care whether a given cell contains voice, video, image, or computer data information. The network cares *only* about the cell header, which contains the route and reassembly instructions.

1.7. A Brief Word about Communication Protocols, Bridges, Routers, Gateways, Datagrams, and Virtual Circuits

Communications protocols are the sets of rules which users have agreed to follow when establishing services and transferring information. In addition to permitting the establishment and management of connections and keeping track of the multitude of connections which may be multiplexed through a common physical network port, protocols are also needed to enable reliable communications in a relatively hostile telecommunication environment characterized by noisy transmission and network congestion. These impairments cause the appearance of random bit errors in delivered data, loss of segments of information from a delivered data packet or stream, and misrouting of information to unintended receivers. To achieve the desired reliability, protocols check for errors in delivered information, search for missing segments, and control the rate of information flow to contain congestion at some acceptably low level.

Unfortunately, not all vendors of information-related equipment have embraced one common agreed-upon set of communication protocols. As a result, the rules followed by one type of terminating equipment may differ from those followed by another. As a result, it is not always possible to establish a connection between any two devices.

A partial remedy for this situation was developed by the International Standards Organization (ISO). A model for communications, known as the Open Systems Interconnect (OSI) reference model, has been developed which describes the functions (but not their implementation) which are provided by communication protocols, and is shown in Figure 1.11. The model contains seven layers, with each layer providing service to the layer above it such that the higher layer can communicate with its "peer" in another termination in a manner which is transparent with regard to the communications processes occurring below it. For example, the *N*th layer in the protocol stack might be responsible for controlling the rate of information flow between the applications running on two terminals. To perform this function, the *N* layers of the two terminals trade real-time information between them concerning the amount of application information which has been sent, the amount received, and the additional amount which can be sent before a pause is needed to enable absorption and processing of the information. The *N*-layer communications assume a certain degree of integrity or quality of service

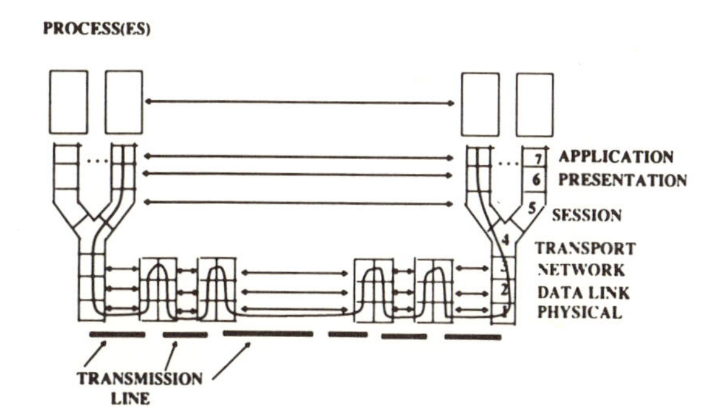

Figure 1.11. A representation of the ISO-OSI seven-layer protocol model.

with regard to the accuracy of the exchanged real-time information. It is the responsibility of layers N-1, N-2, . . . , 2,1 to provide the required accuracy. Moreover, in performing its role, the Nth layer is not interested in data rates on the physical links, voltage levels corresponding to logical zeros and logical ones, or whether the medium is copper, radio, or optical fiber. Again, it is the responsibility of layers N-1, N-2, . . . , 2,1 to work out these details, so that layer N can do its job transparently with regard to that which is transpiring at the lower levels.

As shown in Figure 1.11, the seven-layer OSI model permits multiple applications to be multiplexed over one physical port; such multiplexing may be provided at layer 4 (transport layer). Among other things, it is the responsibility of layer 4 to merge information from different applications, apply headers to enable the applications to be distinguished, and route time-multiplexed arrivals to the correct terminating application. The OSI model admits sublayers for each layer contained in the model.

A much more thorough and complete description of the OSI model can be found in any of several excellent texts.

As is also shown in Figure 1.11, information generated by one transmitting "application" may need to be transported through several subnetworks to reach the intended receiver. These subnetworks may use different physical media, different transmission formats, different routing headers, different flow control strategies, etc. Whenever information leaves one "subnetwork," it may become necessary to "close" its communications protocol to a certain layer in the protocol stack, and to "reopen" the protocol at that layer in the new "subnetwork." For example, if subnetwork A uses a particular type of N-layer flow control protocol, and subnetwork B uses another, then the interface between the two subnetworks must (1) maintain the dialogue with the application on A using protocol A up to and including layer N, while (2) maintaining a dialogue with the application B using protocol B up to and including layer N. It is assumed, here, that the protocols are compatible at layers $(N + 1)$ and higher so that, again, the protocol conversions provided by layers N and below are totally transparent to layers $(N + 1)$ and higher, allowing the $(N + 1)$st and higher layers from the terminating devices to communicate "peer to peer." The layer, N, at which the interface between two subnetworks must "close" the protocol defines the type of interconnecting device required.

A *bridge* is a unit that interconnects two subnetworks that use a common Link Layer Control (LLC) procedure, but may use different Media Access Control (MAC) procedures. The interface between the LLC and the MAC occurs within layer 2 of the OSI model. The LLC sublayer consists of data packets fully formatted with prescribed header fields and data fields for transmission over a network, in need *only* of being placed onto the physical medium. The MAC sublayer is responsible for securing time on a shared transmission facility, and the physical layer (layer 1) is responsible for bit timing, voltage levels, clock recovery, etc. An example of two different MAC procedures, both of which support a common LLC procedure, are the IEEE 802.3 standard for Carrier-Service-Multiple-Access with Collision Detection (CSMA/CD), used on a shared bus LAN, and the IEEE 802.5 standard for a token-based ring LAN; more will be said of these later. Two compatible applications, using identical communication protocols at the LLC sublayer and above, but with one resident on a CSMA/CD bus LAN and the other on a token ring LAN, can communicate if the two LANs are linked by a CSMA/CD-to-token passing bridge. The functions required of a bridge may be sufficiently simple so that real-time implementation in VLSI is possible.

Similarly, a *router* is a unit that interconnects two networks sharing common protocols at the network layer (layer 3) and above, but possibly using different data link layers (layer 2). Thus, for example, the format of the LLCs for the two networks may differ in some portion of the header. The router would then be responsible for closing the protocol at layer 2 on one network and reopening at layer 2 on the other, thereby enabling two compatible applications using a common communication protocol at layer 3 and higher, but resident on networks differing at layer 2 and below, to communicate. A router, being more complicated than a bridge, may require some degree of software processing, thereby presenting more of a bottleneck to the rate of exchange of information or information throughout. Finally, a *gateway* is a unit that interconnects two networks which differ at layer 3 or above. Since the two networks may differ all the way up to and including layer 7, a gateway may be required to close/open the protocol for each network through the application layer (layer 7), a complicated task requiring non-real-time software and often causing significant delay and degraded information throughput.

Among their other virtues, broadband networks seek to avoid some of the issues associated with communication protocols. First, the need for much of the protection provided by communication protocols is reduced since bandwidth is abundant, congestion is less frequent, and low-noise transmission facilities are used. In the nonbroadband environment, these issues are so serious that the same type of protection is often provided via different strategies at different layers of the protocol stack, e.g., error detection may occur at layer 2, guaranteeing a certain (low) probability of undetected error to layer 3, which may have its own error control procedures to further reduce the undetected error probability to layer 4, and so on. By circumventing these impairments, broadband telecommunications may permit very meaningful simplification of communication protocols.

Second, broadband networks are generally defined and standardized below layer 1.5 (the MAC layer arising with OSI layer 2). The broadband network is transparent to the higher-layer protocols which may be in use, and can be used to transparently transport information between two end-point applications which have agreed to use the same protocols at layer 1.5 and above. It is the responsibility of the end-point pairs to agree on use of compatible protocols at layer 1.5 and above, or to use the services of a third end point to provide any necessary protocol conversion at or above layer 1.5; the transport network *per se* is not responsible for bridges; routers, and gateways, although the protocol translation services normally provided by a bridge, router, or gateway module may be offered by connecting an appropriate module to one of the transport network's access ports where the unit would appear to the network like any other terminal. This transparency is achieved by means of an access port adaptation layer, responsible for converting the format of the user-supplied information into the universal format required by the network. There are as many different types of adaptation layers appearing among the network's access ports as there are different types of protocols chosen by the users. More will be said later about the adaptation layer. In this way, each pair of end-users is free to use any protocol at layer 2.0 and above, as long as the needs of these protocols are met by the grade of service provided by the broadband network universal interface. Alternatively, each user may use some standard protocol, and enjoy connectivity with another user which uses a different protocol by requesting a connection via a router or gateway. The router or gateway service might be an offering of the network provider or might be offered by a "fourth party" (the three participants always present are the two users and the network service provider). The participants may all belong to a common enterprise, with network transport provided by a separate enterprise (possibly a public network); the users and router/gateway provider may belong to a common enterprise different from that of the network transport provider (virtual private network); all participants may belong to different enterprises, etc.

One final word about communication protocols. These are often classified by the descriptors "connection-oriented" and "connectionless." Connection-oriented protocols require a call setup procedure, despite the fact that information flows in packet format with header fields containing the routing information. The call setup procedure selects a path or route to be used by all packets associated with a given connection, and the traffic intensity appearing on each physical network link is controlled by limiting the number of connections sharing that link. Path selection statistically spreads the total applied load equitably among all links and packet switching nodes, generally trying to avoid congestion. Since the statistical load to be presented by each requested connection is known by the call processor, the path can be chosen to avoid unacceptable degradation to the quality of service previously guaranteed to earlier connections; if a suitable path cannot be found, the connection can be blocked (similar to a "busy signal" in today's voice network). If a new connection is admitted, a "virtual connection" number is assigned to that connection, and appears in the header field of all packets belonging to that connection. The virtual connection number implicitly identifies both the source and destination for each packet upon call establishment. Each switch along the selected path is informed of the assigned virtual connection number, and is provided with routing instructions to be followed whenever a packet containing that virtual connection number arrives. The connection is termed "virtual" because, unlike synchronous time multiplexing, no network resources (e.g., time slots) are reserved for the exclusive use of that connection. Rather, the user perceives the appearance of a connection, along with many of the services expected from a connection (e.g., packets arrive at the receiver in the sequence generated), but the network resources are, in reality, statistically shared among some multitude of connections. The connection is, therefore, "virtual." Among the services guaranteed by a real connection is bounded latency or delay in the network; a virtual channel cannot guarantee bounded latency since the network load is statistical but, through a combination of admission control, flow control, congestion control, and virtual connection prioritization, it can provide guarantees such as "no more than X% of packets will be delivered with a delay in excess of Y milliseconds." The values for X and Y can be selected by the user at call setup time, depending on willingness to pay the presumed greater charges for higher service quality.

The three phases associated with connection-oriented protocols are (1) call establishment, (2) information transfer, and (3) call release. Call establishment and information transfer procedures are illustrated in Figure 1.12a–c; call release procedures are similar to those used for call establishment and are not illustrated. Shown in Figure 1.12 are generic "access ports" (open circles), switching nodes (closed circles), and interconnecting transmission links. Each access port has a permanent virtual channel number assigned for communication with the call processor, which is attached to the transport network through an ordinary access port and appears to the transport network as any other user or application; call processing may, in fact, be the shared responsibility of processors attached to geographically distributed access ports but, for simplicity, we shall assume the existence of one centralized processor when explaining virtual connection procedures. Using this permanent virtual connection, a user may request a connection to some desired destination; the requested connection might be two-way to enable full duplex operation. The call processor, using a permanent virtual connection from itself to the destination, asks the destination if it wishes to accept the connection. If so, the call processor will attempt to find a path which, when loaded with an additional virtual connection, will not cause the quality of service enjoyed by other connections to fall below the guaranteed minimum level; if the destination refuses to accept the connection, or if a suitable path cannot be found, the connection is blocked and the originator is so informed via a permanent virtual connection from the call processor. If the connection can be established, all switching nodes along the selected path are informed by the call processor of the new virtual connection number, and are provided with appropriate routing instructions; permanent virtual connections from the call processor to the switching nodes are used for this purpose. During the information transfer phase, the call processor drops out of the picture and the enabled access ports exchange informa-

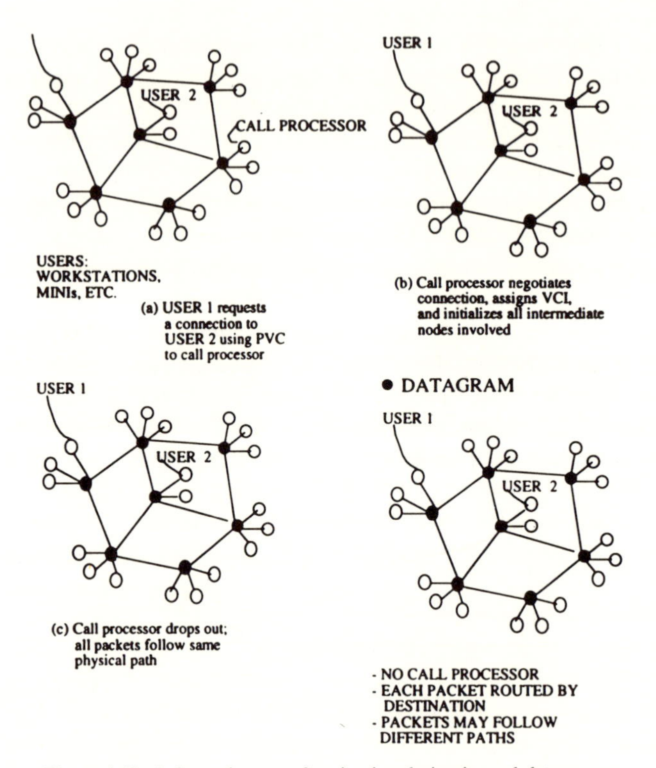

Figure 1.12. Information transfer via virtual circuits and datagrams.

tion over the assigned path (or paths, for for a duplex connection). Each packet contains the virtual connection number in its header, and only the header is processed (in real time, with custom VLSI circuitry) by the switching nodes to effect routing decisions. Since all packets affiliated with a given virtual connection follow the same route, they are delivered in the same sequence in which they were generated. A process similar to call initiation is used when either access port wishes to disconnect; again, permanent virtual connections are used to signal between the access ports and call processor, and between the call processor and the switching nodes.

Connectionless protocols are somewhat more basic. No call setup procedure or call processor is involved, each packet is treated by the transport network as an independent entity, and each packet contains source and destination information within its header. The destination field is used by each encountered switching node to effect a routing decision. However, since more than one switch output port may permit the packet to ultimately arrive at its correct destination, the decision of the switch might be influenced by local congestion conditions. For example, if the switch has routed more packets (which may be associated with different access ports) to one of two outputs acceptable for some given packet, then the switch may choose to select the relatively underutilized output for that given packet. In this fashion, packets will eventually arrive at their intended destinations, but the delivery sequence may be different from that with which the packets were generated by the source since, for a given source–destination pair, not all packets will traverse the same path through the network. Also, since there are no call setup procedures, admission control cannot be used and alternative techniques may be needed to guarantee quality of service.

Transport of information in broadband networks tends to be via connection-oriented protocols since a wide variety of traffic types are to be integrated and carried. Since quality-of-service needs may be different for different traffic types, traffic control is more difficult than for homogeneous data-only types of networks where the simplicity of connectionless protocols can be exploited. In the broadband domain, all traffic control techniques, including admission control at call setup time, must be invoked to permit traffic integration since, for example, the resources needed by a bandwidth-intensive image or full-motion video dialogue between two end points might adversely affect the quality of service expected by hundreds of lower-speed data users. A network manager is needed to ensure fairness in the use of network resources, given the enormous differences among the types of traffic to be integrated onto the broadband network.

The fact that the transport network employs connection-oriented protocols does *not* imply that specific end-users cannot perceive a connectionless service. For example, many connectionless LANs can be interconnected by means of a connection-oriented broadband network. To provide such a service, the LAN access ports to the broadband network might be fully interconnected over the broadband network by means of full-duplex permanent virtual connections. Then, a sequence of time-multiplexed connectionless packets (datagrams) generated by one LAN, each intended for a different receiving LAN, would arrive at the input to the broadband network which serves that LAN. Here, the packets would be demultiplexed and routed to their correct receivers by means of the permanent virtual connections among the LAN access ports. In this way, users attached to connectionless LANs could send datagrams to users attached to remote LANs without first invoking a call setup procedure.

1.8. Contention in Telecommunication Networks

Stripped to its barest essentials, a telecommunication network can be viewed as a multiport system responsible for (1) transmission of information over remote distances, (2) routing of information applied at the various input ports to the correct output ports, and (3) resolution of any contention that may arise when two or more information processes simultaneously seek to use some common facility.

Application of information to a telecommunication network may be viewed as a sequence of random, uncoordinated events. An attempt to initiate a single telephone connection between two users, say, in New York and Los Angeles typically occurs without the coordination or the cooperation of two other users who are also attempting a telephone connection between, say, Philadelphia and San Diego, although both connections may seek to use a common switch in Omaha. Thus, contention may arise at the Omaha switch. Similarly, two computers may simultaneously seek to send a datagram to the same destination computer, in which case contention arises over the use of the network output port serving that destination computer.

These two examples serve to illustrate the two types of contention which may arise in telecommunications. For the first, some multitude of requests may contend for a common or intersecting set of network resources (e.g., transmission, routing, storage facilities). Such contention affects the quality of service seen by the end-user, relative to that which would be experienced if adequate resources were always available to satisfy all requests, since some requested connections may not be satisfied or information delivery may be delayed. The effects of this type of contention on the quality of service experienced by the network user can be controlled by network design: deployment of additional network resources to satisfy the anticipated traffic load will provide quality of service equal to the design objective. However, economic considerations mitigate against deployment of resources sufficient to meet every possible pattern of requests. A network designed, for example, to handle the demand corresponding to *every* telephone simultaneously requesting a connection to *any other* telephone would require a prohibitively large quantity of telecommunication equipment, most of which would sit idle during normal, less stressing conditions. Conversely, a network designed to handle an "average" traffic load might perform poorly when slightly stressed. Thus, the amount of equipment deployed is chosen to provide some desired quality of service under greater-than-expected, but not worst-case, traffic loading conditions.

The second type of contention is more fundamental. Here, multiple information sources simultaneously try to access a common network output port, and quality of service cannot be improved by network design alone (additional telecommunication equipment will not prevent a busy signal when a parent tries to dial home and one family member is a talkative teenager). A goal of network design, then, is to achieve a quality of service no poorer (or only slightly poorer) than that produced by unavoidable "output contention." The network should be designed such that the effects of network-internal congestion are small compared with the effects of output contention, and perceived quality of service should be dominated by output contention.

Although output contention is unavoidable, there are some limited steps that the network might take to offset its effects. Using the "talkative teenager" example once again, the network might, as a service, alert the teenager that a second connection is being attempted; should the teenager wish to respond, the first connection is placed on hold, and the teenager then decides whether or not to accept the parent's call. As a second example, the network can be instructed to deliver contending packets to a common receiver in a sequence which respects the priorities of the contending packets for real-time delivery. While not avoiding output contention, a combination of network design and enhanced services can often soften its effect on service quality.

While network-internal contention can be controlled by design and output port contention is unavoidable, neither type of contention can be ignored, and the network must resolve contention however it may arise. There are only two ways of resolving contention in a telecommunication network. The first technique involves reservation: at the time that a connection is requested, the network "manager" determines whether or not network resources exist to be assigned for the exclusive use of that connection, and whether or not the output port can accept the connection. For example, the network manager may seek a circuit on the various synchronously time-multiplexed transmission systems separating the input and output ports (both directions for a full-duplex connection). Among the many paths that the connection might take, the network manager needs to find only one with the required resources available on each link. If the resources are available, and the output port can accept the connection, then the network manager will make an allocation of the identified network resources for the *exclusive* use of the new connection for the entire duration of the connection. Since resources have been allocated on an exclusive basis, no subsequent requests can interfere with the newly established connection. If the output port cannot accept the connection, or if available network resources cannot be allocated for the exclusive use of the requested connection, then the requesting party will be informed that the connection cannot be made (in telephony, this would correspond to a busy signal). This first approach, in which network resources are *reserved* for the exclusive use of the new connection, is referred to as circuit switching, and quality of service is typically defined by the blocking probability, i.e., the likelihood that a newly requested connection cannot be made.

For a second approach, the network does not seek to allocate resources on an exclusive basis. Rather, the network accepts all applied input signals, temporarily storing or delaying their delivery until the resources required to do so become available, or until the output port becomes free. This second approach is dependent on having sufficient storage capability within the network to "smooth" over the statistical variations in arrivals from the various input ports. Here, the network has been designed with adequate resources such that under stressful (but again, not worst case) traffic loading, the typical delay suffered by applied messages is acceptably small most of the time, and the fraction of messages "lost" by the network whenever the information to be stored exceeds the storage capacity is acceptably low. This type of contention resolution is known as packet switching, and the quality of service is typically defined by several parameters: average message delay, fraction of messages with delay greater than some extreme limit of acceptability, and probability of message loss. For connectionless (datagram) packet switching service, network scheduling such as is needed for circuit switching is avoided since all applied traffic is accepted, and quality of service is maintained through a combination of network design and internal management of congestion (i.e., the network manager, sensing that the instantaneous demand for an internal link is too high, may begin to alternately route some messages to their respective destinations over different paths). For connection-oriented (virtual connection) packet switching service, a scheduler is reintroduced, not to reserve resources on an exclusive-use basis but, rather, to determine whether the additional traffic represented by some newly requested connection will statistically overload the network, thereby causing a degradation in the quality of service enjoyed by previously established connections. In a sense, within a virtual connection-oriented network, the network resources and the right to be connected to the desired output port are reserved on a statistical rather than on an exclusive-use basis: the network transmission links, switching nodes, and smoothing buffers are statistically shared among all admitted connections on an instantaneous, as-needed packet-by-packet basis.

1.9. Problems

P1.1. A new transmission format uses a frame length of 1 msec which is subdivided into 100 time slots. Each time slot contains 100 bits. (a) What is the data rate of the transmission line? (b) If each time slot corresponds to one fixed-rate channel, what is the data rate of each such channel? (c) How many time slots must be assigned to a connection requiring a data rate of 2 Mbits/sec?

P1.2. A set of black-and-white photographs is to be stored and delivered electronically. Each raster scan contains 100 horizontal lines, and each line is sampled 1000 times. Each sample is quantized to one of 256 gray levels. (a) How many bits of information are contained in each uncompressed image? (b) Assuming a browsing rate of 2 images/sec, a transmission rate of 1 Mbit/sec, and negligible time for image compression and expansion, by what factor must each image be compressed?

P1.3. The packet format for some particular data network contains a connection number field of length equal to 20 bits. (a) Assuming that each packet is to be delivered to a single receiver, to how many distinct destinations can each transmitter address its packets? (b) Repeat part a, assuming that ten connection numbers are reserved for multicasting, and each multicast connection can involve, at most, 100 receivers.

P1.4. A Time Slot Interchange (TSI) stores the information contained in each transmission frame arriving at its input such that an arbitrary permutation of the frame's time slots can be produced at the TSI's output. For example, if the frame contains three time slots sequentially numbered 1,2,3, then, at the output of the TSI, it is possible to reorder the information contained within the time slots such that the output time slots contain, in time sequence, information from input time slots 3, 2, 1. TSIs can be used to resolve output port contention in time-multiplexed circuit switches. Consider a 3-input, 3-output switch. Each of the three input transmission frames contains 10 time slots. (a) Suppose the time slots for inputs 1, 2, and 3 are sequentially addressed to switch outputs as follows:

Input 1: 1, 2, 1, 1, 3, 3, 2, 3, 2, 2
Input 2: 2, 2, 3, 3, 2, 1, 1, 2, 3, 3
Input 3: 1, 2, 1, 1, 3, 3, 1, 1, 2, 3

Note that conflicts arise whenever two or more concurrent input time slots are to be switched to the same output. By means of three TSIs, resequence each of the inputs such that all conflicts are eliminated. (b) Next, consider the following three input sequences:

Input 1: 2, 3, 3, 2, 2, 3, 2, 1, 1, 2
Input 2: 3, 3, 2, 2, 3, 1, 1, 2, 2, 2
Input 3: 1, 3, 2, 2, 3, 1, 2, 3, 3, 2

Can these be resequenced to eliminate conflicts? If not, why not?

P1.5. A data source produces packets, each with an average length of 1 kbyte. The source is active, on average, 10% of the time and, when active, delivers information at a peak rate of 10 Mbit/sec. (a) What is the average time duration of each packet? (b) What is the average data rate of the source? (c) How many sources can be statistically multiplexed onto a 5 Mbit/sec transmission link without causing the average traffic demand on that link to exceed its capacity?

P1.6. A Metropolitan Area Network is used to interconnect several Local Area Networks. Traffic appearing on each LAN consists of a data packet containing 2 kbyte every 10 msec, and 20% of each LAN's locally generated traffic terminates at a remote LAN. Traffic patterns are uniform. Find the minimum data rate needed on the MAN such that 20 such LANs can be interconnected.

P1.7. A statistical concentrator time-multiplexes packets arriving from three sources onto a common output link. At the concentrator, each source terminates at its own dedicated buffer. The three buffers are cyclically served in the sequence 1–2–3, and a buffer is skipped over if and only if it is empty. All packets are of fixed length, and packet arrivals from all sources have time-aligned boundaries. Packet arrivals are as shown in the following table.

Packet present on:

Time slot No.	Input 1	Input 2	Input 3
1	yes	no	yes
2	no	yes	yes

(cont.)

Packet present on:

Time slot No.	Input 1	Input 2	Input 3
3	no	no	yes
4	no	yes	no
5	no	no	no
6	yes	no	no
7	no	no	no
8	no	no	no
9	yes	yes	no
10	no	no	no
11	no	no	no

Draw a time sequence showing the packets appearing on the output link, with each packet labeled by its source number. In the first time slot, the cyclical server is positioned to accept a packet, if present, from source number 1.

P1.8. A retransmission protocol requires the receiver to acknowledge each arriving packet determined to be error free. The link bit error rate is 10_{-6}, each packet contains 1000 bits (including the error-detecting code bits), and the error-detecting code can detect all possible bit error patterns which may occur among the 1000 bits. In addition, the probability that an applied packet is lost by the network (e.g., buffer overflow) is 10_{-4}. Any packet not acknowledged within three transmission attempts is abandoned. Find the probability of abandonment.

P1.9. A data communication protocol uses window flow control: each received packet is acknowledged and, for each packet source, at most five unacknowledged packets can be outstanding. The round-trip network propagation delay (elapsed time from packet generation to receipt of its acknowledgment) is 40 msec, each packet of length 100 bytes, and the transmission data rate is 56 kbit/sec. Each source immediately generates a new packet upon receipt of an acknowledgment. (a) Draw a timing diagram illustrating the time sequence of events (packet generation and acknowledgment) corresponding to the first 20 packets generated by a given source. (b) What is the average source throughput, that is, the average data rate in bits per second at which information is supplied by a given source? (c) What is the effective utilization efficiency of the transmission link? (d) Repeat c, but for a window size of 100 packets.

P1.10. Repeat Problem 1.9a–d except that, now, let the transmission data rate be 800 Mbit/sec. (e) What is the window size needed to produce a utilization efficiency of 100%? (f) If forward-link congestion should develop at the receiver under the conditions of part (e), how much new information will enter the network before the interruption in acknowledgments is detected by the transmitter? (g) What conclusion do you draw concerning the usefulness of window flow control for broadband networks?

P1.11. Consider an integrated services network in which voice and data connections share a common transmission link of capacity 45 Mbit/sec. Each voice connection requires a dedicated capacity of 64 kbit/sec and each data connection generates fixed-length 1-kbyte packets at an average rate of 100 packets/sec. To achieve its quality-of-service objective, each data connection requires an "equivalent" capacity which is three times as high as its average data rate. Write an inequality defining the transmission link capacity region, that is, defining the set of permissible combinations of voice and data connections which may simultaneously share the transmission link.

P1.12. In some hypothetical integrated bandwidth-on-demand environment, each voice connection requires a capacity of 50 kbit/sec, each video connection requires 5 Mbit/sec, and each data connection produces a 100-msec burst of information at a peak data rate of 10 Mbit/sec every second. Plot the peak bandwidth needed between t = 0 and t = 10 seconds for an "information outlet" which supports the following traffic profile:

Time (sec)	# voice connections	# video connections	# data connections
0	0	0	0
1	1	0	0
2	2	0	0
3	2	1	0
4	2	1	1
5	2	1	1
6	2	2	1
7	1	2	1
8	1	1	1
9	0	0	1
10	0	0	0

P1.13. A high-speed network containing 500,000 user access ports has a per-port capacity of 1 Mbit/sec and an aggregate capacity of 100 Gbit/sec. To support the network fault diagnostic routine, each port generates a 1000-bit word every second, receipt of which must be acknowledged by the intended receiver by means of a 1000-bit status reply. Assuming normal operation, what fraction of aggregate network capacity is committed to fault diagnostics?

P1.14. A data link operating at a rate of 56 kbit/sec exhibits a bit error probability of 10_{-6}. Each data packet contains 1 kbyte, including an error-detecting field capable of detecting any pattern of bit errors. Each time a packet is received containing a detected error, the protocol requires that it be retransmitted along with each of the nine prior packets (Go-back-10 retransmission scheme). What is the maximum rate of information transfer across the data link?

P1.15. An entertainment video signal is developed by means of a raster scan containing 500 scan lines for each of three primary colors. The entire scan is repeated every 10 msec, and each scan line is sampled 500 times. Each sample is quantized to one of 256 voltage levels. Assuming a video codec which removes signal redundancy and achieves a compression factor of four, find the data rate required to transmit the resulting digital video signal (ignore the associated audio signals).

P1.16. Consider the video signal of Problem 1.15, except that, now, the codec achieves a compression factor of 100. A video conference call uses such a codec to compress the signal generated at each of 10 sites, and each site is equipped with nine monitors such that it can concurrently view all the remote sites. (a) What is the total traffic demand, in bits per second, generated by the video conference? (b) What is the required receiving capacity per conference site (again, ignore the associated audio signals)?

P1.17. Consider a multiprocessor containing 100 computing elements. When active, each computing element generates information at a peak rate of 100 Mbit/sec. Suppose that synchronous time division-multiplexed circuit switching is used to fully interconnect the processing elements, and that each circuit operates at the per-computing element peak rate of information generation. (a) What is the required data rate of the transmission link serving each computing element? (b) Assuming that each computing element sends to only one other at any given time, what is the maximum link utilization efficiency?

References

1. R. W. Lucky, *Silicon Dreams: Information, Man, and Machine*, St. Martin's Press, New York, 1989.
2. R. L. Geiger, P. E. Allen, and N. R. Stoader, *VLSI Design Techniques for Analog and Digital Circuitry*, McGraw–Hill, New York. 1990.
3. D. A. Hodges and H. G. Jackson, *Analysis and Design of Digital Integrated Circuits*, 2nd ed., McGraw–Hill, New York, 1988.
4. O. Wing, *Gallium Arsenide Digital Circuits*, Kluwer Academic, 1990.
5. A Tanenbaum, *Computer Networks*, 2nd ed., Prentice–Hall, Englewood Cliffs, N.J., 1988.
6. T. Li, Advances in optical fiber communications: An historical perspective, *IEEE J. Selected Areas Commun.* **SAC-1**, April 1983.
7. P. S. Henry, Introduction to lightwave transmission, *IEEE Commun. Mag.* **23**(5), May 1985.
8. S. E. Miller and I. Kaminow, *Optical Fiber Telecommunications II*, Academic Press, New York, 1988.
9. E. Desurvire, Lightwave communications: The fifth generation, *Sci. Am.* Jan. 1992.
10. C. A. Brackett, Dense wavelength division multiplexing networks: Principles and applications, *IEEE J. Selected Areas Commun.* **SAC-8**(6), Aug. 1990.
11. K. Murakami and M. Katoh, Control architecture for next generation communication networks based on distributed databases, *IEEE J. Selected Areas Commun.* **SAC-7**(3), April 1989.
12. J. J. Garrahan, P. A. Russo, K., Kitami, and R. Kung, Intelligent network overview, *IEEE Commun. Mag.* **31**(3), March 1993.
13. S. R. Schach, *Software Engineering*, Aksen Associates.
14. F. Halsall, *Data Communications, Computer Networks, and OSI*, Addison–Wesley, Reading, Mass., 1988.
15. Members of the Technical Staff, Bell Telephone Laboratories, *Transmission Systems for Communications*, Bell Telephone Laboratories, Inc., 1982.
16. *IEEE Trans. Commun.*, special issue on Digital Switching, **COM-27**(7), July 1979.
17. *AT&T Tech. J.*, special issue on the 5ESS™ Switch, **64**(6, Part 2), July–Aug. 1985.

18. A. R. Modarressi and R. A. Skoog, Signaling system No. 7: A tutorial, *IEEE Commun. Mag.* **28**(7), July 1990.
19. J.-Y. Le Boudec, The asynchronous transfer mode: A tutorial, *Comput. Net. ISDN Syst.* **24**, 1992.
20. A. S. Acampora and M. J. Karol, An overview of lightwave packet networks, *IEEE Net. Mag.* **3**(1), Jan. 1989.
21. C. D. Tsao, A local area network architecture overview, *IEEE Commun. Mag.* **22**(8), Aug. 1984.
22. *IEEE Commun. Mag.*, special issue on Metropolitan Area Networks, April 1988.
23. *IEEE Commun. Mag.*, special issue on Multimedia Communications, **30**(5), May 1992.
24. *IEEE J. Selected Areas Commun.*, issue on Medical Communications, **10**(7), Sept. 1992.
25. *IEEE J. Selected Areas Commun.*, issue on Multimedia Communications, **8**(3), April 1990.
26. *IEEE J. Selected Areas Commun.*, issues on Packet Speech and Video **7**(5), June 1989.
27. J. Walrand, *Communication Networks: A First Course*, Aksen Associates, 1991.
28. W. Stallings, *Data and Computer Communications*, 2nd ed., Macmillan Co., New York, 1988.
29. M. DePrycker, ATM switching on demand, *IEEE Net. Mag.* **6**(2), March 1992.
30. *Switching Systems*, American Telephone and Telegraph Co., 1961.
31. J. Auerbach, TACT: A protocol conversion toolkit, *IEEE J. Selected Areas Commun.* **SAG-8**(1), Jan. 1990.
32. E. Ball, N. Linge, P. Kammar, and R. Tasker, Local area network bridges, *Comput. Commun.* **11**(3), June 1988.
33. C. A. Sunshine, Network interconnection and gateways, *IEEE J. Selected Areas Commun.* **SAC-8**(1), Jan. 1990.
34. C. V. Bachman and P. Mondain-Monval, Design principles for communication gateways, *IEEE J. Selected Areas Commun.* **SAC-8**(1), Jan. 1990.
35. L. Roberts, The evolution of packet switching, *Proc. IEEE* **6**(11), 1978.
36. 36. J. D. Day and H. Zimmerman, The OSI reference model, *Proc. IEEE* **71**(12), Dec. 1983.
37. W. F. Emmons and A. S. Chandler, OSI session layer: Services and protocols, *Proc. IEEE* **71**(12), Dec. 1983.
38. L. L. Hollis, OSI presentation layer activities, *Proc. IEEE* **71**(12), Dec. 1983.
39. J. C. McDonald, ed., *Fundamentals of Digital Switching*, Plenum Press, New York, 1983.
40. M. M. Buchner, Jr., M. Iwama, T. J. Herr, R. E. Staekler, and L. J. Gitten, Evolution to a universal information services environment, ISSLS '86, Tokyo.
41. M. Decina, Evolution toward digital wideband communication networks, 1987 Int. Switching Symp., Phoenix.
42. P. E. White, The role of the broadband integrated services digital network, *IEEE Commun. Mag.* **29**(3), March 1991.

2

Review of Local Area Networks

2.1. Functional Elements

Many of the functional components, attributes, and operating principles of broadband networks are also exhibited by premises-based Local Area Networks (LANs) Among these are the notions of a shared high-speed resource, real-time packet switching via custom VLSI, "universal" access ports, and multiple simultaneous connections supported through each port. However, LANs do not have the enormous capacity of a broadband network and are traditionally intended to support homogeneous bursty data traffic rather than heterogeneous multimedia traffic.

As shown in Figure 2.1, functional components of a LAN are:

1. A shared transmission medium—All messages transferred among the interconnected users are transported on some time-interleaved basis over a common transmission medium. The protocol used to detect and mediate contention among users for message transport time on the medium is known as the medium access protocol.
2. The network access stations—These are logically dispersed over the transmission medium (physically, the access stations may be geographically distributed or colocated; more about this later), and each presents a "universal" port through which a user presents and accepts information. The speed of the universal port may be the same as, or may be different from, that of the medium. An access station is responsible for implementing the medium access protocol, placing information onto the medium, inspecting the header field of each packet transported over the medium to select those intended for local reception, and implementing lower-level error and flow control protocols. The access stations also contain the buffers to store packets awaiting transmission onto the medium while any contention is being resolved, and to store information selected from the medium prior to delivery to the output port (if the data rate of the shared medium exceeds that of the output port.)
3. Network controller—In a connection-oriented packet-switched network, a

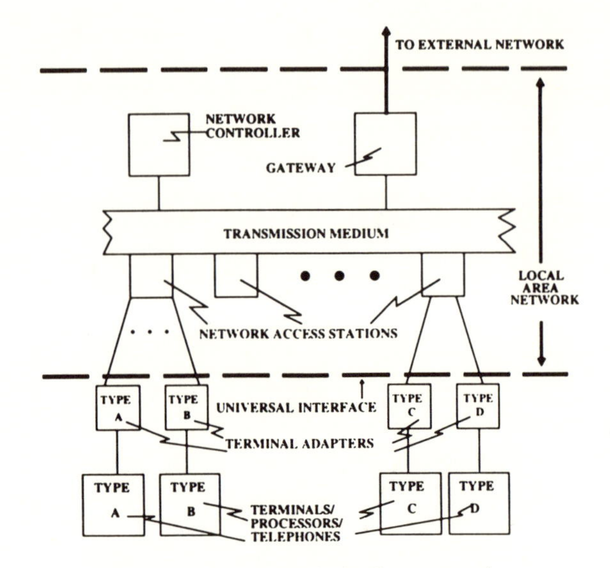

Figure 2.1. Components of a local area network.

network controller is always needed to perform, at a minimum, such functions as admission control and call processing. It is an optional component of a connectionless network. The network controller, itself, might be a centralized unit which connects through an access station and communicates with other access stations over the shared medium. Alternatively, the network controller may itself be a distributed processor, the elements of which communicate over the LAN via their own access stations. In the limit, the network control responsibility might be distributed among processors residing within the access stations themselves.

4. A gateway—This is needed to allow users connected to the LAN to communicate with external users.

Not part of the LAN, but necessary to enable devices to connect to the LAN, are several types of Terminal Interface Units of Terminal Adapters. These are needed to transform the information signals presented by a particular type of device into the universal format acceptable to the access station, and to transform signals arriving from the network back into a format acceptable to the terminating device. There are as many Terminal Adapters as there are distinct types of devices to be connected over the LAN. One type of Terminal Adapter might be responsible for connecting slow asynchronous data terminals, another might provide high-speed multiplexed access for a host computer. An extreme type of Terminal

Adapter might be responsible for connecting an analog telephone, in which case the Terminal Adapter would need A/D and D/A converters, packet assemblers, disassemblers, and clock recover circuitry.

The data rate of the LAN shared medium might be as low as 100 kbit/sec, or as high as 100 Mbit/sec, and the geographical service region may be characterized by a linear dimension somewhere between a few meters and a few kilometers, appropriate for private, single-owner, premises-based data networking within a building or throughout a small campus. Typical environments might include offices, universities, factories, laboratories, and hospitals.

2.2. Functional and Physical Topologies

A telecommunication network has both physical and functional (or logical) characteristics. The functional characteristics are defined by the communication mechanisms among the network elements. The physical characteristics are defined by geographical locations of transmission, switching, processing, and storage elements.

A LAN, being a subset of all possible telecommunication networks, generally assumes one of only two types of functional topologies: bus or ring. A functional bus, shown in Figure 2.2, contains a broadcast medium to which each access station is passively attached. Packets placed onto the medium by a particular access station can propagate either (1) bidirectionally, in which case each access station taps the medium only once, and that tap is responsible for both transmission and reception (writing and reading), or (2) unidirectionally, in which the medium is tapped twice by each access station, once to write messages which propagate toward a head end and once to read the messages as they are returned from the head end. Variations on this basic theme are possible: for a unidirectional system, the head end might regenerate packets and actively separate the transmit from the broadcast bus, or each access station might actively intersect the bus to regenerate signals as they propagate. The important point is that, once placed onto the medium, a packet cannot be removed or intentionally preempted by another access station; the packet is broadcast past the read tap of each access station until

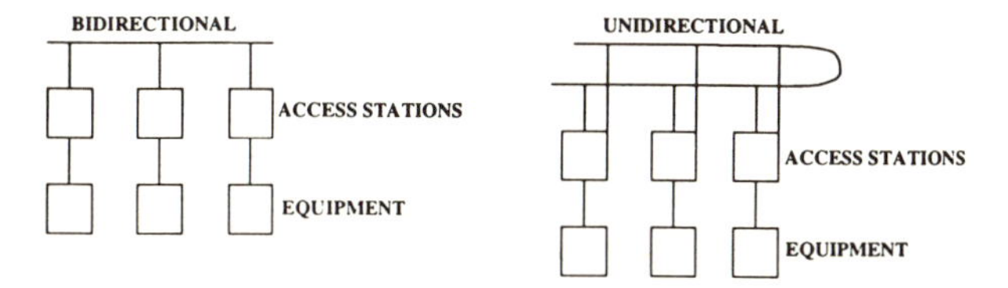

Figure 2.2. A broadcast bus type of functional topology.

it reaches the end of the bus, which is terminated in some dummy load (e.g., a resistive termination).

The access stations of a functional ring, as shown in Figure 2.3, are arranged in a closed loop, with the write tap of each access station feeding the read tap of the next NIU around the circle. These access station pairs are interconnected by unidirectional point-to-point (as opposed to multiple access/broadcast) transmission facilities. Each access station actively regenerates all packets propagating around the ring, and each access station may either relay or preempt an arriving packet, temporarily holding that packet while a locally generated packet is inserted. Each packet written onto the ring must be removed by one of the access stations or else it would recirculate forever. Typically, the packet is removed by the originating rather than the destination access station so that the sender can verify delivery. The packet propagates fully around the ring, and is copied by the intended receiver as it passes that access station. The receiver then modifies a special status field contained somewhere within the packet, thereby allowing the sender to verify that the packet was indeed received.

As shown in Figure 2.4a, a functional bus might take the physical form of a long, linear transmission medium with access stations geographically distributed along its length and with each terminal device locally attached to its respective access station via a short dedicated transmission facility. An alternative physical topology, depicted in Figure 2.4b, might be that of two physically identical rooted trees, one transmit and one broadcast, traversing the same paths (only one such tree appears in Figure 2.4b). Each tree branches out multiple times from the head end; the branches themselves sprout branches, which sprout smaller branches, etc. Each access station is passively connected to a "leaf" of each tree. All signals injected onto the transmit tree propagate unidirectionally toward the head end, successively entering larger and larger branches until they enter the tree trunk at the head end. The head-end trunk of the transmit tree is also the head-end trunk of the broadcast tree, and all signals arriving at the head end are simply reinjected onto the broadcast tree, splitting and flowing down each branch and eventually passing each leaf upon which the read tap of an access station is passively connected.

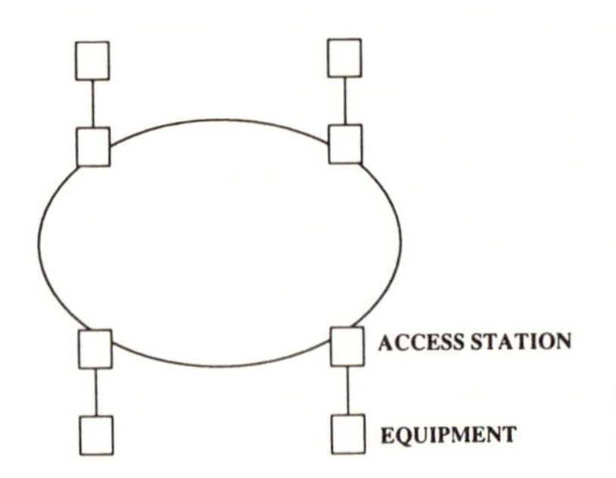

Figure 2.3. A point-to-point ring type of functional topology.

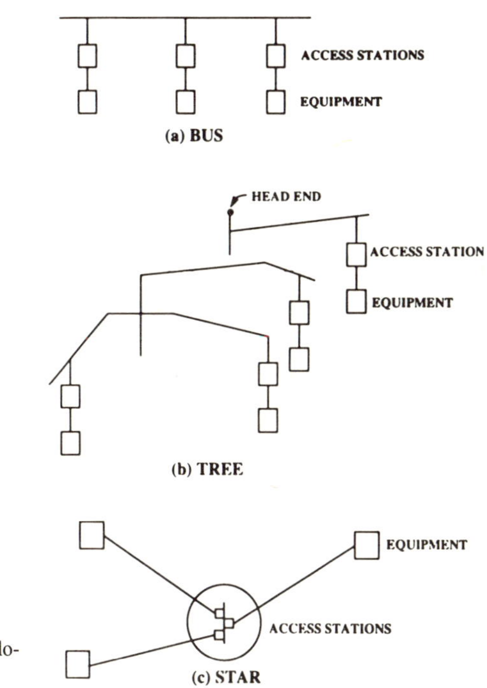

Figure 2.4. Three possible physical topologies for a functional bus LAN.

Finally, as shown in Figure 2.4c, the physical topology of a functional bus might be that of a star. Again, the access stations are distributed along either a bidirectional or two unidirectional transmission media, and are passively attached, but, physically, the length of the medium is short (the broadcast bus might be the backplane wiring of an equipment rack into which the access stations are plugged). Each terminating device is connected to its respective access station via long, *dedicated* (contention free) access links.

A functional ring may take the shape of any physical topology, as long as the point-to-point transmission links connecting the access stations form a closed loop.

2.3. Random Access for a Functional Bus LAN

Among the simplest-to-implement strategies for sharing a functional bus among the various access stations is the so-called random access approach. Here, an access station having a fully formatted packet ready to be placed onto the transmit bus simply writes that packet onto the bus; contention arises when two or more access stations attempt to access the bus such that, after accounting for the nonzero propagation delay along a physically distributed medium, two or more of the packets partially overlap in time at some point along the bus. Since, from that point, both packets are broadcast over the same path, subsequent separation is not possible, and all access station read taps will sense the partially overlapping packets. This medium access approach, originally conceived by researchers at the University of Hawaii, has come to be known as "Aloha." In its simplest (and most common) form, the informational content of any overlapping segments are unintelligible. Overlapping packets are said to have suffered a "collision," and the data formats (including the error detecting codes) are such that loss of any segment of information destroys the entire packet. Thus, any packet having suffered a collision is destroyed. Since the transmission medium is randomly time-shared among access stations, time lost to collisions is unavailable to carry successfully delivered information, and the medium's utilization efficiency will be less than 100%. The throughput of the network, or successfully carried traffic, is the utilization efficiency multiplied by the data rate on the shared medium.

Since the medium is broadcast in nature, each transmitting access station, in addition to selecting broadcasted messages intended for local reception, also listens for its own transmissions. If an access station does not detect one of its own transmissions, neither does the destination access station. The originating access station assumes that a collision has occurred and will reattempt to place that packet onto the medium. However, if all parties to a multiway collision immediately attempt retransmission, there is a strong likelihood that a second collision will occur. To prevent a pattern of successive collision–retry–collision from developing, each retransmitting access station waits some random amount of time before its reattempt. Then, a second collision will occur only if two or more reattempting access stations are unfortunate enough to have chosen the same random time (or, of course, if some other access station, not a party to the original collision, should attempt to access the medium at the same time as one of the reattempts). Each time a collision occurs, the contending access stations enter their backup mode, independently choosing some random point in time at which to reattempt the access. Eventually, each of the contending packets to a collision event will successfully access the medium. For a given packet, the delay from when that packet initially tries to access the medium until the access is successful is a random variable, the statistics of which are dependent on the intensity of traffic loading and the backoff strategy employed. The utilization efficiency is also dependent on traffic loading and backoff strategy. Several different backoff strategies have been proposed and studied, some working better than others over some limited offered traffic intensity regime.

Slotted Aloha is a variant of the basic theme. Here, time on the medium is divided into repetitive fixed-length slots; users can transmit only fixed-length packets of duration equal to one time slot, and the users must synchronize their transmissions to arrive within the boundaries of one time slot.

Under these conditions, it is readily shown that the maximum bus utilization efficiency is 36%, that is, using an optimum strategy for access attempts, at most 36% of the time slots will be successfully accessed by exactly one packet; the remaining 64% of the time slots either remain idle or contain two or more colliding packets. If two or more stations simultaneously attempt to access a given time slot, the information content of all are destroyed and the time slot is effectively wasted. For an *unslotted* Aloha system (access stations do not need to synchronize their transmissions to arrive within the boundaries of fixed-length time slots), the maximum bus utilization efficiency falls to 18%.

The inefficiency of a simply Aloha access strategy, and the availability of custom VLSI circuitry capable of rapidly executing more sophisticated and efficient access algorithms, have resulted in the adoption of different medium access standards for LANs. Accordingly, our further discussion of Aloha is of historic and instructive interest only.

2.4. Throughput Analysis of a Slotted Aloha System*

Consider the system shown in Figure 2.5, in which time is divided into contiguous fixed-length intervals called time slots, each of duration T seconds. Each of N users is connected through its respective station to each of two unidirectional buses, the transit and broadcast buses, which are connected at the head end. Traffic offered by each user consists of fixed-length packets. The length of each packet, in bits, and the data rate of the bus are such that the length of each packet, in time, is *exactly* equal to the duration of one time slot T. Each user knows its propagation delay from the head end of the bus, and whenever it seeks to place a packet onto the transmit bus, it synchronizes its transmission such that, at the head end, the packet falls precisely into one time slot; for no user does a packet overlap two time slots as seen at the head end. These attributes define the slotted Aloha system.

Each access station connects to the transmit bus via a packet buffer which temporarily stores each packet generated by the corresponding user while it awaits placement onto the transmit bus. These buffers also contain packets which must retry to gain access as a result of some prior collision. To avoid repeated collisions among the same set of users, each access station having a packet ready for transmission waits some random time interval before attempting to access the transmit bus. Each medium access controller detects collisions by monitoring the broadcast bus, recognizes successfully received packets intended for its respective receiver, and decides when an access should be attempted.

Let p be the probability that a given user attempts to access some given time slot, that is, p is the probability that a given user has at least one packet ready for transmission and attempts to transmit in that given time slot. Users having at least

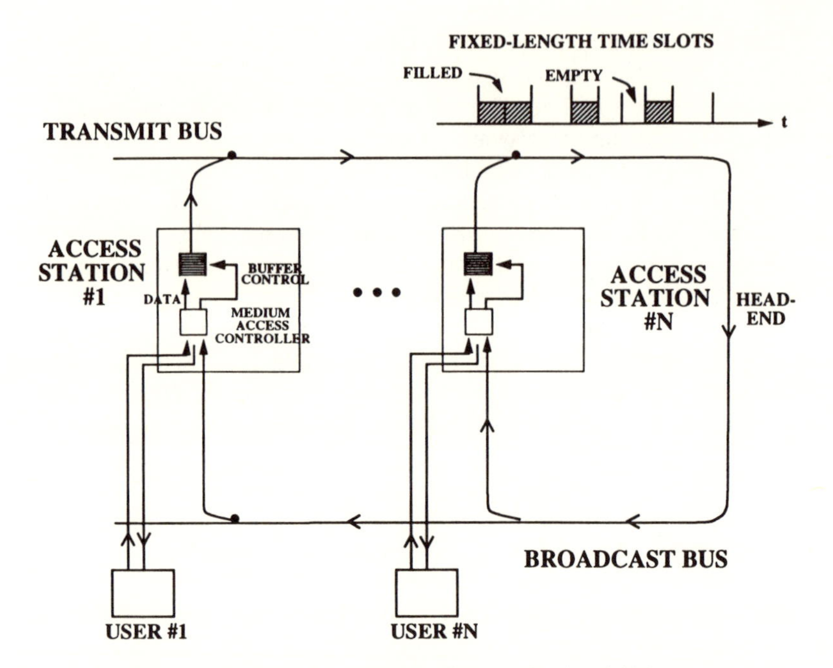

Figure 2.5. Model used to study the efficiency of a slotted Aloha system.

one packet awaiting transmission attempt to access the transmit bus independently of other users. We will assume that the system *overload* conditions apply, that is, the average rate at which packets arrive from the users to the access stations is greater than the rate at which they are successfully placed onto the bus, so that each of the N access stations always has an ample supply of packets awaiting transmission in its buffer. The objective now is to compute the saturation throughput under such overload conditions, that is, the maximum rate at which information can be delivered successfully over the bus. Then, in an operational system, the total offered load must be less than the saturation throughput; if the offered load should exceed the saturation throughput, then information would arrive at a rate greater than that at which it departs, and the number of packets stored in each input buffer (assumed to be infinite in length) would continually grow with time, causing an infinite amount of delay for each new arrival. The number of users attempting to access some given time slot, Q, is therefore a binomially distributed random variable:

$$P(Q = k) = \binom{N}{k} p^k (1 - p)^{N-k} \tag{2.4.1}$$

A successful event is one for which exactly one user attempts to access a given time slot (if no attempt is made, that time slot is lost; if two or more attempts are made, that time slot is again lost to a collision). Thus,

$$P_s = P(\text{success}) = P(Q = 1) = \binom{N}{1} p(1 - p)^{N-1} = Np(1 - p)^{N-1} \quad (2.4.2)$$

The attempt probability, p, which maximizes the success probability is found by differentiating P_s with respect to p:

$$\frac{dP_s}{dp} = -N(N - 1)p(1 - p)^{N-2} + N(1 - p)^{N-1} = 0 \quad (2.4.3)$$

$$\Rightarrow p = 1/N \quad (2.4.4)$$

Thus, the maximum success probability is given by:

$$P_s^{(\text{MAX})} = N(1/N)(1 - 1/N)^{N-1} = (1 - 1/N)^{N-1} \quad (2.4.5)$$

Finally, for large N,

$$\lim_{N\to\infty} P_s^{(\text{MAX})} = \lim_{N\to\infty} (1 - 1/N)^{N-1} = 1/e = 36\% \quad (2.4.6)$$

The result (2.4.6) tells us that, for a slotted Aloha system, at most 36% of the available time slots can be effectively used.

2.5. Throughput Analysis of an Unslotted Aloha System*

For slotted Aloha, a collision can occur if and only if two or more packets arrive in the same time slot at the head end of the bus. Consider now an unslotted Aloha system. Again, users access with fixed-length packets, each of length T, but no attempt is made to time the transmissions so that packets arrive within non-overlapping T-second windows at the head end, as was done for slotted Aloha; although fixed-length packets are again used, the bus is unslotted.

We seek to find the saturation throughput, that is, the maximum rate at which information can successfully flow over the bus assuming that all inputs always have a nonzero supply of packets awaiting transmission.

Let p be the probability that a given user attempts to begin accessing the bus anywhere within an arbitrarily chosen window of width T. Clearly, if two or more packets begin access within the window, some packet overlap will occur. A successful transmission will occur within the T-second window beginning with the start of one packet if that is the only packet beginning to attempt access within the window and no other packet attempted to start access within the preceding T-second window. Access attempts are assumed to be independent.

An arbitrarily chosen T-second test window will see the start of a successful transmission if three conditions are satisfied:

1. One of the users begins access within that window.
2. Once that first user to access begins transmission, no other user transmits within the T-second window needed for the initiating user to complete its transmission.
3. No user attempted to begin access within the T-second window preceding the initiating event within the test window.

These three conditions are independent, with probabilities P_1, P_2, and P_3 given, respectively, by:

$$P_1 = P\,[\{\text{user 1 initiates the successful event}\} \text{ or } \{\text{user 2 initiates the successful event}\} \text{ or} \ldots \text{or } \{\text{user } N \text{ initiates the successful event}\}] \qquad (2.5.1)$$

that is:

$$P_1 = P\,[\bigcup_{i=1}^{N} \{\text{user } i \text{ initiates the successful event}\}] \qquad (2.5.2)$$

$$P_2 = (1 - p)^{N-1} \qquad (2.5.3)$$

$$P_3 = (1 - p)^{N-1} \qquad (2.5.4)$$

The events in the union of (2.5.2) are mutually exclusive, that is, only one user can initiate the successful event. Furthermore,

$$P\,\{\text{user } i \text{ initiates the successful event}\} = p,\; i = 1, 2, \ldots, N \qquad (2.5.5)$$

Since the events in (2.5.2) are mutually exclusive, the probability of their union is equal to the sum of their individual probabilities, and

$$P_1 = \sum_{i=1}^{N} P\,\{\text{user } i \text{ initiates the successful event}\} \qquad (2.5.6)$$

$$= \sum_{i=1}^{N} p = Np \qquad (2.5.7)$$

Finally, the probability that the test window contains the beginning of a successful transmission is given by

$$P = P\,\{\text{success}\} = P_1 P_2 P_3 \qquad (2.5.8)$$

$$= Np(1 - p)^{N-1} (1 - p)^{N-1} = Np(1 - p)^{2N-2} \tag{2.5.9}$$

The optimizing value of p is found from

$$\frac{dP_s}{dp} = -N(2N - 1)\, p(1 - p)^{2N-3} + N(1 - p)^{2N-2} = 0 \tag{2.5.10}$$

$$\Rightarrow p = \tfrac{1}{2}N \tag{2.5.11}$$

$$P_s^{(MAX)} = \tfrac{1}{2}(1 - 1/2N)^{2N-2} \tag{2.5.12}$$

$$\lim_{N\to\infty} P_s^{(MAX)} = \frac{1}{2e} \cong 18\% \tag{2.5.13}$$

Thus, for unslotted Aloha, the probability that a randomly chosen time slot of duration T contains the start of a successful transmission is, at most, 18%, and, on average, at most 18% of the bus time can be effectively used.

2.6. Carrier-Sense-Multiple-Access (CSMA)

The environment originally envisioned for Aloha and slotted Aloha was random access over a broadcast radio channel (terrestrial or satellite). Consider the satellite channel. Here, the round-trip propagation delay (to and from the satellite) is about 0.25 sec, a time interval long compared with the time duration of a typical data packet (e.g., a 1-kbyte data packet transmitted at a data rate of 1 Mbit/sec corresponds to a time duration of 8 msec). Although random access Aloha is certainly simple to implement, its throughput performance is entirely independent of packet duration and propagation delay. Other schemes, nearly as simple to implement, can significantly improve upon the Aloha throughput performance by exploiting the much shorter propagation delay (relative to packet duration) typically found in the LAN environment.

CSMA is one such medium access protocol which has gained prominence for broadcast bus functional topologies. CSMA, when used in conjunction with a procedure called collision detection (to be described), serves as the basis for a widely used LAN standard developed by a committee of the Institute of Electrical and Electronic Engineers (IEEE 802.3 committee). The scheme is equally applicable to both unidirectional and bidirectional buses, and will be illustrated for a unidirectional bus such as shown in Figure 2.3.

Unlike Aloha, CSMA is a random access protocol in which any access station having a packet awaiting transmission first listens to activity on the medium (using its read tap on the broadcast bus). If that user senses any activity, it infers that some other user is currently transmitting a packet, and refrains from accessing the transmit bus. Only if the user senses an idle bus (no "carrier" present) can it attempt to transmit its packet.

Thus, unlike Aloha, which involves absolutely no coordination among accessing users, CSMA involves some loose coordination: transmission of any newly generated packet is deferred to any packet already occupying the medium. The intent is to reduce the frequency of collisions, and thereby improve the bus utilization efficiency.

Ideally, CSMA should completely avoid collisions: before attempting a transmission, the presence or absence of another earlier-arriving packet is first established, and transmission can proceed only if the bus is clear. Hence, it would appear as if no collisions can occur. However, on closer examination, we see that the nonzero propagation delay along the bus can indeed give rise to collisions. Suppose some access station has a newly generated packet to transmit. Following the medium access protocol, that station first listens with its read tap to the broadcast bus. If that bus is found to be idle, the station begins its transmission. However, some other access station may have earlier placed a packet onto the medium, the leading edge of which has not yet propagated past the read tap of the station having the newly generated packet. That second packet is therefore transmitted onto the medium, and may collide with the earlier packet. Notice, however, that once the earlier packet has existed, collision free, on the medium for a length of time greater than the longest propagation delay from any write tap to any read tap (henceforth called the round-trip propagation delay of the bus), no further collision is possible, since the presence of that packet will then be known to all potential contenders; unlike Aloha, whose access stations transmit irrespective of the presence of other packets, the stations of a CSMA system politely defer to an existing transmission. Thus, with CSMA, a newly generated packet is vulnerable to collision only during the time period when its presence is being made known (the round-trip propagation delay); once its presence is known, it has effectively seized the bus and can transmit collision-free for an arbitrarily long period. In fact, the bus utilization efficiency monotonically increases as the ratio of the packet duration to round-trip propagation delay increases, approaching 100% for very large ratios. Conversely, if the packet duration is short compared with the round-trip propagation delay, then carrier sensing is of little value since the information is dated (by the time a packet's presence is felt on the broadcast bus, its existence on the transmit bus may already be over), and, unless care is taken, throughput performance may actually be lower than that of Aloha.

If a collision should occur, again the contents of all packets involved in the collision are destroyed. However, rather than discharging the entire contents of packets involved in a collision onto the medium, each transmitting access station can continue listening to the broadcast bus and, within one round-trip propagation delay, sense the presence of any collision. Transmission can thereby be immediately terminated to prevent further loss of time because of a known collision.

This scheme is known as Collision Detection, and the medium access protocol is known by the acronym CSMA/CD. Upon detection of a collision, all transmitting users immediately terminate their transmission, wait some random period of time (to avoid repeatedly colliding within the same set of access stations), and attempt retransmission.

CSMA/CD would be of very limited utility on a satellite channel because of the long propagation delays involved. However, for a LAN, reasonably high throughput can be achieved. By way of example, consider a bus of length equal to 2000 feet. The propagation speed of an electromagnetic signal in free space is approximately 1 foot/nsec; let us assume a propagation speed of $\frac{1}{2}$ foot/nsec on the medium. The round-trip propagation delay is therefore $\tau = 4000$ feet/($\frac{1}{2}$ foot/nsec) = 8 μsec. At a data rate of 1 MBit/sec, a 1-kbyte packet lasts for $T = 8$ msec. Thus, under these conditions, the ratio of packet size to round-trip propagation delay is 1000 which (as will be shown in Section 2.8) produces a bus utilization efficiency of nearly 100%.

However, for broadband systems, the channel speed is typically greater than 100 MBit/sec, the geographical coverage of a single bus system may range up to several tens of miles, and at 100 Mbit/sec, a hypothetical 100-byte packet lasts for 8 μsec (remember that in broadband networks, all types of information are formatted into short, fixed-length cells). Assuming a bus length of 20,000 feet, the round-trip propagation time is approximately 80 μsec, the ratio of packet length to propagation delay is only 0.1, and the bus utilization efficiency is under 10% (it would actually be better to use unslotted Aloha). Thus, while CSMA/CD is easy to implement and is attractive for LANs, it is not particularly suited for broadband networks spanning metropolitan or wider area service regions.

A plot of the CSMA/CD channel utilization efficiency as a function of the ratio of packet size to round-trip propagation delay ($T/2\tau$) appears in Figure 2.6. We note some slight sensitivity to the number of system users, N and that if $T/2\tau$ is too small, it is better to use Aloha. The curves of Figure 2.6 are derived in the following section.

2.7. Throughput Analysis of Carrier-Sense-Multiple-Access with Collision Detection*

Consider a unidirectional bus as shown in Figure 2.7a. All packets are transmitted toward the head end on a write bus and at the head end, the write bus is connected to a read bus which broadcasts all arriving signals onto the read bus. Each of N network access stations is connected to both the read and write buses.

To determine the throughput of such a bus-type system where the CSMA/CD protocol is used, we assume that both buses are of equal length and that the total round-trip propagation delay from the beginning of the write bus to the end of the

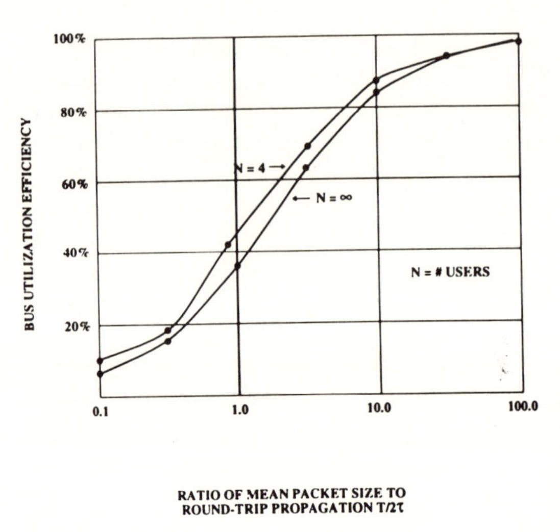

Figure 2.6. Bus utilization efficiency of a CSMA/CD system, plotted as a function of the key determining factor: packet length relative to round-trip propagation delay ($T/2\tau$).

read bus is 2τ sec. To simplify the analysis, we assume that all access stations are located at the end of the dual buses farthest from the head end, i.e., each experiences a round-trip propagation delay for its own transmission of 2τ. Further, we shall assume a time-slotted system in which the slot duration is 2τ, and we shall also assume that all data packets are of the same, but otherwise arbitrary length T sec, where T is totally independent of the slot time 2τ. Finally, since we are interested in computing the maximum amount of traffic that can be carried by the network (saturation throughput), we shall assume that all access stations always have at least one packet awaiting transmission.

Time slots are defined as seen at the read tap of each access station, and the access scheme is as follows. At the beginning of each time slot, each read tap determines (in zero time) whether the beginning of that slot is active (i.e., "carrier present"). If carrier is sensed, then for the duration of that slot, no station will attempt to access the write bus. If carrier is absent, then with probability p, each of the N stations will attempt to access the write bus. If two or more stations attempt to access the same time slot, then one slot later (i.e., 2τ sec later, at the beginning of the next slot), each read tap will instantaneously sense the collision and, if T was greater than 2τ, all parties to the collision will immediately terminate transmission (if T was less than 2τ, the transmissions would have been completed before the collision was detected). If a collision occurred, then each station contends for the next slot on the write bus with probability p. Similarly, if $T < 2\tau$ and a successful transmission was detected, or if no carrier was present, each station contends with probability p. However, if a successful transmission is detected and $T > 2\tau$, then all stations will defer contending until the entire packet

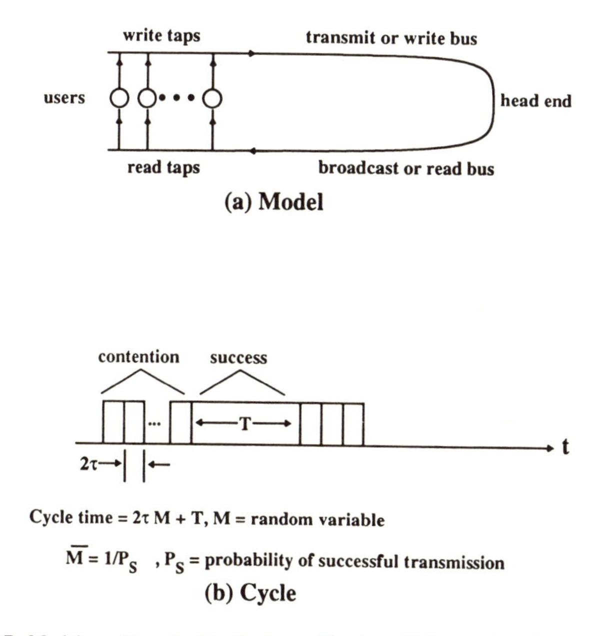

Figure 2.7. Model used to calculate the bus utilization efficiency of a CSMA/CD system.

has been transmitted (i.e., after 2τ sec has elapsed, a successful packet seizes the bus for the duration of its transmission). Since all packets are of duration T (known to all access stations), contention (along with redefinition of time slot boundaries) can resume 2τ sec before the end-of-packet occurs as seen on the read bus.

At the beginning of a time slot that is available for contention (i.e., no ongoing successful transmission is using that slot on the write bus), the probability that a successful transmission is initiated is given by the probability that exactly one out of N stations attempts access:

$$P_s = Np(1 - p)^{N-1} \tag{2.7.1}$$

Let the integer random variable M be the number of time slots which elapse before a successful transmission is initiated (see Figure 2.7b). Since attempts occur independently on successive time slots until a successful outcome occurs, M has a geometric probability distribution:

$$p(M = k) = P_s(1 - P_s)^{k-1}, \qquad k \geq 1 \tag{2.7.2}$$

The total elapsed time F from completion of the last successfully transmitted packet to completion of the next successfully transmitted packet is therefore:

$$\mathcal{T} = 2\tau M + T \tag{2.7.3}$$

The average time taken to successfully transmit a packet, $\langle \mathcal{T} \rangle$, is therefore:

$$\langle \mathcal{T} \rangle = 2\tau \langle M \rangle + T \tag{2.7.4}$$

where

$$\begin{aligned} \langle M \rangle &= \sum_{k=1}^{\infty} kp(M = k) \\ &= P_s \sum_{k=1}^{\infty} k(1 - P_s)^{k-1} \\ &= \frac{P_s}{[1 - (1 - P_s)]^2} = \frac{1}{P_s} \end{aligned} \tag{2.7.5}$$

Thus,

$$\langle \mathcal{T} \rangle = \frac{2\tau}{P_s} + T \tag{2.7.6}$$

Clearly, $\langle \mathcal{T} \rangle$ is minimized when P_s is maximized. From (2.7.1), P_s is maximized when $p = 1/N$, and

$$P_s^{(MAX)} = (1 - 1/N)^{N-1} \tag{2.7.7}$$

$$\Rightarrow \lim_{N \to \infty} P_s^{(MAX)} = 1/e \tag{2.7.8}$$

Finally, we see that for large N, and the optimizing value for p,

$$\langle \mathcal{T} \rangle = 2\tau e + T \tag{2.7.9}$$

The channel efficiency, then, defined as the ratio between the packet duration (the minimum bus occupancy time of a packet in the absence of any collision as measured at the head end) and the average of the actual amount of bus time effectively consumed by a packet (which includes the wasted time arising from collisions prior to successful delivery of a packet) is given by:

$$\eta = \frac{T}{T + 2\tau(e - 1)} = \frac{1}{1 + (e - 1)/a} \tag{2.7.10}$$

where $a = T/2\tau$, the ratio of the packet duration to round-trip propagation time.

From the result (2.7.10), which is plotted in Figure 2.6, we see that the channel utilization efficiency increases toward unit as the ratio of packet time to round-trip propagation delay becomes large. Also, we note that for $T = 2\tau$ and N very large, the efficiency equals $1/e$ (slotted Aloha result), and that for $T < 2\tau$, the efficiency is less than that for slotted Aloha, falling toward zero as $T/2\tau$ becomes very small. The reason for this last effect is that for $T < 2\tau$, after an access attempt, each station must wait 2τ sec to try again, despite the fact that a successful transmission takes a shorter time interval. Clearly, if it is known that the packet duration is T, then each station should contend with probability p every T sec, not every 2τ sec; waiting to sense carrier for 2τ sec when it is known *a priori* that a successful transmission is completed in $T < 2\tau$ sec is equivalent to resolving contention with outdated, irrelevant information. The better strategy is to shift to slotted Aloha whenever $T < 2\tau$. Thus, for $T/2\tau$ small, channel utilization efficiency will not fall below $1/e$.

Note that if the packet length T is a random variable that might assume a value greater than 2τ, then the shift to the slotted Aloha protocol cannot be made since it is not known *a priori* that every time slot is available for contention; those packets for which $T > 2\tau$ and which had successfully contended for the write bus will have "reserved" time into subsequent time slots and all subsequent transmission attempts must be deferred. In such a case, (2.7.10) is the channel utilization efficiency, conditioned on T, and the unconditioned efficiency is given by:

$$\langle\eta\rangle = E\left\{\frac{1}{1 + (e - 1)\, 2\tau/T}\right\} \tag{2.7.11}$$

where the random variable over which the expectation is taken is the packet duration T. For probability distributions yielding an efficiency below that of unslotted Aloha, it would be better to abandon CSMA/CD in favor of the unslotted Aloha protocol.

2.8. Token Rings

An architectural alternative to a functional bus is the token ring. Access to the ring is controlled by a token, some unique pattern of bits which propagates around the ring as a short-length packet, being regenerated by each access station along the way (see Figure 2.3). A user having a packet to transmit must await arrival of the token prior to placing its packet onto the ring. Upon receipt of the token, a user removes the token from the ring, and inserts its packet. After that packet circulates once around the ring (being copied as it passes its receiving station), it eventually arrives back at its source. The generating user then removes the packet from the ring, reinserts the token, and ships this to the next user around the loop. If that user has a packet awaiting transmission, it repeats the above process; if not, it simply relays the token to the next user.

Some simple modifications to this medium access protocol are possible. With the above scheme, only one packet is outstanding on the ring at any given moment. Rather than relaying each transmitted packet all the way around the ring and back to its source for removal, thereby wasting valuable time on the ring, it is possible for the receiving station to remove the packet, generate a new token, and send the new token to the next station along the ring. A potential problem with this approach is "token starving" of some ring access stations. Suppose user K_1 has just received a token, and has a packet to transmit. It removes the token and destroys it, since a new token will be generated by station K_2, the intended receiver for station K_1's packet. When station K_2 receives the packet, it removes that packet from the ring and sends a new token to station $K_2 + 1$. Suppose station $K_2 + 1$ removes the token and sends a packet to station $K_1 - 1$, which removes the packet and delivers a new token to station K_1, which restarts the process. In this fashion, only stations K_1 and $K_2 + 1$ will ever get to see the token.

Another alternative involves having a station which has just transmitted a packet immediately reinserting an idle token, rather than waiting for the packet to travel completely around the loop. At successive stations around the loop, the packet is relayed to the next station, being locally copied at the receiver along the way, and the trailing token is also relayed until it arrives at a station having a packet awaiting transmission. Here, the token is removed, the new packet inserted, and the token immediately reinserted following the new packet. Again, only a source can strip a circulating packet from the ring. This scheme has the virtue of permitting multiple packets to simultaneously exist on the ring, thereby improving utilization efficiency, while avoiding the deadlock of the first alternative scheme described above. This second alternative is the basis of the IEEE 802.5 medium access standard for LANs.

The medium access controller located in each access station is responsible for recognizing the token as it circulates around the ring. Accordingly, each access station introduces a fixed processing delay to the circulating packets as they are regenerated and inspected for the unique combination of bits corresponding to the token. The utilization efficiency of a Token Ring system, defined as the fraction of the ring data rate actually available to carry user traffic, is dependent on both the access station processing delay and the total propagation delay incurred on the transmission links around the ring. If the number of access stations is small, then the effects of propagation delay ultimately limit utilization efficiency (which, like CSMA/CD, diminishes as the ratio of packet length to propagation delay diminishes). However, for a fixed-length ring, the effects of processing delay begins to dominate as the number of access stations increases, with the utilization efficiency actually improving as more stations are added. When the number of access stations is very large, the utilization efficiency is dependent only on the ratio of the packet size relative to the processing delay, decreasing as this ratio decreases. This asymptomatic effect for large N is plotted in Figure 2.8, and the inefficiency mechanism is further described and analytically studied in the following section.

2.9. Throughput Analysis of Token Ring*

Consider a ring containing N regenerative access stations. Each access station presents a processing delay of D_H sec, and the overall loop delay D is given by

$$D = ND_H + D_p \tag{2.9.1}$$

where $D_p = \sum_{i=1}^{N-1} D_{i,\,i+1} + D_{N,1}$ is the total propagation delay on the transmission links connecting the access stations into a closed ring; $D_{i,i+1}$ $(1 \le i \le N - 1)$ is the propagation delay along the transmission link between stations i and $i + 1$, and $D_{N,1}$ is the propagation delay along the transmission link between stations N and 1.

Suppose that each access station always has at least one packet of duration T awaiting transmission whenever the token arrives. Further, suppose that station 1 is in receipt of the token. Referring to Figure 2.9, we see that at time $t = 0$, station 1 begins to transmit a packet (which we shall call packet No. 1) toward station 2, and, at time $t = T$, that packet transmission has been completed. Station 1 then immediately transmits the token to station 2.

At time $t = D_{1,2}$, the leading edge of packet No. 1 arrives at station 2, and at time $t = D_{1,2} + D_H$, packet No. 1 has left station 2, and station 2 begins to transmit

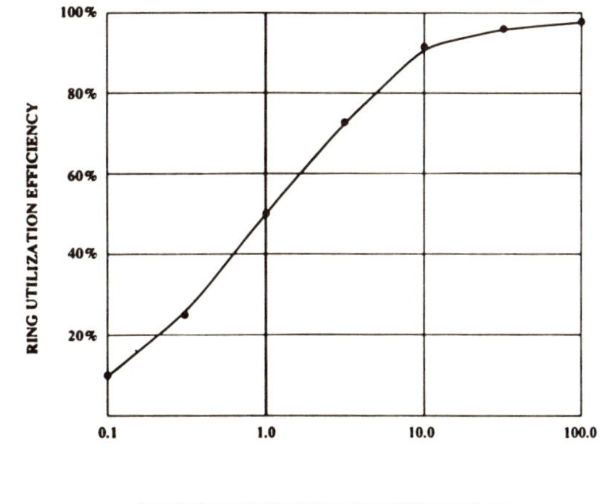

Figure 2.8. Utilization efficiency of a Token Ring system plotted as a function of packet length relative to round-trip delay.

its own packet (which we shall call packet No. 2). The trailing edge of packet No. 2 departs from station 2 at time $2T + D_{1,2} + D_H$, followed by the token.

At time $t = D_{1,2} + D_{2,3} + D_H$, the leading edge of packet No. 1 has arrived at station 3; at time $t = T + D_{1,2} + D_{2,3} + 2D_H$, the trailing edge of packet No. 1 has left station 3; and at time $t = 2T + D_{1,2} + D_{2,3} + 2D_H$, the trailing edge of packet 2 has left station 3. Station 3 then begins to transmit its own packet (which we shall call packet No. 3) toward station 4, and at time $t = 3T + D_{1,2} + D_{2,3} + 2D_H$ the trailing edge of packet 3 has left station 3, followed by the token.

Continuing in this fashion, we see that the leading edge of packet 1 arrives back at the input to station 1 at time $t = D_p + (N - 1)D_H$. At time $t = T + D_p + (N - 1)D_H$, packet 1 has been stripped from the ring and at time $t = T + D_p + ND_H$, the leading edge of packet 2 is leaving station 1 for station 2, where it will be stripped from the ring. At time $t = NT + D_p + ND_H$, the trailing edge of packet N leaves station 1, followed by a newly generated packet from station 1, which we call packet 1a.

We note from the above that the time interval between the leading edges of departure for packets generated by station 1, T_R, is therefore

$$T_R = NT + D_p + ND_H \tag{2.9.2}$$

During this time, a total of N packets were transferred onto the ring (one from each of the N stations). The minimum amount of time needed to transfer N packets onto the ring, corresponding to $D_H = D_p = 0$, is NT sec. Thus, the ring utilization efficiency, N, is given by:

$$\eta = \frac{NT}{NT + ND_H + D_p} \tag{2.9.3}$$

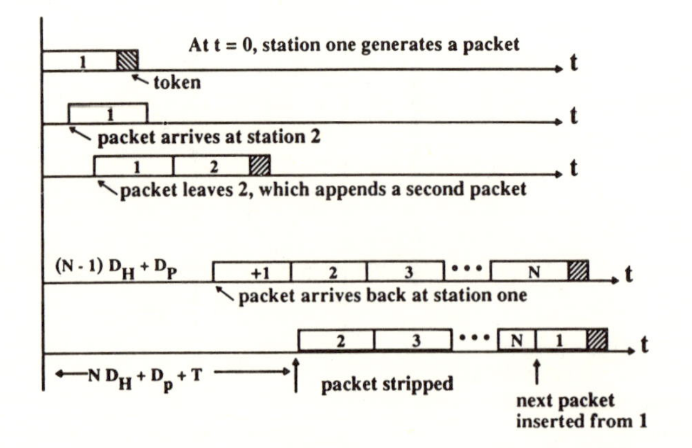

Figure 2.9. Model used to calculate the utilization efficiency of a Token Ring system.

$$= \frac{N}{N + D/T} \tag{2.9.4}$$

where $D = ND_H + D_p$ is the total round-trip delay of the ring. This is the result that is plotted in Figure 2.8. We note that, just as with CSMA/CD, the utilization efficiency for fixed N diminishes as the ratio of packet duration to round-trip delay, T/D, becomes small. However, unlike CSMA/CD, the round-trip delay is dependent on N, the number of access stations, since each regenerative access station introduces processing delay D_H (this can be fairly large, since D_H must be at least sufficient to allow an access station to recognize the token). In fact, the utilization actually improves as N increases, despite the fact that increasing N also increases the round-trip delay as more access stations and processing delay are added. We note from (2.9.4) that the ring's maximum utilization efficiency in the large N asymptomatic regime is given by:

$$\eta_{MAX} = \lim_{N\to\infty} \eta = \frac{1}{1 + D_H/T} \tag{2.9.5}$$

Thus, unlike CSMA/CD, the utilization efficiency cannot be made arbitrarily close to one by shrinking the physical size of the network (forcing D_p toward zero).

2.10. Short Bus Architecture

The short bus architecture, shown in Figure 2.10, is an attempt to improve on the performances of the distributed bus CSMA/CD and the token ring systems. Because the bus is physically short, end-to-end propagation delay does not degrade utilization efficiency. Also, because the short bus is passively tapped by each access station, the processing delay inefficiency arising with the regenerative access stations of the token ring is also absent. The short bus, in fact, exhibits a feature known as "perfect capture": if one or more access stations has a packet to transmit, one packet is guaranteed collision-free access to the bus. Thus, the bus utilization efficiency under heavy load is 100%.

Access to the bus exploits the short propagation delay to allow cooperation among contending stations in resolving their conflicts without loss of transmission time. As shown in Figure 2.10, the bus is unidirectional, and time is divided into fixed-length slots, each capable of handling one fixed-length packet. Access stations having packets awaiting transmission begin contending at the leading edge of each time slot, and the bus is so short that its propagation delay is small compared with one bit interval. Each fixed-length packet contains a header, which is used to resolve contention, followed by a data payload. Each of N access stations is assigned a module number, and the packet header contains the module number written in binary.

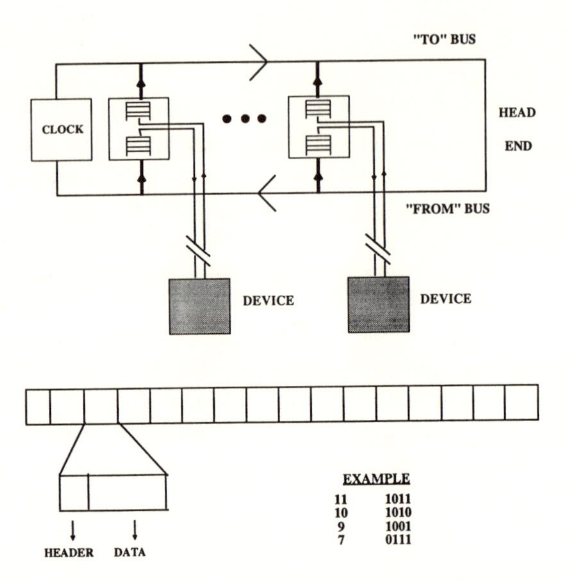

Figure 2.10. Functional architecture of a short bus LAN and an illustration of contention resolution.

The contention bus essentially operates as a wired OR gate: if, within any bit interval, any of the modules writes a logical “1” onto the contention bus, then each contending module will sense a logical “1,” independent of what it has written.

Contention begins at the start of a new time slot. Each contending module begins to write its contention header onto the transmit bus. The contention rule is as follows. If, while contending, any module should write a logical “0” into a particular bit interval, but a logical “1” is received, then that module must stop contending immediately, without so much as transmitting even one additional bit. A given module continues to contend only as long as, on a bit-by-bit basis, the contents of the transmit bus agree with that module’s contention header.

As an example, suppose that for a particular time slot, modules 7, 9, 10, and 11 have packets awaiting transmission. The modules, and their contention headers, are as follows:

Module No.	Contention header
11	1011
10	1010
9	1001
7	0111

During the first bit slot of the contention interval, modules 9, 10, and 11 each write a logical "1" onto the transmit bus; module 7 writes a logical "0," but detects the presence of a logical "1." Module 7 therefore immediately drops out of contention. In the second bit slot, each of the remaining contenders (9, 10, and 11) writes a logical "0," but detects a logical "1" and therefore drops out of contention. Accordingly, 9, 10 and 11 remain in contention. In the third bit slot, modules 10 and 11 each write a logical "1"; module 9 writes a logical "0," but detects a logical "1" and therefore drops out of contention. Accordingly, only 10 and 11 remain in contention; during the fourth bit slot, module 11 writes a logical "1," and module 10 writes a logical "0" but detects the presence of a logical "1." Module 10 therefore drops out of contention, and module 11 has won the right to place its packet onto the transmit bus.

Clearly, this contention scheme gives priority to high-numbered modules. In fact, the highest-numbered module can win contention for every time slot, thereby denying access to all other modules. To correct this inequity, one additional bit can be placed at the beginning of the contention header, which now consists of this so-called "pseudo round-robin" bit followed by the usual module address.

The contention rule is as follows. If, during the first bit slot of a contention interval, a module which was contending with lead (pseudo round-robin) bit set at "0" should detect a logical "1" on the transmit bus, it must immediately cease contending, but should contend for the next time slot with its lead bit again set at logical "0." If a module contending with its lead bit set at logical "0" should detect a logical "0" on the transmit bus during the first bit interval, it should remain in contention using the remaining bits (module address) in the contention field; under these conditions, the normal contention rules apply, and if a module should write a logical "0" but sense a logical "1," it must immediately drop out of contention. However, if a module contending with its lead bit set at logical "0" should subsequently lose contention to another module with lead bit also set at logical "0," that module should contend in subsequent time slots with lead bit set at logical "1" until it successfully accesses the bus. After the first successful access, that module will next contend (whenever it has its next packet available for transmission) with its lead bit set again at logical "0."

The effect of these contention rules is to create variable-length "cycles" of time slots. Modules which are in contention at the beginning of a cycle will have priority for the bus over any modules which attempt to access during the cycle but which were not in the contention at the beginning of the cycle. Thus, a lower-numbered module in contention at the beginning of a cycle will have priority, during that cycle, over a higher-numbered module which attempts to access after the cycle has begun. Cycle formation and contention resolution within a cycle is illustrated by the following example, as depicted in Figure 2.11.

Suppose that, for some particular time slot, four modules have packets awaiting transmission. Let us assume that all are contending with lead bit set at zero; at the end of a cycle containing four time slots (one for each contending module), another group of modules will contend, each with lead bit set at zero, signaling the start of the next cycle.

Now, of the four contending modules, the one with highest module number wins the first time slot. The remaining three, having contended with lead bit set at zero but having lost the initial contention to a module which also contended with lead bit set at zero, continue to contend for successive time slots, but with lead bit set at "1." Beyond the first time slot, contention among the remaining three will be resolved exclusively on the basis of module number. Meanwhile, any other modules which may generate a packet for transmission while the original four are resolving their contention will contend with lead bit set at "0," and, independent of module number, will continually lose contention until all of the original four have been served, since the new arrivals contend with lead bit set at "0" but are in competition with modules contending with lead bit set at "1." Thus, the new arrivals are denied access until each of the original group of four has transmitted one packet. After each of the original group of four has been served, the modules which generated new packets during the four time slots needed to serve the original group new comprise a new group, all members of which contend for the fifth time slot with lead bit set at zero. This brings us to the beginning of a new cycle, which contains six time slots, one for each of the six packets which arrived while the original group of four were resolving their conflicts. Using the pseudo round-robin contention algorithm, each member of this new group of six will place one packet onto the bus before any new packets which may originate while the new group is being served. The length of the new cycle, in time slots, is equal to the number of members in the new group. This process repeats continuously and,

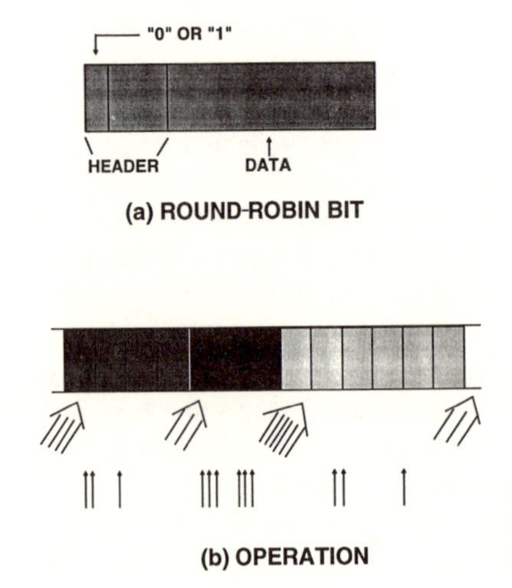

Figure 2.11. Example of variable-length cycle formation for a short bus LAN with pseudo round-robin medium access.

during any cycle, the group being served has priority access to the bus over any new arrivals, independent of module number.

Thus, inclusion of one additional bit in the contention header, to complement the module address, removes the inherent unfairness of module number-based contention. Furthermore, if at the beginning of any time slot, one or more modules have a packet awaiting transmission, one packet will be successfully transmitted during that slot. No time is lost to collisions or propagation delay, the bus is "perfect capture," and the saturation utilization efficiency is 100%. Under conditions of heavy load (all modules always have a supply of packets awaiting transmission), if the number of modules is equal to N, then each module will transmit every Nth time slot, and all time slots will be filled with valid user information.

2.11. Delay Performance*

The analysis contained in Sections 2.7 and 2.8, and the discussion of Section 2.10 allow us to find the maximum level of traffic that can be handled by a CSMA bus, token ring, and short bus LAN, respectively. To produce these results, the assumption was made that the offered load is much greater than the network capacity, and the carried load is then computed. As long as the load actually offered to the network is less than the maximum load which can be carried, then all of the applied traffic will eventually emerge from the network. The issue then becomes one of characterizing the delay experienced from the time that a new block of data is generated by a source until that block is delivered to the intended receiver. In general, the latency, or length of time that a block of data is resident in the network, consists of three components: transmission time, queuing delay, and processing delay.

In this section, we study the latency issue in LANs and present appropriate analytical methodologies. Although the treatment is focused on LANs, the latency phenomenon arises in all broadband network settings, and the discussion of this section is intended to enhance the reader's understanding of these effects in whatever context they may arise.

The transmission delay itself contains two components. The first is the generation time, or the amount of time that it takes to clock a given block of data, bit by bit, onto a network link. (If the block contains R bits and the clock speed of the medium is D bits/sec, then the generation time is simply R/D sec.) The second component of a transmission delay is the propagation delay, caused by the fact that the speed of light is not infinite, and that it takes a nonzero amount of time to move a block of data over some nonzero extent of the medium.

The processing delay corresponds to the amount of time that it takes to prepare a block of data for transmission over the medium, and to verify that the block was correctly received. In addition, for some types of traffic and communication protocols, failure to correctly receive a block of data will initiate a request for a retransmission, adding a nondeterminate component to the processing delay.

Queuing delay is caused by network congestion and arises when two or more service requests concurrently demand some common resource. In such an event, the requests will be sequentially served in accordance with some service discipline, with those requests awaiting service temporarily stored in a buffer or queue. The number of queued requests is, in general, a random variable since requests, in general, are randomly generated. Furthermore, the number of queued requests generally increases as the rate of generation of requests increases (greater traffic intensity). When the traffic intensity increases beyond the maximum load that can be carried by the network, the length of the queues grows without bound.

If all of the packets flowing between a given source–destination pair follow the same physical path through the network (as would be the case, for example, for virtual circuits), then the propagation delay for that pair is a deterministic constant. If different packets associated with some given connection may take different paths from source to destination (involving links of different distances), then the propagation delay becomes random. As already noted, the processing delay may also be fixed or random. In either case, packet generation time, propagation delay, and processing delay are determined by the size of the packets, the transmission speeds, the physical layout of the network, and the processing protocols employed; all can be controlled by design. The queuing delay, however, arises from the statistical nature of multiuser communication networks, and our study of LAN delay performance will be limited to the queuing delay only. Even here, our treatment will be semiquantitative. Analysis of queuing delay is dependent on the models chosen for traffic statistics and, furthermore, for some types of media access protocols, computation of the average queuing delay (let alone the moments of the delay) is analytically intractable.

Slotted Aloha

Let us begin with slotted Aloha. We will qualitatively assess the media access or contention delay, which arises when fixed-length packets randomly arrive at access stations and are stored awaiting successful transmission onto the medium. The model appears in Figure 2.12. There are N access stations and the probability of a new arrival in any time slot at any access station is λ. Arriving packets are stored in input buffers which are served in a first-in-first-out fashion. Each time slot, the head-of-queue packets at all access stations having packets to send attempt to access the bus with probability p.

Suppose that at a given time there are $n < N$ access stations having queued data packets. From (2.4.2), since p is fixed and n buffers are contending, the probability of a successful transmission, P_s, is given by:

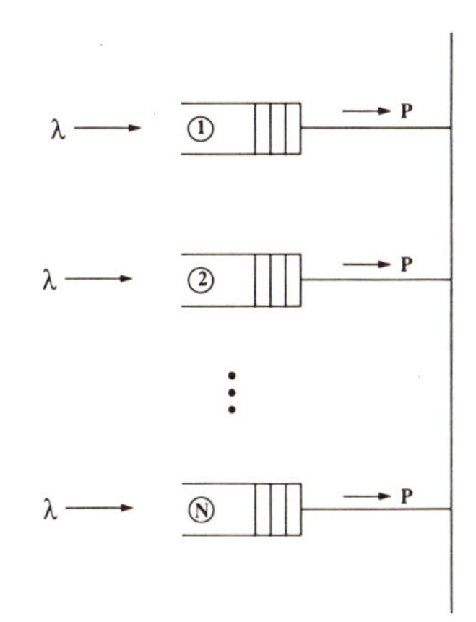

Figure 2.12. Model used to study queuing delay in a slotted Aloha system.

$$P_s = np(1 - p)^{n-1} \qquad (2.11.1)$$

Equation (2.11.1) is plotted in Figure 2.13 and is drawn for the case $N > 1/p$. We note that the maximum value of P_s is approximately $1/e$, and that the value of n which maximizes P_s is approximately $1/p$. Also shown in Figure 2.13 is $N\lambda$, the rate at which new packets enter the system per time slot.

The slotted Aloha medium access scheme exhibits the phenomena of bistability. Referring to Figure 2.13, we note that over the domain of n for which $N\lambda > np(1 - p)^{n-1}$, the probability of a successful transmission in a given time slot (2.11.1) is less than the probability of an arrival in a time slot. Under this condition, the total number of packets stored in the queues will tend to increase (since we are dealing with random events, the number of stored packets at the end of any particular time interval might actually diminish, but the system is clearly biased toward an increase). Thus, the average delay experienced by a newly generated packet will tend to increase. However, over the domain for which $N\lambda < np(1 - p)^{n-1}$, the rate at which new packets arrive is less than the success probability, so that the number of stored packets tends to decrease. The system therefore tends to operate at a value $n = n_1$ such that $N\lambda = n_1 p(1 - p)^{n_1-1}$, with n sometimes growing beyond this value (which tends to force n to diminish) or with n diminishing below this value (which tends to force an increase in n). The system may continue to operate around this value of n for an extended period.

Suppose, however, that the instantaneous value of n should increase beyond

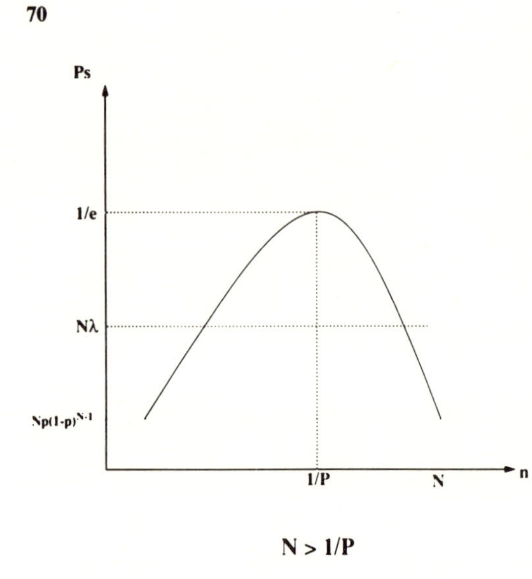

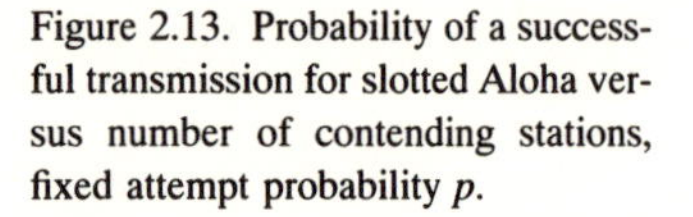
Figure 2.13. Probability of a successful transmission for slotted Aloha versus number of contending stations, fixed attempt probability p.

that which causes P_s to peak. Then, if an additional packet should arrive at an empty access station before the next departure, the success probability will actually diminish, causing the backlog of packets awaiting transmission to diminish less slowly. In this regime, the success probability per time slot remains greater than the expected number of arrivals per time slot, and, eventually, the system restabilizes around the point at which $N\lambda = n_1 p(1 - p)^{n_1 - 1}$. However, the amount of time during which the congestion in the network exceeds its typical value may be quite long. A newly arriving packet therefore experiences one of two generic "conditions." If n is less than the value which maximizes P_s, then the average delay will be relatively short. However, if n is greater than this value, then the average delay may become quite long. The length of time for which the system typically remains in this long-delay "state" is dependent on the difference between $N\lambda$ and $Np(1 - p)^{N-1}$, the latter corresponding to the success probability if all N buffers have at least one packet awaiting transmission, and becomes longer as this difference diminishes. In fact, if $N\lambda > Np(1 - p)^{N-1}$, as shown in Figure 2.13, then the system may never return from the high-delay state since, if n should grow beyond the point for which the success probability is smaller than the arrival rate, the bias will be toward increasing n up to its maximum value of N. Under these conditions, the average delay grows to infinity; the system will start off tending to stabilize at a point for which $n_1 p(1 - p)^{n_1 - 1}$. If n should ever increase beyond the value n_2 determined by $N\lambda = n_2 p(1 - p)^{n_2 - 1}$, then there is a strong likelihood that n will stabilize at the new value $n = N$. Under these conditions, the rate of arrivals exceeds the rate of departures, and the queue lengths will grow without bound.

Finally, if p is reduced to a value such that $np(1 - p)^{n-1}$ monotonically increases over the domain $0 < n < N$, as depicted in Figure 2.13, then the bistability disappears and the queuing delay will be finite provided $N\lambda < Np(1-p)^{n-1}$.

We conclude from this discussion that the delay experienced in a slotted Aloha system is critically dependent on the relationships among N, p, and λ. The critical domains can be summarized as follows:

a. $p < 1/N$ — stable, finite delay
   $N\lambda < Np(1-p)^{N-1}$
b. $p < 1/N$ — delay goes to infinity
   $N\lambda > Np(1-p)^{N-1}$
c. $p > 1/N$ — bistable, finite delay
   $N\lambda < Np(1-p)^{N-1}$
d. $p > 1/N$ — bistable, delay goes to infinity
   $N\lambda > Np(1-p)^{N-1}$

From these observations, we can draw the following qualitative conclusions concerning the average queuing delay as a function of offered load for a fixed number of access stations N. As shown in Figure 2.14, for large p, the average delay at first slowly builds as the network is loaded. Since the load is small and p is large, there is a good likelihood that a newly arriving packet will see an empty network and will be quickly served (p is large) before the next arrival, thereby avoiding contention. However, for the same value of p, as the load increases, congestion develops along with bistability (case c above), causing the average delay to rise sharply. When λ is such that case d applies, the delay goes to infinity.

For a smaller value of p, the average delay under light load increases since, again, a newly generated packet is unlikely to find a prior packet to be present but

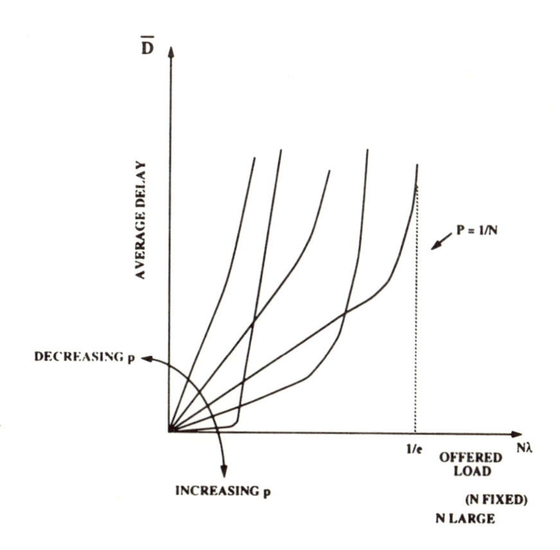

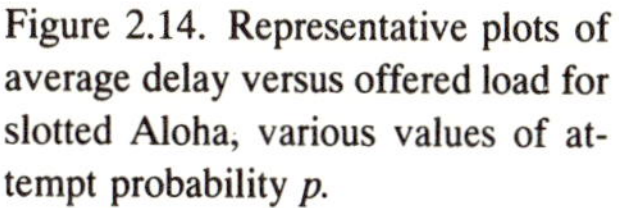
Figure 2.14. Representative plots of average delay versus offered load for slotted Aloha, various values of attempt probability p.

since p is smaller, that packet will wait longer before it is placed onto the bus. As the load increases, congestion again develops along with bistability, causing the delay to rise sharply. Again, when λ is such that case d applies, the delay goes to infinity.

As the value of p is reduced further, we eventually reach the value $p = 1/N$, at which time the network can accept the maximum load, corresponding to $N\lambda = (1 - 1/N)^{N-1}$. For this value of p, we see that the average delay experienced for light load is more than that experienced for larger values of p, but that the increase of delay with load is more gradual since the system is not bistable.

As p is reduced still further (case a), the load at which the delay goes to infinity begins to diminish since the maximum success probability is reduced, and the delay at light load is substantial since a newly generated packet may sit in a buffer, allowing empty time slots to pass contention-free since p is small. Whereas there may be some benefit to be enjoyed by operating the network at a value $p>1/N$ (very low latency at very light load), there is never any benefit associated with operation at $p<1/N$.

CSMA/CD

Qualitatively, the mean delay versus load characteristics for CSMA/CD are similar to those for Aloha. For CSMA/CD with N and p fixed, the maximum rate at which packets can be offered is given by:

$$N\lambda = \frac{1}{T + 2\tau/P_s} \tag{2.11.2}$$

where P_s is given by (2.7.1). Equation (2.11.2) reflects the fact that the average rate of arrivals cannot exceed the average rate of departures. If the arrival rate should exceed that given by (2.11.2), then the queue lengths will grow without bounds. From (2.11.2), we see that the offered load must be bounded by:

$$N\lambda T < \frac{T}{T + 2\tau/P_s} = \frac{1}{1 + \frac{2\tau}{T}(1/P_s)} \tag{2.11.3}$$

which is simply the bus utilization efficiency. We note from (2.11.3) that the discharge rate, for fixed p, is again a function of the number n of access stations having packets awaiting transmission:

$$N\lambda < \frac{1}{T}\left\{\frac{1}{1 + \frac{2\tau}{T}\left[\frac{1}{np(1-p)^{n-1}}\right]}\right\} \tag{2.11.4}$$

and if we plot the right-hand side of (2.11.4) as a function of n, CSMA/CD can also exhibit either stable or bistable behavior. However, unlike slotted Aloha, the domain of p and n over which CSMA/CD is bistable is strongly affected by the ratio of round-trip propagation delay to packet size, $2\tau/T$. If $2\tau/T$ is large, the right-hand-side of (2.11.4) can be approximated as

$$N\lambda \le np\,(1 - p)^{n-1}/2\tau \tag{2.11.5}$$

and the regimes of bistability are scale replicas of those for slotted Aloha, with a resulting similarity in delay versus loading. However, when $2\tau/T$ is small, then the "instantaneous throughput," given by the right-hand-side of (2.11.4), becomes increasingly flat, with peak approaching unity (as shown in Figure 2.7). Thus, when $2\tau/T$ is small, the delay versus load curves appear qualitatively as shown in Figure 2.15. Although p is a factor, its importance is greatly diminished, at least for $p > 1/N$, and under these conditions, the offered load approaches 100%. As $2\tau/T$ grows, the importance of p becomes more pronounced and, for $2\tau = T$, the delay versus load curves appear qualitatively as shown in Figure 2.16, drawn for N large.

Short Bus

Unlike Aloha and CSMA/CD, the delay performance of the short bus architecture can readily be quantitatively modeled and studied. We shall model the packet arrivals as a Poisson point process; that is, the number of packets, K, arriving in any time slot of duration T, is a discrete random variable with probability distribution

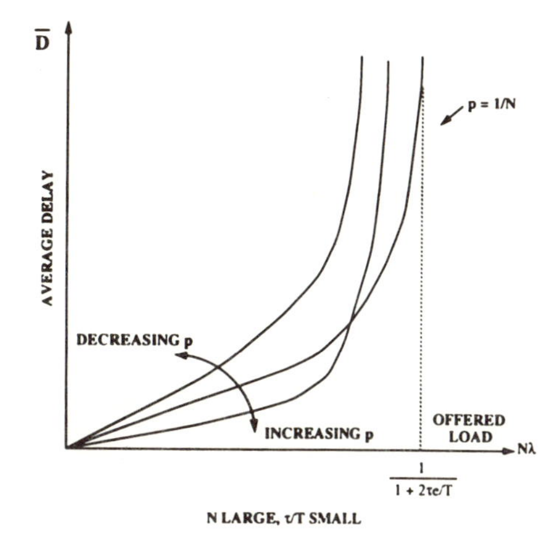

Figure 2.15. Representative plots of average delay versus offered load for CSMA/CD, various values of attempt probability p, propagation delay short relative to packet length.

$$P(K = k) = e^{-\lambda T}(\lambda T)^k/k! \tag{2.11.6}$$

where λ is a parameter known as the packet arrival rate. Packets which arrive during a time slot enter the queue associated with their respective source, as shown in Figure 2.10. Since there are N sources, the arrival rate per source is λ/N. During any time slot, one fixed-length packet, selected at random from the head of one of the queues having packets awaiting transmission, is chosen to be sent onto the channel. Our analysis will seek to find the statistics of the total number of packets awaiting transmission at the end of each time slot. From this, we can find the average amount of time that the typical packet must wait prior to service. We note in advance that the mean delay is independent of the service discipline; that is, a randomly arriving packet at one of the access stations, will, on average, wait the same amount of time, independent of access station number and independent of the order in which waiting packets are served (Little's formula, derived later).

The delay analysis is actually somewhat more general than that needed for the short bus with fixed-length packets in that we will assume the lengths of the packets are random variables, each drawn independently from an arbitrary continuous probability distribution $m(t)$. This allows us to demonstrate the analytical methodology for solving problems of this type in its greatest generality.

Since the arriving packets are Poisson distributed, the number of arrivals during the service time of any packet is independent of the number of arrivals during the service interval of any other packet. Furthermore, the number of packets

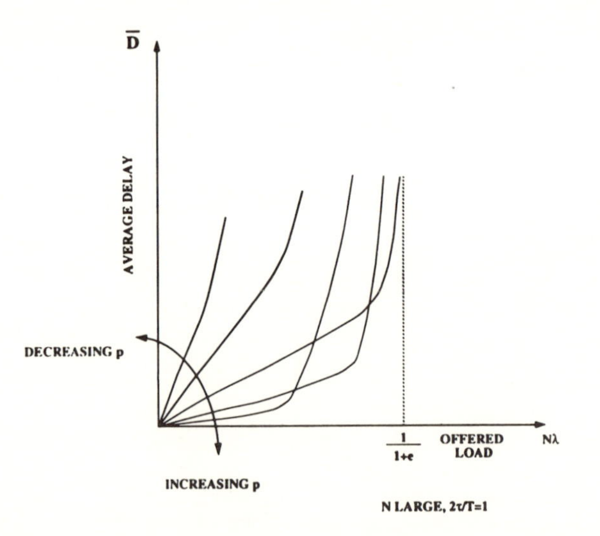

Figure 2.16. Representative plots of average delay versus offered load for CSMA/CD, various values of attempt probability, propagation delay long relative to packet length.

awaiting transmission at the end of any current service interval is dependent only on the number of packets awaiting transmission at the end of the prior service interval and the number of new arrivals during the current interval. We say that the number of packets awaiting transmission forms an embedded Markov chain.

Let N_i be the number of packets awaiting transmission at the end of the ith service interval, $i = 1,2,\ldots$. Then, if $N_i > 0$, exactly one packet will be served during the $(i + 1)$st interval. Therefore, we can write:

$$n_{i+1} = n_i - s_i + a_{i+1} \tag{2.11.7}$$

where a_{i+1} is the number of arrivals during epoch $i + 1$ and $S_i = 1$ if $n_i < 0$; this represents the departure of the packet served during the $(i + 1)$st epoch. Clearly, $S_i = 0$ if $N_i = 0$ since, then, there is no packet to be served during the $(i + 1)$st interval.

Let us define the generating function $\psi_b(x)$ of any nonnegative discrete random variable, b, to be:

$$\psi_b(x) \mathrel{\hat{=}} E\{x^b\} = \sum_{b=0}^{\infty} x^b p_b \tag{2.11.8}$$

where p_b is the (discrete) probability distribution of b, and $E\{\ \}$ is the expectation operator. We note from (2.11.8) that

$$\psi_b(1) = \sum_{b=0}^{\infty} p_b = 1 \tag{2.11.9}$$

and

$$\psi_b'(1) = \sum_{b=1}^{\infty} b x^{b-1} p_b \Bigg|_{b=1} \tag{2.11.10}$$

$$= \sum_{b=1}^{\infty} b p_b = E\{b\}$$

where P_0 is the probability that $b = 0$ and $\psi'(x) = d\psi/dx$. Applying the definition (2.11.8) to (2.11.7), we obtain

$$\psi_{n_{i+1}}(x) = E\{x^{n_i - S_i + a_{i+1}}\} \tag{2.11.11}$$

$$= E\{x^{a_{i+1}}\}\ E\{x^{n_i - S_i}\}$$

$$= \psi_{a_{i+1}}(x)\ E\{x^{n_i - S_i}\}$$

Since the random variables $a_1, a_2, \ldots, a_i, \ldots$ are independent and identically distributed, the index on a_{i+1} may be dropped in (2.11.11). Also, the probability distribution for the number of packets awaiting transmission at the end of any given interval is independent of the sequential number of that interval, and we can drop the index on n_i, S_i, and n_{i+1}. Thus,

$$\psi_n(x) = \psi_a(x) E\{x^{n-S}\} \tag{2.11.12}$$

Now,

$$\begin{aligned} E\{x^{n-S}\} &= \sum_{n=0}^{\infty} x^{n-S} p_n \\ &= p_0 + \sum_{n=1}^{\infty} x^{n-1} p_n \\ &= p_0 + 1/x[\psi_n(x) - p_0] \end{aligned} \tag{2.11.13}$$

Thus,

$$\psi_n(x) = \psi_a(x)[p_0 + 1/x\{\psi_n(x) - p_0\}] \tag{2.11.14}$$

yielding

$$\psi_n(x) = p_0 \frac{x\psi_a(x) - \psi_a(x)}{x - \psi_a(x)} \tag{2.11.15}$$

From (2.11.10),

$$\begin{aligned} E\{n\} &= \psi'(1) \\ &= p_0 \left. \frac{[x - \psi_a(x)][\psi_a(x) + x\psi_a'(x) - \psi_a'(x)]}{[x - \psi_a(x)]^2} \right|_{x=1} \\ &\quad - p_0 \left. \frac{[x\psi_a(x) - \psi_a(x)][1 - \psi_a'(x)]}{[x - \psi_a(x)]^2} \right|_{x=1} \\ &= p_0 \left. \frac{\psi_a(x) - \psi_a^2(x) - x(x-1)\,\psi_a'(x)}{[x - \psi_a(x)]^2} \right|_{x=1} \end{aligned} \tag{2.11.16}$$

We note from (2.11.10) that $\psi_a(1) = 1$ and that the expression (2.11.16) is indeterminate (the numerator and denominator are both zero at $x = 1$). We therefore apply L'Hopital's rule (see Appendix 1) to obtain the result:

$$E\{n\} = p_0 \frac{\psi_a'(1)}{1 - \psi_a'(1)} + p_0 \frac{\psi_a''(1)}{2[1 - \psi_a'(1)]^2} \tag{2.11.17}$$

Next we examine $\psi_a(x)$ in greater detail. Conditioned on a particular service time t, since the number of arrivals in $(0,t)$ is Poisson,

$$\psi_a(x|t) = E\{x^a|t\} = \sum_{a=0}^{\infty} e^{-\lambda t} \frac{(\lambda t)^a x^a}{a!} \tag{2.11.18}$$

Thus, removing the conditioning,

$$\begin{aligned}\psi_a(x) &= \int_0^{\infty} \sum_{a=0}^{\infty} e^{-\lambda t} \frac{(x\lambda t)^a}{a!} m(t)dt \\ &= \Phi[(1-x)\lambda]\end{aligned} \tag{2.11.19}$$

where $\Phi(y)$ is the moment generating formation of the random variable t:

$$\Phi(y) = E\{e^{-yt}\} = \int_0^{\infty} e^{-yt}\, m\,(t)dt$$

From (2.11.19),

$$\psi_a'(x)\Big|_{x=1} = -\lambda\Phi'(0) = \lambda m = E\{a\} \tag{2.11.20}$$

where $m = E\{t\}$ is the mean service time. Also,

$$\psi_a''(x)\Big|_{x=1} = \lambda^2\Phi''(0) = \lambda^2 E\{t^2\} \tag{2.11.21}$$

where $E\{t^2\}$ is the second moment of the service time distribution. Note in (2.11.20) that λm is the expected number of arrivals in the average packet time, which we define to be the offered load, a number which must be less than one since we can serve, at most, only one packet per service time:

$$\Lambda \triangleq \lambda m \tag{2.11.22}$$

Finally, returning momentarily to (2.11.7), we note that

$$E\{n_{i+1}\} = E\{n_i\} - E\{S_i\} + E\{a_{i+1}\} \tag{2.11.23}$$

As before, we can suppress the indices i and i+1 since the system statistics at the time of a departure are not dependent on the index number of that departure, yielding:

$$E\{S\} = E\{a\} \tag{2.11.24}$$

But $S = 1$ only if $n > 0$, implying that

$$E\{S\} = \Sigma p(n) = 1 - P_0 = \lambda m \tag{2.11.25}$$

$$P_0 = 1 - \lambda m \tag{2.11.26}$$

Substituting (2.11.20), (2.11.21), (2.11.22), and (2.11.26) into (2.11.17), we obtain the desired relationship:

$$E\{n\} = (1 - \Lambda)\left[\frac{\Lambda}{1 - \Lambda} + \frac{\lambda^2 E\{t^2\}}{2(1 - \Lambda)^2}\right] \tag{2.11.27}$$

$$= \Lambda + \frac{\lambda^2 E\{t^2\}}{2(1 - \Lambda)}$$

Equation (2.11.27) relates the average number of packets in all queues (plus the one being served) to the offered load Ω and to parameters which characterize the Poisson arrival process and the service time distribution. It was derived only under the assumptions of Poisson arrivals and independent service times. For fixed-length packets of duration, $T, E\{t\} = T$ and $E\{t^2\} = T^2$, and the average number of packets in the system, $\bar{n}_F$, is given by:

$$\bar{n}_F = \Lambda + \frac{\Lambda^2}{2(1 - \Lambda)} \tag{2.11.28}$$

where the subscript F denotes "fixed" or "deterministic" service time. We see for fixed-length packets, the average number of packets in the system is dependent only on the average load = λt = average number of arrivals per time slot [see (2.11.22)].

From (2.11.28), we can now find the average message delay by applying Little's formula, valid for any system:

$$\bar{n} = \bar{\lambda}\bar{D} \tag{2.11.29}$$

Little's formula states that, for any system, the average number of packets in the system, N, is given simply by the product of the average arrival rate λ and the average time spent in the system by a representative packet, D. Before applying Little's formula, we will give a simple graphical "proof." In Figure 2.17, we plot the number of arrivals, N_A, and the number of departures, N_D, for an arbitrary system, as functions of time. Note that N_A and N_D can assume only integer values, that N_A is incremented by unity whenever there is an arrival, that N_D is incremented by unity whenever there is a departure, and that, clearly, $N_D < N_A$. The difference,

$N_A - N_D$, is the number in the system at any moment. Over a long time base $\mathcal{T}$, the average number in the system is given by

$$\bar{N} = \frac{1}{\mathcal{T}} \int_0^{\mathcal{T}} [N_A(t) - N_D(t)]dt \qquad (2.11.30)$$

Note that the integral in (2.11.30) is simply the area, in Figure 2.17, between $N_A(t)$ and $N_D(t)$ over the range 0 and $\mathcal{T}$. Note further that, from Figure 2.17, we can see, for the ith arrival, the time actually spent in the system t_i (this is simply the horizontal distance between the point in time when N_A jumps to the value $N_A = i$ and that point in time when N_D jumps to the value $N_D = i$). Let us assume that over the long time base $\mathcal{T}$, there was a very large number M of arrivals and departures. We see from Figure 2.17 that we can also compute the area between N_A and N_D by adding the time spent in the system for each of the M arrivals and departures:

$$\text{Area} = \sum_{i=1}^{M} t_i = \text{M} \frac{\sum_{i=1}^{M} t_i}{M} = M\bar{D} \qquad (2.11.31)$$

In (2.11.31) we recognize that summing the time spent in the system by each packet, and dividing this by the number of packets served, yields the average time spent in the system. Comparing (2.11.30) and (2.11.31), we conclude that

$$\bar{N}\mathcal{T} = \int_0^{\mathcal{T}} [N_A(t) - N_D(t)]dt = M\bar{D} \qquad (2.11.32)$$

$$\bar{N} = \frac{M}{\mathcal{T}}\bar{D} \qquad (2.11.33)$$

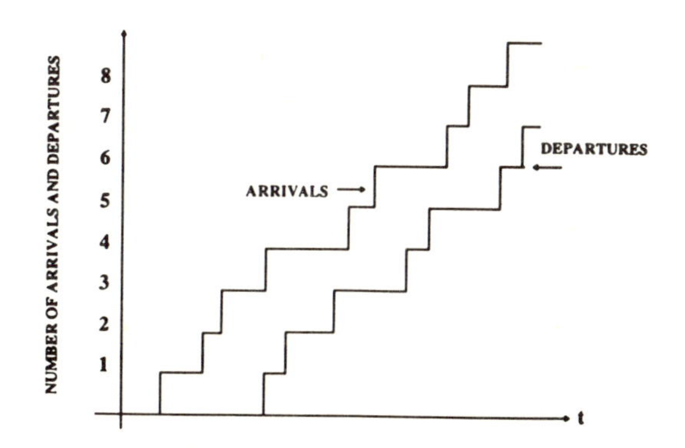

Figure 2.17. A graphical derivation of Little's formula.

But $M/\mathcal{T}$ is the number of arrivals over the large time base $\mathcal{T}$, or the average arrival rate $\bar{\lambda}$, producing the desired result (2.11.29).

For the short bus traffic model, $\bar{\lambda} = \lambda$. Thus, applying Little's formula to (2.11.28), we obtain

$$\bar{D} = \frac{1}{\lambda}\left[\Lambda + \frac{\Lambda^2}{2(1-\Lambda)}\right] \tag{2.11.34}$$

$$= T\left[1 + \frac{\Lambda}{2(1-\Lambda)}\right]$$

Equation (2.11.34) is plotted in Figure 2.18, where the average delay is normalized by the packet size T. We see that, for the short bus, the average delay consists of the time needed to transmit the packet, plus time spent waiting in a queue. As the offered load approaches unity, the average delay approaches infinity as expected. Since the short bus is "perfect capture," no system can provide a lower average delay at any load, i.e., the delay versus load curves for Aloha and CSMA/CD must be greater than those for the short bus.

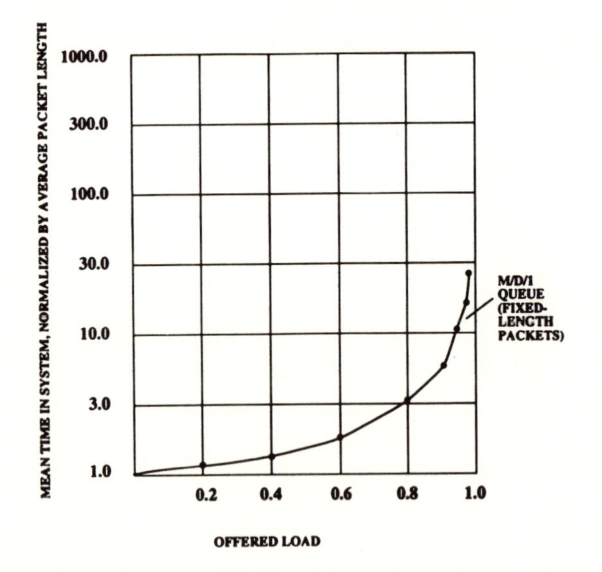

Figure 2.18. Quantitative plot of mean time spent in the system (waiting time plus service time) for the perfect capture short bus.

2.12. Appendix

In this appendix, we apply L'Hopital's rule to the indeterminant form (2.11.16) to produce the result (2.11.17). L'Hopital's rule states that if $f(a) = g(a) = 0$, then

$$\lim_{x \to a} \frac{f(x)}{g(x)} = \lim_{x \to a} \frac{df(x)/dx}{dg(x)/dx} \tag{A2.1}$$

Applying (A2.1) to (2.11.16), we obtain:

$$E\{n\} = p_0 \left. \frac{\frac{d}{dx}[\psi_a(x - \psi_a^2(x) - x(1 - x)\psi_a'(x)]}{\frac{d}{dx}[x - \psi_a(x)]^2} \right|_{x=1} \tag{A2.2}$$

$$= p_0 \left. \frac{\psi_a'(x) - 2\psi_a(x)\psi_a'(x) - (x - x^2)\psi_a''(x) - (1 - 2x)\psi_a'(x)}{2[x - \psi_a(x)][1 - \psi_a'(x)]} \right|_{x=1} \tag{A2.3}$$

$$= p_0 \left. \frac{2x\psi_a'(x) - 2\psi_a(x)\psi_a'(x) - (x - x^2)\psi_a''(x)}{2[x - \psi_a(x)][1 - \psi_a'(x)]} \right|_{x=1} \tag{A2.4}$$

Again, (A2.4) is indeterminate. Applying L'Hopital's rule a second time:

$$E\{n\} = p_0 \left. \frac{\frac{d}{dx}[2x\psi_a'(x) - 2\psi_a(x)\psi_a'(x) - (x - x^2)\psi_a''(x)}{2\frac{d}{dx}[\{x - \psi_a(x)\}\{1 - \psi_a'(x)\}]} \right|_{x=1} \tag{A2.5}$$

$$= p_0 \left. \frac{2\psi_a''(x)[x - \psi_a(x)] + [2 - 2\psi_a'(x)]\psi_a'(x) - (x - x^2)\psi_a'''(x) - (1 - 2x)\psi_a''(x)}{2[x - \psi_a(x)][-\psi_a''(x)] + 2[1 - \psi_a'(x)][1 - \psi_a'(x)]} \right|_{x=1} \tag{A2.6}$$

Finally, remembering that $\psi_a(1) = 1$, we obtain the result (2.11.17).

2.13. Problems

P2.1. For a slotted Aloha system containing 10 access stations: (a) Find the transmission probability per user per time slot which maximizes the overall throughput. (b) For the transmission probability in part a, find the throughput for 5 access stations and for 20 access stations. (c) For 10 access stations, find the throughput for a transmission probability of 0.2 and 0.4.

P2.2. Consider a slotted Aloha system in which a time slot is successfully accessed if occupied by a single user or if occupied by two users (a collision involving two users can somehow be tolerated, but a collision involving three or more packets destroys the information content of all). (a) For arbitrary N, find the transmission probability per user per time slot which maximizes the throughput. (b) Write an expression for the throughput. (c) Find the asymptotic maximum throughput as the number of access stations grows very large.

P2.3. Repeat Problem 2.1, but for an unslotted Aloha system.

P2.4. Consider an unslotted Aloha system containing N identical users, wherein each packet is of duration T seconds. Let p be the probability that a given user generates a packet within any T-second interval. A generated packet immediately contends for the bus, and any packet suffering a collision is abandoned. (a) Find the percentage of packets which are successfully transmitted. (b) For $N = 10$, plot the percentage of successfully transmitted packets as a function of p.

P2.5. For a CSMA/CD Local Area Network, let each packet contain precisely 10,000 bits, the bus length be 100 m, the clock frequency be 10 MHz, and the propagation delay be 10^7 m/sec. (a) Using the asymptomatic results for large N, find the ratio of packet duration to round-trip propagation delay and the maximum utilization efficiency. (b) Repeat part a, except that the LAN is now a Metropolitan Area Network of length 100 km operating at a clock frequency of 100 MHz.

P2.6. Consider a CSMA/CD system containing $N = 10$ stations, each of which always has a supply of fixed-length packets awaiting transmission. Let $T/2\tau = 0.5$, and at the beginning of a contention interval, each station attempts to access the bus with probability $p_1 = 1/2$. If a collision occurs in the kth minislot, all stations contend in the subsequent minislot with probability $p_{k+1} = (1/2)p_k$, $k = 1,2,\ldots$. (a) Find the utilization efficiency. (b) Find the value of p_1 which maximizes the utilization efficiency. (c) Repeat a and b for $N = 100$. (d) For $p_1 = 1/2$, find the asymptotic utilization efficiency as N becomes large.

P2.7. At the beginning of a particular contention interval, each of three stations of a CSMA/CD system has one packet awaiting transmission. Let $T/2\tau = 0.5$ and let no new arrivals be allowed to contend until each of the original three packets has been successfully transmitted. Let the access stations contend for the first minislot of any contention interval with probability $p_1 = 1/2$, and if a collision occurs in the kth minislot, each contends in the subsequent minislot with probability $p_{k+1} = (1/2)p_k$, $k = 1,2,\ldots$. Find the expected time that the first, second, and third packet leaves the system.

P2.8. Consider a CSMA/CD system containing N access stations, each of which always has a supply of fixed-length packets awaiting transmission. Let $T/2\tau = 0.5$ and, at the beginning of a contention interval, let each station contend with probability p_1 for the first k_1 minislots, and with probability p^2 thereafter until a successful transmission takes place. Find an expression for the utilization efficiency.

P2.9. (a) Draw timing diagrams for an $N = 4$ station token ring system wherein each station always has a supply of packets awaiting transmission. Only the sender can remove a packet from the ring. All packets are of 1-sec duration, the token processing time is zero, and the propagation delay between adjacent stations is ¼ sec. The timing diagrams should show events occurring at each station until a total of nine packets have been placed onto the ring. Let the system start at $t = 0$ with station 1 in possession of the token. (b) For the system of part a, what is the ring utilization efficiency?

P2.10. Consider an N-station token ring system in which the propagation delay and token processing time are both zero, all packets are of duration T seconds, and each station always has a supply of packets awaiting transmission. (a) Suppose that only the sender can remove a packet from the ring. Find the total rate at which the ring can transfer packets among its stations (that is, the average packet transfer rate per station, multiplied by the number of stations). (b) Repeat part a, assuming now that the receiver can remove a packet from the ring and that each packet is equally likely to be destined for any receiver. In this case, arrival of the leading edge of a packet at its intended destination is interpreted by the receiving station as an implied token, and a new packet is immediately transmitted. (c) Qualitatively interpret the difference between results a and b above.

P2.11. For a token ring system, plot the utilization efficiency as a function of the ratio of the packet duration to token handling time (T/D_H) for a fixed ratio of propagation delay to token handling time (D_p/D_H) equal to 10 and for (a) $N = 2$; (b) $N = 5$; (c) $N = 20$; (d) $N = 100$.

P2.12. Consider a token ring system containing N equally spaced access stations. Let the propagation delay between adjacent stations be denoted by D_S and the station processing delay be denoted by D_H. As N increases the total ring round-trip delay $D = ND_H + ND_S$ increases proportionately. Does the overall utilization efficiency diminish as more stations are added? Provide an interpretation for your answer.

P2.13. For a token ring system, suppose that the ratio of total delay to packet duration is zero. All packets are of fixed length T. The number of new packets generated by each source within any T-second interval is a Poisson random variable, with average value equal to 1/10. Each time a station captures the token, it transmits all packets currently contained within its buffer, but not any new packets which arrive while that station is holding the token. (a) Plot the average delay experienced by any packet while it awaits transmission as a function of the number of access stations N. (b) Is there a limit to the number of stations which this system can support? If so, what is that limit?

P2.14. Consider a deterministic service time queuing system, with service rendered synchronously within T-second service intervals. At the beginning of a service interval, the server will begin serving the next request in the queue; if none is present, the server remains idle for that service interval. Within any T-second service interval, the number of newly generated request may be 0, 1, or 2, with the following probabilities:

$$P(0 \text{ requests}) = \tfrac{1}{2}$$
$$P(1 \text{ request}) = \tfrac{1}{4}$$
$$P(2 \text{ requests}) = \tfrac{1}{4}$$

(a) Find the expected number of requests in the system at the end of a service interval. (b) Find the average time that a request spends in the system, in units of T.

P2.15. Consider an N-station Token Ring network in which the round-trip delay=0 (propagation delay = processing time = 0). Packets are generated at random at each access station. The arrivals are governed by Poisson statistics, and the average rate of arrivals at each station is λ packets per second. The packet lengths are independent and have exponential probability distribution with mean value T seconds. Each access station has a buffer which can store an arbitrarily large number of packets awaiting transmittal on the ring. Let M = total number of packets in the system, including the one (if any) currently on the ring. Find an expression for the probability that $M = 5$, assuming that $N\lambda T < 1$.

P2.16. Consider three parallel queuing systems. Arrivals to each are Poisson with arrival rate $\lambda/3$, and each has constant service time $3T$. Find an expression for the average number of queued requests, summed over all three queues.

P2.17. Consider a constant time system containing three queues and one server. Upon completing a service request, the server always chooses to serve a new request from that queue having the greatest number of queued requests (in the event of a tie, the server will choose at random from those queues having the greatest number of queued requests). Arrivals to each queue are Poisson with arrival rate $\lambda/3$, and the constant service interval is T. Find an expression for the average number of queued requests, summed over all three queues.

P2.18. Compare the results of Problems 2.16 and 2.17 and explain the difference, if any.

P2.19. Consider a short bus LAN. Each user generates packets in accordance with a Poisson process with average arrival rate per user λ packets per second. The access station uses a priority-based access mechanism: the highest-numbered user having a queued packet gains access to the bus. All packets are of constant length. (a) Find the average delay for a packet generated at random by any one of the users (assume $N\lambda T < 1$). (b) Find the average delay for a packet randomly generated by one of the following set of users:

$$\#N/2 + 1, N/2 + 2, \ldots, N$$

(assume N is even)

(c) Find the average delay for a packet generated by user N.

P2.20. Consider a system containing four queues and one server. Arrivals to each queue are Poisson with arrival rate $\lambda/4$. Arriving requests are given "time stamps," and the server picks the "oldest" request in any queue for service. All requests require a constant service time T. After a request has been served, the server picks the next request. (a) Find an expression for the expected number of requests in the system, total, including the one being served, for $\lambda T < 1$. (b) Find an expression for the expected number of requests in the system produced by the first source.

P2.21. A short bus LAN operates at a clock rate of 10 MHz, and 10% of each packet is required for medium access control. Users present continuous bit rate traffic, each at a rate of 100 kbyte/sec. How many users can be simultaneously served?

P2.22. Consider a short bus LAN containing three access stations. All packets are of length T, and the probability that a user generates a new packet within any given T-second service interval is p. Bus access is priority-based, with priority given in any time slot so that station with the highest module address. Assuming that $3pT$

< 1, find expressions for the average number of packets awaiting transmission to stations 3, 2, and 1, respectively.

P2.23. A slotted Aloha system contains 10 access stations. In each time slot, an access station having one or more packets awaiting transmission generates a new packet with probability λ, and transmits with probability $p = ½$. Find the range of values for λ such that the system is bistable with finite delay.

P2.24. For the slotted Aloha system of Problem 2.23, let $\lambda = 1/20$. Suppose that, at the beginning of a time slot, there are packets in $n = 2$ access stations. (a) Find the probability that a departure occurs before the next packet arrives at the system. (b) Repeat part a for $n = 8$. (c) Is the average delay for this system bounded? Why?

P2.25. For fixed N and $T/2\tau$, derive an expression for the utilization efficiency of a CSMA/CD system as a function of the probability p that a station attempts to access the bus during a contention interval minislot.

P2.26. Plot the utilization efficiency of a CSMA/CD system as a function of the probability p with which each station attempts to access the bus during the contention interval, for (a) $2\tau/T = 0.01$, (b) $2\tau/T = 0.1$, and (c) $2\tau/T = 1.0$.

P2.27. Consider a CSMA/CD system with N access stations and $2\tau/T = 1/10$. Suppose that packet arrivals to each access station are Poisson with arrival rate λ, and let p be the probability that an access station contends for a minislot during the contention interval. Find the range of p and λ over which the system (a) is stable with finite average delay; (b) is bistable with finite average delay; (c) has unbounded delay; and (d) is bistable with unbounded delay.

P2.28. Consider a short bus LAN with fixed packet length T. Each of N access stations generates a new packet with probability λ during each interval T. Derive an expression for the average delay experienced by a packet, including its service time.

P2.29. Consider an N-station short bus LAN for which each packet receives a time stamp when it is generated. Access to the bus is priority based, with the oldest packet served first. Arrivals at each station are Poisson with arrival rate λ, and the fixed packet length is T. (a) Find an expression for the average delay experienced by a packet, including its service time. (b) Repeat part a, except that, now, the most recently generated among all stored packets has priority access.

P2.30. N access stations are connected to a short bus. The fixed-length packet interval is T. Access stations are either active or passive. Each active station generates packets characterized by Poisson arrivals with rate λ. Passive stations generate no new packets. Before going active, each station must request permission to join the population of already-active stations. How many stations can be admitted to the active group such that the average delay (including service time) is less than $10T$? The answer may be expressed in terms of λ and T.

References

1. D. D. Clark, K. T. Pogran, and D. P. Reed, An introduction to local area networks, *Proc. IEEE* **66**(11), 1978.

2. K. Kuemmerle and M. Reiser, Local area communication networks—An overview, *J. Telecommun. Net.* **1**(4), 1982.
3. IEEE Project 802 Local Area Network Standards, 802.2 Logical Link Control, 802.3 CSMA/CD Access Method and Physical Level Specification, 802.4 Token Passing Bus Access Method and Physical Level Specification, 802.5 Token Passing Ring Access Method and Physical Level Specification, IEEE Computer Society.
4. Local Area Networks (CSMA/CD Baseband), Coaxial Cable System, Standard ECMA-80; Local Area Networks, Physical Layer, Standard ECMA-81; Local Area Networks, Link Layer, Standard ECMA-82, Eur. Comput. Manuf. Assoc., Geneva, Sept. 1982.
5. Local Area Networks, Token Ring, Standard ECMA-89; Local Area Networks, Token Bus (Broadband), Standard ECMA-90, Eur. Comput. Manuf. Assoc., Geneva, Sept. 1983.
6. N. Abramson, The ALOHA System—Another Alternative for Computer Communications, 1970 Fall Joint Comput. Conf., AFIPS Conf. Proc., Vol. 37, AFIPS Press, Montvale, N.J., 1970.
7. L. Kleinrock and S. S. Lam, Packet Switching in a Slotted Satellite Channel, 1973 Natl. Comput. Conf., AFIPS Conf. Proc., Vol. 42, AFIPS Press, Montvale, N.J. 1973.
8. L. Kleinrock and S. S. Lam, Packet switching in a multiaccess broadcast channel: Performance evaluation, *IEEE Trans. Commun.* **COM-23**(4), April 1975.
9. S. S. Lam and L. Kleinrock, Packet switching in a multiaccess broadcast channel: Dynamic control procedures, *IEEE Trans. Commun.* **COM-23**(9), Sept. 1975.
10. R. M. Metcalfe and D. R. Boggs, Ethernet: Distributed packet switching for local computer networks, *Commun. ACM* **19**(7), July 1976.
11. L. Kleinrock and F. A. Tobagi, Packet switching in radio channels: Part 1—Carrier sense multiple access modes and their throughput-delay characteristics, *IEEE Trans. Commun.* **COM-23**(12), Dec. 1975.
12. S. S. Lam, A carrier sense multiple access protocol for local networks, *Comput. Net.* **4**, 1980.
13. W. Bux *et al.*, A reliable token-ring system for local area communication, 1981 IEEE Natl. Telecommun. Conf., Conf. Rec., New Orleans.
14. W. Bux, Local area subnetworks: A performance comparison, *IEEE Trans. Commun.* **COM-29**(10), Oct. 1981.
15. A. K. Agrawala, J. R. Agre, and K. D. Gordon, The slotted vs. the token-controlled ring: A comparative evaluation, Proc. COMPSAC 1978, Chicago.
16. W. Bux, F. Closs, K. Kummerle, H. Keller, and H. R. Muller, Architecture and design considerations for a reliable token ring network, *IEEE J. Selected Areas Commun.* **SAC-1**(5), Nov. 1983.
17. A. S. Acampora, M. G. Hluchyj, and C. D. Tsao, A centralized bus architecture for local area networks, IEEE ICC Conf. Rec. Boston, June 1983.
18. A. G. Fraser, A virtual channel network, *Datamation* **21**, Feb. 1975.
19. A. G. Fraser, Towards a universal data transport system, *IEEE J. Selected Areas Commun.* **SAC-1**(5), Nov. 1983.
20. F. A. Tobagi, F. Borgonovo, and L. Fratta, Expressnet: A high performance integrated services local area network, *IEEE J. Selected Areas Commun.* **SAC-1**(5), Nov. 1983.
21. *IEEE Commun. Mag.*, special issue on Architecture of Local Area Networks, **22**(8), Aug. 1984.
22. L. Kleinrock, *Queueing Systems*, Vols. 1 and 2, Wiley, New York, 1976.
23. J. D. C. Little, A proof of the queueing formula $L = \lambda W$, *Oper. Res.* **9**(3), 1961.
24. W. Stallings, Local networks, *Comput. Surv.* **16**(1), March 1984.
25. J. F. Hayes, *Modeling and Analysis of Computer Communication Networks*, Plenum Press, New York, 1984.
26. G. B. Thomas, Jr., *Calculus and Analytical Geometry*, Addison–Wesley, Reading, Mass., 1962.
27. M. Schwartz, *Telecommunication Networks, Protocols, Modeling, and Analysis,* Addison–Wesley, Reading, Mass., 1987.

3

Packet Switch Interconnection Fabrics

3.1. Switch Functionality

The appeal of broadband networks which carry all types of information in some common fixed-length packet format stems in large measure from the feasibility of implementing hardware-based, real-time packet switching fabrics. Such switches route arriving packets on the basis of information contained in each packet header. Because of the high data rates involved, the reading of (and decision-making reaction to) the arriving packets must be implemented entirely in hardware; software processing is incapable of keeping pace with the high rate of packet arrivals. In general, such a hardware-based self-routing switch contains some plurality N of input and output ports and must therefore be capable of processing N packets in parallel. To maintain a physically compact design, custom-designed VLSI circuitry is essential.

At a minimum, two functions must be supported by the N-input, N-output switch (so-called $N \times N$ switch): (1) routing of each applied, fixed-length packet (which we shall refer to as a cell) to its correct destination port, and (2) resolving the contention which arises if two or more simultaneously arriving cells both seek access to a common output port. Since packet switching is involved, output port contention is resolved by means of store-and-forward buffering: arriving packets which cannot be immediately delivered because of contention are stored in one or more buffers within the switch and are delivered on a deferred basis in accordance with some prescribed buffer servicing discipline. The purpose of the buffers is to "smooth out" the statistical fluctuation in the arrival patterns so that as the instantaneous arrival rate of packets destined for some common output fluctuates around its mean, the output is a smooth continuum of cells. Different switch architectures are differentiated on the basis of the sequence in which the two functions are performed, and, for some architectures, the functions are provided by multiple sequential stages with a portion of each function provided by each stage.

A generic example of a time-slotted packet switch appears in Figure 3.1. Shown are the N-input lines and N-output lines. (We assume that, for each input, some input circuitry, not shown, has time-aligned the slot boundaries appearing on the lines leading to the switch.) Also shown are the fixed-length time slots on each line. Numbered time slots contain active cells, each labeled by its destination port. We notice that in the first time slot, cells simultaneously arrive on inputs 1 and 2, both destined for output 1. The switch must route both cells to output 1 as well as write them sequentially on output line 1. We note further that each cell appearing on an output line contains a destination number corresponding to that line, implying that the switch has properly sorted the arriving cells.

A switch of this type is at the core of a broadband network. Such a switch might be used to interconnect multimedia traffic that has already been aggregated by some other type of network or apparatus. Alternatively, a single device might receive its own dedicated port on such a switch by virtue of its high rate of cell generation. For example, referring to Figure 3.2, the traffic presented to one port of the $N \times N$ packet switch might have been produced by a statistical multiplexer, the job of which is to gather low-bandwidth cells generated by the actual terminal devices (workstations, data bases, telephones, etc.) and to multiplex these into a single high-bandwidth stream (remember that the peak data rate on each link feeding into the statistical multiplexer is the same as that of its single output; the only distinction between a low-bandwidth and a high-bandwidth stream is the percentage of link time slots which are filled with information-bearing cells). Other inputs to the $N \times N$ switch might carry cells arriving from Local Area Networks or Metropolitan Area Subnetworks, which then act like distributed ring or bus-type statistical multiplexers, gathering cells generated by terminal devices dispersed over some geographical domain and supplying these time-multiplexed cell streams to the switch for subsequent routing. Still other inputs might carry cells produced by a single device (e.g., high-resolution video camera, supercomputer). The presence of the $N \times N$ cell switch allows (1) low-bandwidth

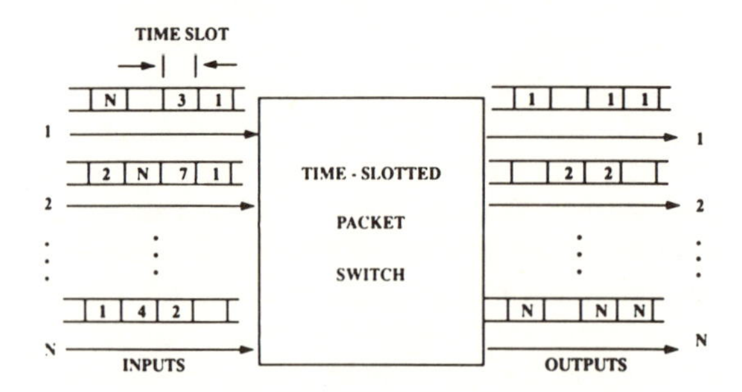

Figure 3.1. A generic time-slotted packet switch.

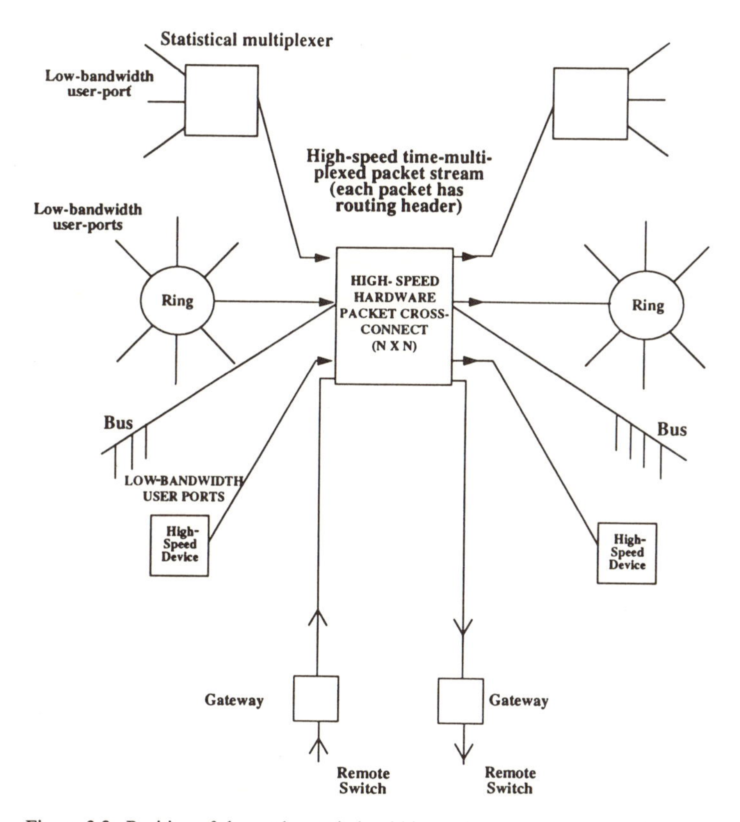

Figure 3.2. Position of the packet switch within a telecommunication environment.

devices connected to one statistical gatherer to communicate with low-bandwidth devices connected to other statistical gathers (the subnetworks appearing at the switch output ports serve to distribute the routed cells to their correct terminating devices, the inverse function of the multiplexing done at the switch inputs); (2) low-bandwidth devices connected via a statistical gatherer to communicate with high-bandwidth devices; and (3) high-bandwidth devices to communicate with other high-bandwidth devices. Some ports on the switch may be connected to other remote switches, thereby enabling broadband communications over wide-area service regions as cells generated from devices connected to the multiplexers feeding into one switch are routed through multiple intermediate switches to devices served by the distribution networks connected to output ports of some remote switches.

We note in passing that since the traffic presented to each input port is already highly aggregated from many separate sources, traditional methods of source flow control may be inappropriate to manage congestion within the $N \times N$ cell switch since source flow control information would need to be sent not only to the affected switch inputs but also through the multiplexing/distribution networks to the appropriate terminating devices which are generating the traffic to be controlled. Another problem associated with flow control concerns the propagation delay associated with sending a congestion-indicating control signal back to the source of traffic and the volume of new information having entered the network at broadband data rates while that controlling signal was being sent. Clearly, problems of this type would be magnified when many $N \times N$ switches are interconnected into a wide-area network, and alternatives to source flow control must be sought. Among the possibilities are open-loop admission control (applicable to virtual circuit networks) wherein requests for new connections are blocked if the expected traffic load presented by that new connection might cause unacceptable congestion to develop for any existing connections. Another possibility, if alternate routes exist from originating source to ultimate destination, would be to dynamically alter the routes taken by the cells corresponding to each given source–destination pair in an attempt to "spread the load" and avoid any "overworked" portion of the network. Yet another possibility is to manage the buffer space in each switch in such a way as to prioritize cell delivery based on latency or lost cell requirements. More will be said about these traffic control and performance management issues in Chapter 6.

3.2. The Need for Queuing in a Space Division Packet Switch

The statistical smoothing needed to resolve switch output port contention can be done immediately as the cells arrive at the switch, or after the cells have been routed to their appropriate destination ports, or within the various stages of a multistage design. We shall subsequently consider the two extremes, input and output queuing, and also consider queuing in a multistage design, but first we shall investigate the need for any type of queuing.

Consider an N-input, N-output time-slotted switch. We shall model the traffic applied to such a switch as follows. For any input, the probability that a given time slot contains an active cell is given by p and, if a cell is present, it is equally likely to be destined for any switch output. Appearance of an active cell in any time slot of a given input is an event which is statistically independent from the appearance of a cell within any other time slot on that or on any other input. The probability, p, is the offered load on any link; $p = 1$ corresponds to each and every input time slot being filled.

Suppose that the switch does not contain any buffers. Then, a cell will be lost if two cells should simultaneously arrive (on different inputs, of course), both bound for the same output; in general, k-1 cells will be lost if k cells simultaneously arrive, anywhere among the N inputs, all bound for the same output. If the fill factor or offered load, p, is small, then the likelihood of two or more concurrent cell arrivals, all bound for the same output, becomes small, and the percentage of cells lost to output port contention will be low. Conversely, as the fill factor p approaches unity, an increasing fraction of cells will be lost to output port contention.

This effect is illustrated in Figures 3.3 and 3.4; the analysis underlying these curves is presented in Section 3.3 Figure 3.3 shows the carried load as a function of the input fill factor p for switches of various dimensionality N. Here, carried load is defined to be the fraction of output time slots actually carrying cells. Ideally, the fraction of filled time slots at the switch output should equal the fraction of filled time slots at the input (all applied cells are carried by the switch; none are lost). We see that, except for the trivial case $N = 1$ (a one-input one-output switch which is, of course, not a switch at all but merely a direct link from input to output), output port contention prevents the output fill factor from equaling the input fill factor. We also note an asymptotic effect: as the dimensionality of the switch increases, a large switch asymptote is rapidly approached. A representative result is as follows. Suppose the input fill factor of a large switch is $p = 1$. Then, the output fill factor is about 63%, that is, if no smoothing buffers are used to compensate for output port contention, then, under the above conditions, about 37% of the applied cells will be lost.

Figure 3.4 contains, essentially, the same information of Figure 3.3 but is plotted to show the fraction of applied packets which are lost (the so-called lost packet rate). Curves of this type are usually more appropriate since they allow one to determine the amount of traffic which can safely be offered to the switch while

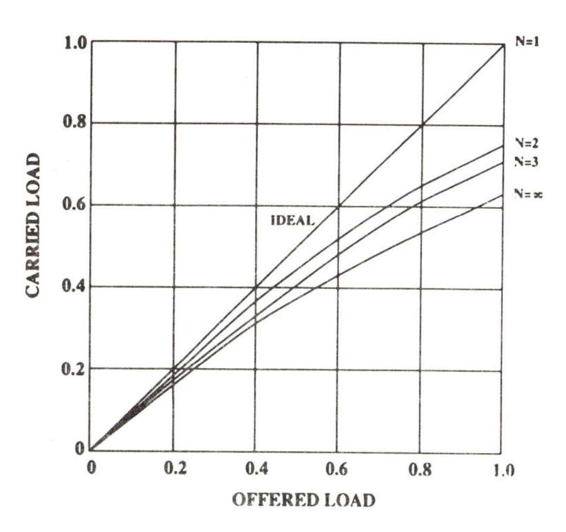

Figure 3.3. Carried load versus offered load for a packet switch without smoothing buffers.

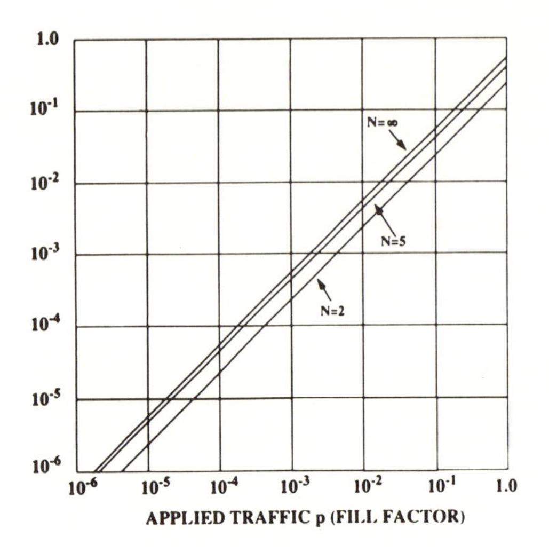

Figure 3.4. Lost packet rate versus offered load for a packet switch without smoothing buffers.

satisfying some lost packet rate objective. For example, we see that if a lost packet rate of 1% is required, then the load offered to a large switch must be limited to about 2%. In a connection-oriented network, limiting the applied load to conform with this permissible fill factor is the responsibility of the admission controller; knowing the volume of traffic which can be accommodated, and knowing the volume of traffic generated by a given virtual connection, the admission controller must appropriately block requests for new connections in order to maintain a volume of traffic not in excess of the permissible fill factor. We see that, without smoothing buffers, the maximum fill factor permitted such that a low lost packet rate (say, less than 10^{-5}) is maintained is very small, implying that the capacity of the input and output links, along with the high-speed circuitry of the switch itself, are grossly underutilized. Smoothing buffers are obviously a necessity for a space division packet switch.

3.3. Analysis of Lost Packet Performance in the Absence of Smoothing Buffers*

As before, let the input fill factor be p, the destination of each arriving cell equally likely be any of the N switch outputs, and arrivals totally uncorrelated in time and across inputs. Under these conditions, we see that the probability that a given time slot on a given input contains a cell destined for a particular output is equal to p/N (if a cell is present, it is equally likely to be bound for any of the N

outputs). Because cell arrivals are uncorrelated among the various inputs, the number of cells, K, bound for a particular output having all arrived within the same time slot among the various inputs is a discrete random variable between 0 and N with a binomial probability distribution; that is,

$$P(K = k) = \binom{N}{k}(p/N)^k(1 - p/N)^{N-k}, \qquad k = 0, 1, \ldots, N \tag{3.3.1}$$

Since the switch is assumed to be buffer-free, cells will be lost if two or more cells should simultaneously arrive bound for a common output; the switch can deliver only one such competing cell, with the others being lost within the switch. One cell will be lost if there are two such arrivals, two cells will be lost if there are three such arrivals, etc. Thus, the average number of lost packets, $\langle L \rangle$, intended for a given output is given by:

$$\langle L \rangle = \sum_{k-2}^{N} (\mathrm{k} - 1)\binom{N}{k}(p/N)^k(1 - p/N)^{N-k} \tag{3.3.2}$$

We recognize that the sum in (3.3.2) is simply the sum for k running between 0 and N, minus the $k = 0$ and $k = 1$ terms. Thus,

$$\langle L \rangle = \sum_{k-0}^{N} (k - 1)\binom{N}{k}(p/N)^k(1 - p/N)^{N-k} + (1 - p/N)^N \tag{3.3.3}$$

We further recognize that the sum in (3.3.3) is equal to the average value of the random variable K, minus the sum of all of the probabilities in the binomial expression (unity). Thus,

$$\begin{aligned} \langle L \rangle &= N(p/N) - 1 + (1 - p/N)^N \\ &= p + (1 - p/N)^N - 1 \end{aligned} \tag{3.3.4}$$

If no packets were lost by the switch, then the fraction of time slots filled on any output line would be the same as the fraction filled on any input, p. Without buffering, the average number of packets lost to any output per time slot is <L>. Thus, the fill factor, F, for any output is simply

$$F = p - \langle L \rangle = 1 - (1 - p/N)^N \tag{3.3.5}$$

The fill factor corresponds to the average amount of traffic which is carried by any output of the switch per time slot, and, again, p is the offered load per input line. From (3.3.5), we can therefore plot the carried load as a function of the offered load for various size switches, as was done in Figure 3.3. We note that, for $N = 1$ (no switch), $F = p$ and the carried load is equal to the offered load (this is the ideal case). For $N>1$, not all of the applied cells are delivered by the switch.

As N increases, the deviation from the ideal becomes more pronounced. Furthermore, the throughput rapidly approaches the large switch asymptote:

$$\lim_{N\to\infty} F = \lim_{N\to\infty} [1 - (1 - p/N)^N] = 1 - e^{-p} \tag{3.3.6}$$

The difference between the applied traffic and the carried traffic is the lost traffic. This is plotted in Figure 3.4, normalized by the applied traffic p. Thus, Figure 3.4 shows the packet loss rate for the switch, i.e., the fraction of applied cells that is lost by the switch:

$$F_L = \langle L \rangle / p = 1 - \frac{1}{p} + \frac{1}{p}(1 - p/N)^N \tag{3.3.7}$$

The plots of Figure 3.4 clearly demonstrate that without buffering, the lost packet rate is unacceptably high except for extremely light loading. Thus, the switch can carry very little traffic if smoothing buffers are absent.

3.4. Input Queuing

Having thus illustrated the need for smoothing, the next question which arises concerns the actual placement of the buffers. The two extremes are (1) smooth at the input and (2) smooth at the output. Input smoothing is illustrated in Figure 3.5. Here, the cells which arrive on each of N input lines are initially placed into smoothing buffers, prior to placement on the correct output lines. Operation is as follows. In any given time slot, the leading cells in those input buffers which are all destined for a common output contend for that output, using some suitable contention scheme such as the short bus pseudo round-robin scheme described in Chapter 2. For example, if the cell at the head of queue No. 1 is intended for output

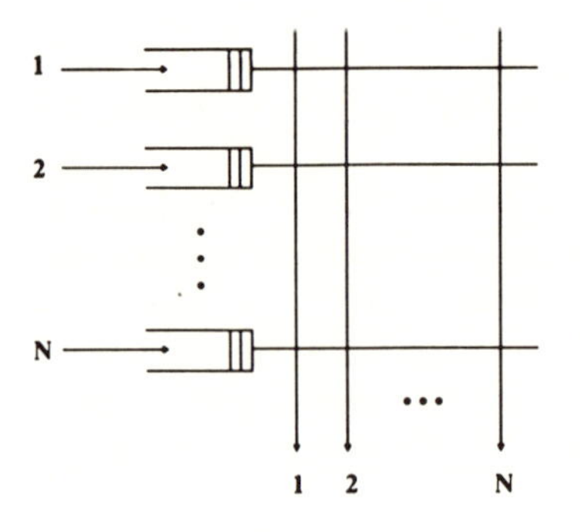

Figure 3.5. Architecture of an input queued space division packet switch.

No. 1, then the switch connects the bus leading from buffer No. 1 to contention circuitry for output No. 1; simultaneously, the bus leading from each other buffer whose head-of-queue cell is intended for output No. 1 is also connected to the corresponding contention circuitry for output No. 1. At the same time, all head-of-queue cells intended for output No. 2 contend for output No. 2, and so forth. Thus, the output lines appear as N short buses, all operating in parallel. Those head-of-queue cells which lose the contention will try again in the very next time slot.

Although this switch requires N^2 packet routing elements and N^2 contention elements, these can readily be implemented in VLSI circuitry, and a large number of such elements can be synthesized on each "chip." Although the buffers can also be implemented in VLSI, it is generally true that each buffer requires its own dedicated chip or chip set (this is because RAM technology generally permits the contents of only one memory location to be written or read at a time; the switch requires that in each time slot, cells arriving on as many as N input lines be simultaneously stored and the lead cells from as many as N buffer locations be simultaneously accessed). In addition, each RAM chip generally requires some RAM control circuitry (also implemented in VLSI) to implement the queue service discipline this will usually be first-in first-out, but not necessarily so. If we measure the complexity of the switch by the number of VLSI chips that are needed in its implementation, this will generally be dominated by buffers and buffer controllers, rather than by routing and contention elements, since the former are much more complex than the latter. A desirable trait of the input-queued switch is that, by this measure, the switch complexity scales linearly with N (one buffer/controller needed for each input). Unfortunately, the traffic-related performance of the input-queued switch suffers from a phenomenon known as "head-of-queue blocking." In essence, a "test" cell in some given queue seeking access to some particular output link may be denied such access by a lead cell seeking access to some other output, even though no other cell in the system is seeking access to the same output as that of the "test" cell. If the output sought by the lead packet is heavily contended for, then the "test" cell may need to wait for a prolonged period before it can be served (in particular, the test cell must wait until the lead packet has successfully contended before its own contention can begin). For example, suppose that within some given time slot, the lead packet in each queue is addressed to the same output and, at the same time, the second cell in queue No. 1 is addressed to a different contention-free output. Then, despite the fact that there is a cell in the system which is destined for that other output, a time slot on that other output will be unfilled; contention among "head of queue" packets has blocked access to unused outputs by non-head-of-queue cells. Because of this effect, we see that some time slots on the output links will unavoidably remain empty, even if cells are already in the system which could have accessed those empty time slots. Since cells cannot arrive on the input lines at a rate greater than that with which they are placed onto the output lines, we conclude that the fraction of input time slots which can contain active cells is less than 100%; the effectiveness or traffic-bearing capability of an input-queued switch is therefore less than 100%. It is shown in the following section that the effective traffic-bearing capability of a large input-queued switch is limited to 63% if arriving cells are served on a random basis; a simpler first-in-first-out service discipline causes the traffic-bearing capability to fall to 58%.

3.5. Traffic-Handling Capability of an Input-Queued Packet Switch*

An exact study of the traffic-handling capability of a first-in first-out input-queued switch is quite tedious and, in the following, we present a much simplified analysis which adequately demonstrates the head-of-queue effect while avoiding excessive mathematical complexity. Then, we explain the difference in throughput predicted by the two analyses.

We are interested in computing the maximum amount of traffic which actually departs from the switch under overload conditions (saturation throughput); that is, we assume that the arrival rate at each switch input is greater than 100% (clearly, not a physically meaningful situation, but one which readily allows us to compute the maximum rate of departure, or saturation throughput; this was done in Chapter 2 when the saturation throughputs of various LAN architectures were found). Under these conditions, each input buffer is always filled with a large number of cells. Consider now one particular output, say output j. In any given time slot, output j will carry a cell if the lead packet in at least one buffer is destined for output j (we assume that contention for all output lines is "perfect capture," that is, a cell will always leave on a given output if at least one lead packet is destined for that output; use of the short bus contention scheme ensures "perfect capture"). We assume that in a given time slot, each lead packet is equally likely to be destined for any output, and that the output addresses for each of the N lead packets are statistically independent (it is here that our simplified analysis is, at best, an approximation; more will be said of this later). Thus, the "fill factor" for output j is simply the probability, P_F, that one or more out of N lead packets is destined for output j:

$$P_F = \sum_{k=1}^{N} \binom{N}{k} \left(\frac{1}{N}\right)^k \left(1 - \frac{1}{N}\right)^{N-k} = 1 - \left(1 - \frac{1}{N}\right)^N \tag{3.5.1}$$

For large N,

$$\lim_{N\to\infty} P_F = 1 - 1/e = 63\% \tag{3.5.2}$$

Thus, for large N, only about 63% of the output time slots are filled, and we conclude that this is the maximum load that can be applied to the switch (at most, 63% of the input time slots can be filled with active cells, or else the rate of departures will be less than the rate of arrivals and the queues will continually grow in length).

Now, in fact, we note that the maximum load which can be applied to the switch is only 58%, under the conditions that consecutive arrivals on a given input line are independent, equally likely to be destined for any output, and are uncorrelated with cells arriving on other inputs. The difference (63% versus 58%) is qualitatively explained on the basis of the following observation: in a given time slot, the destinations of the lead packets in the input buffers are *not* statistically independent. Rather, if in the prior slot, contention was experienced on one output, then in the subsequent slot, those cells which did not depart on that output remain in the system and will contend in the subsequent slot. Correlation among the destinations of the lead packets aggravates the head-of-queue blocking effect and actually reduces the likelihood that one or more lead cells will be destined for some other output, thereby lowering the likelihood that the other output will contain an active cell.

We conclude our discussion of input queuing with the observation that, if in each time slot, a candidate cell for departure is selected *randomly* from any position within each queue (as opposed to serving the queues on a first-in first-out basis, then the approximate analysis (predicting 63% maximum throughput) becomes exact since, in each time slot, the destinations of the candidate cells chosen from the input queues become statistically independent. It is questionable as to whether or not the additional throughput (63% versus 58%) justifies the added complexity required to implement random selection.

3.6. Output Queuing

Output queuing is illustrated in Figure 3.6. Here, cells which arrive on a given input line are broadcast along an input bus, and their headers are interrogated by each member of a set of N switching elements, one associated with each output port. If the destination field of a given cell coincides with the output port of a given switching element, then that switching element will close for the duration of that cell, allowing the cell to enter the associated queue.

Note that, unlike input queuing, the routing is done first, followed by the queuing, and that instead of N queues, now N^2 queues are needed. The Jth column of queues contains only presorted cells which are destined for output No. J, $J = 1, 2, \ldots, N$. Cells contained in this set of N queues then contend for output J, using, for example, the short bus contention scheme. As with input queuing, there are N^2 switching elements and N^2 contention elements. However, since queued cells are presorted, head-of-queue blocking cannot occur: a cell bound for output

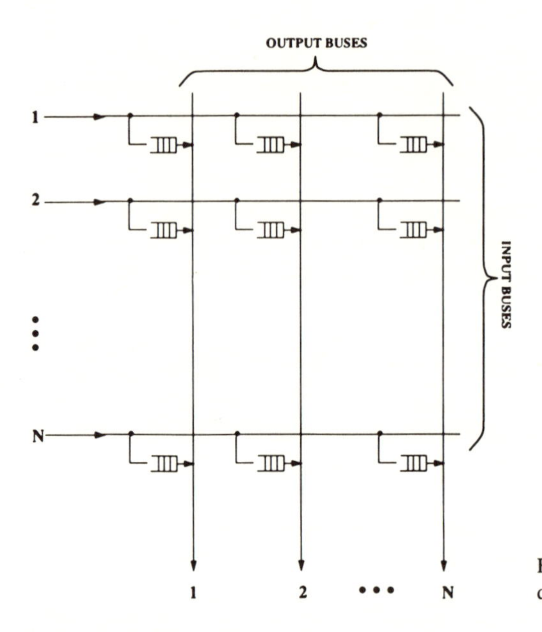

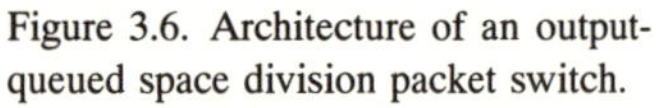
Figure 3.6. Architecture of an output-queued space division packet switch.

J cannot be denied access to output J by the presence of another cell bound for a different output. Thus, no output time slot will be idle if a cell is present within the system bound for that output. Since the short bus contention scheme is perfect capture, the switch may be loaded at an effective utilization of 100%: as long as the average load presented to each output is less than one per time slot, then each time slot of the input links may carry an active cell. The column of N queues containing presorted cells awaiting transmission on a particular output line comprises the "output queue" serving that line.

Unfortunately, the price that must be paid to avoid head-of-queue blocking and increase maximum throughput from 58% to 100% is that the required number of buffers is now N^2. If we measure switch complexity by the required number of buffers, we see that the complexity of a switch with output queuing grows quadratically with the switch dimensionality N, rather than linearly as was the case with input queuing. The switch architectures to be considered in the remaining sections of this chapter represent an attempt to resolve the following dilemma: How might the throughput performance of output queuing be realized with the implementation simplicity of input queuing?

Before considering alternative architectures, the delay versus loading performance of an output buffered switch will be analytically studied; although related curves for input queuing will also be presented, the interested reader is referred to the literature for their derivation. Results of this analysis appear in

Figure 3.7, where T is the time duration of a single cell. These curves are derived by focusing attention on one particular output and computing the expected number of cells that are awaiting service within the N queues serving that output. As before, arriving traffic is modeled by assuming that a cell is present on an input line with probability p (the fill factor or applied load). If present, a cell is equally likely to be destined for any output. Cell arrivals in different time slots and/or on different inputs are assumed to be statistically independent. We note that the delay experienced with output queuing becomes large as the applied load approaches 100%. However, for a large switch, the delay experienced with input queueing becomes large as the applied load approaches 58%, reflecting the effects of head-of-queue blocking.

One final note. From their architectural designs (see Figures 3.5 and 3.6), both the input- and output-queued switches can support broadcast and/or multicast services. For either of these services, the destination field of each cell could contain a broadcast or multicast address. Since all cells are broadcast along their respective input buses, the switching elements (one per output line, for each input line), when checking the destination field in each header, would simply pass all cells having the appropriate single destination address along with those cells containing multicast or broadcast destination addresses to which the switching element has been preprogrammed to accept. Each switching element would, however, now need a binary lookup table containing a "one" in those memory locations corresponding to the multicast address which the element must accept, and switch complexity will grow accordingly.

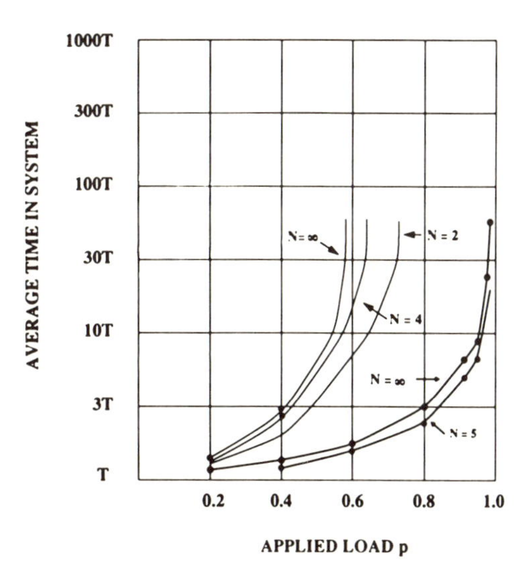

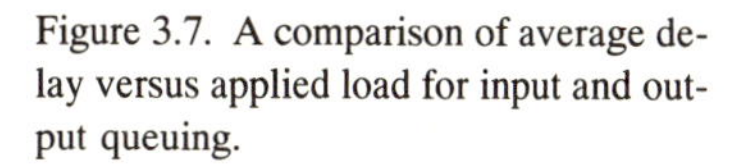
Figure 3.7. A comparison of average delay versus applied load for input and output queuing.

3.7. Mean Delay for a Packet Switch with Output Queuing*

Let us focus attention on any one particular output, say output J. We note that up to N cells may simultaneously arrive for output J within any given time slot (within each time slot, each input may contain a cell bound for output J). For the traffic model presented in the prior section, the number of cells simultaneously arriving for output J within any one time slot is a binomially distributed random variable:

$$P(K = k) = \binom{N}{k}\left(\frac{p}{N}\right)^k (1 - p/N)^{N-k} \tag{3.7.1}$$

Following the analysis of Section 2.11, we wish to compute the expected number of cells in the column of queues awaiting service on output J at the end of a service interval. Let n_i be the number of queued cells at the end of the ith service epoch, let K_i be the number of cell arrivals during the ith service epoch, and let S_i be the number of departures during ith service epoch. Then,

$$n_{i+1} = n_i - S_i + K_i \tag{3.7.2}$$

Since there can be at most one departure per time slot, $S_i = 1$ if $n_i > 0$; $S_i = 0$ if $n_i = 0$. Taking expectations,

$$E\{n_{i+1}\} = E\{n_i\} - E\{S_i\} + E\{K_i\} \tag{3.7.3}$$

and since the expectations in (3.7.3) are independent of the service epoch index i, we can drop the subscript and obtain

$$E\{S_i\} + E\{K_i\} = p \tag{3.7.4}$$

Next, repeating (2.11.17), we write:

$$E\{n\} = p_0 \frac{\psi_K'(1)}{1 - \psi_K'(1)} + p_0 \frac{\psi_K''(1)}{2[1 - \psi_K'(1)]^2} \tag{3.7.5}$$

where

$$\psi_K(x) = E\{x^k\} \tag{3.7.6}$$

K is the discrete random variable with binomial probability distribution (3.7.1), and p_0 is the probability that there are no cells in the system. From (3.7.6),

$$\psi_K(x) = \sum_{k=0}^{N} x^k \binom{N}{k} (p/N)^k (1 - p/N)^{N-k} \tag{3.7.7}$$

$$\psi_k'(x) = \sum_{k=1}^{N} kx^k \binom{N}{k} (p/N)^k (1 - p/N)^{N-k} \tag{3.7.8}$$

$$\psi''(x) = \sum_{k=2}^{N} k(k-1)\, x^{k-2} \binom{N}{k} (p/N)^k (1 - p/N)^{N-k} \tag{3.7.9}$$

Thus,

$$\psi_K'(x) = E\{K\} = p \tag{3.7.10}$$

$$\psi_K''(x) = E\{K^2\} - E\{K\} \tag{3.7.11}$$

Since K is binomially distributed,

$$\sigma_K^2 = E\{K^2\} - E^2\{K\} \tag{3.7.12}$$

$$E\{K^2\} = \sigma_K^2 + E^2\{K\} = N\frac{p}{N}\left(1 - \frac{p}{N}\right) + p^2 \tag{3.7.13}$$

and

$$\psi_k''(1) = p(1 - p/N) + p^2 - p = p^2(1 - 1/N) \tag{3.7.14}$$

Next, since $S_i = 1$ if $n_i > 0$ and $S_i = 0$ if $n_i = 0$,

$$E\{S_i\} = \sum_{i=1}^{\infty} p(n_i) = 1 - p_0 \tag{3.7.15}$$

and, from (3.7.4),

$$1 - p_0 = p \Rightarrow p_0 = 1 - p \tag{3.7.16}$$

Substituting (3.7.10), (3.7.14), and (3.7.16) into (3.7.5), we obtain

$$\begin{aligned} E\{n\} &= (1 - p)\frac{p}{1 - p} + (1 - p)\left[\frac{p^2(1 - 1/N)}{2(1 - p)^2}\right] \\ &= p + \frac{p^2(1 - 1/N)}{2(1 - p)} \end{aligned} \tag{3.7.17}$$

We note that for $N = 1$,

$$E\{n\} = p \tag{3.7.18}$$

as it should be since under these conditions, there can be at most one arrival per time slot, and that arrival will always be served during the next time slot. The number of cells awaiting service can therefore be only zero (if there was no arrival during the time slot) or one (if there was an arrival during the time slot). Equations (3.7.18) follows directly.

Finally, applying Little's theorem [see (2.11.29)], and recognizing that the average rate of arrival to output J is simply p/T, where T is the time slot length, the average delay $\bar{D}$ is given by:

$$\bar{D} = T\left[1 + \frac{p(1 - 1/N)}{2(1 - p)}\right] \tag{3.7.19}$$

Again, for $N = 1$, we notice that the average delay is always equal to the time slot duration, independent of the load p; for $N = 1$, no congestion can occur since there can be at most one arrival per time slot for any given output j, and that arrival will always be served in the next time slot. We also note that for $N > 1$, the average delay is always finite for p less than 1, but becomes unbounded as p approaches unity. Also, we see that the delay is an increasing function of N, but rapidly approaches the large N asymptote.

This result is plotted in Figure 3.7 for various values of N. For completeness, we also show the average delay as a function of offered load for input queuing with various values of N, borrowing from previously published results. For large N, the average delay remains bounded only for $p < 0.58$, a result of head-of-queue blocking as previously discussed.

3.8. Switching Elements

As previously mentioned, the two functions which must be provided by a packet switch are (1) routing of cells from their various inputs to their correct outputs, and (2) resolving contention by means of store-and-forward buffering. The routing function is synthesized by a network of 2 × 2 switching elements, often referred to as β elements. The contention resolution function is synthesized by smoothing buffers which can be located at the switch input, switch output, or interspersed between the β elements in the cell routing network.

The generic β element appears in Figure 3.8. It consists of decision circuitry which operates on the cell headers, a latch to hold the result of that decision for the duration of the cell, shift registers to delay the cells while their headers are being processed, and a 2 × 2 cross connect which is set in either the "bar" state (left input-to-left output, right input-to-right output) or the "cross" state (left input-to-right output, right input-to-left output). Also shown is a schematic of the 2 × 2 cross connect. We note that, by itself, the 2 × 2 cross connect consists of only six gates. When the control signal is logically high, AND gates 1 and 4 are enabled, setting the "bar" state; when the control signal is logically low, AND gates 2 and 3 are enabled (note the inverters at their control signal inputs, represented by the open ovals), and the switch is set in the "cross" state. The complexity of the β element is typically determined by the functionality required of the decision circuitry. For some switch architectures, the decision functionality is quite simple: the state of the switch is determined exclusively by the activity bit on the left input (bar state if left input is active, cross state otherwise). For other switch architectures, the decision might be dependent on the destinations of each input cell as inferred from the virtual connection number contained in the cell header. In this case, a lookup table is often needed, the pointers to which are the virtual connection numbers, and the output of which is the one-bit decision used

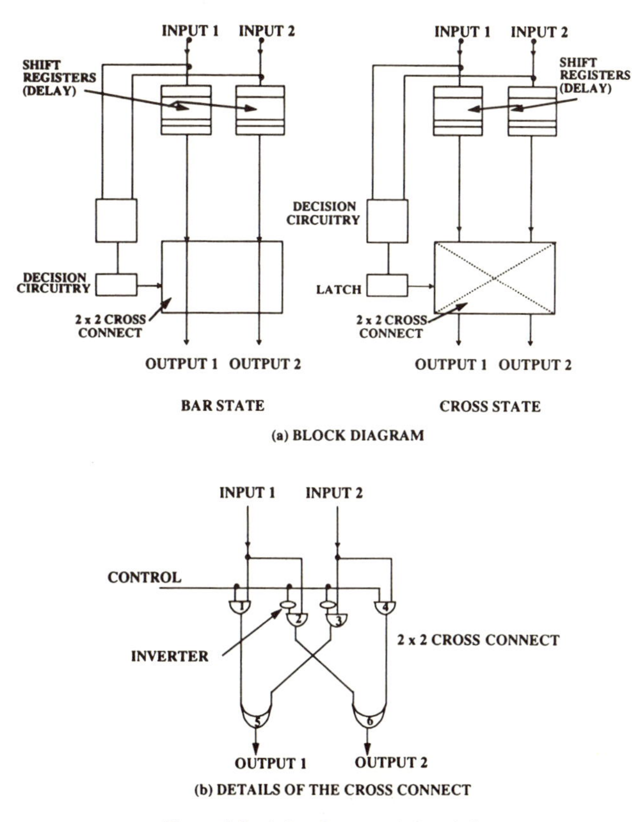

Figure 3.8. A 2 × 2 contention switch.

to set the state of the β element. We note that in all cases, the β element is implemented entirely in custom-design VLSI (implying real-time speed, or the ability to keep up with the cell stream), and that the β element is sufficiently simple that many such elements can be synthesized on a single VLSI chip.

How many β elements are needed to synthesize an $N \times N$ packet switch? In general, the answer to this question is dependent on a multitude of factors, including physical placement of the buffers, expected traffic intensity, lost packet rate objective, and delay objective. Another consideration might be β element complexity, although as already mentioned, this is usually not a significant consideration in a custom VLSI-based design. However, if the β elements were to be implemented with discrete transistors (or, worse yet, with vacuum tubes, as might have been the case in the not-so-distant past), then the number of β elements needed would be a major design-influencing factor.

3.9. A Modified Input-Queued Switch and the Required Number of β Elements

It is instructive to consider the number of β elements which are needed to synthesize the cell routing function, not only for historical purposes, but also because of the additional insight provided. Let us consider a major variant of an input-buffered switch as depicted in Figure 3.9. Here, input cells are first sorted by destination and, for each input, N buffers are provided, each containing packets intended for one common output. The instantaneous contents of all N^2 buffers are known by the controller. In each time slot, the controller selects a packet from no more than one buffer from each group associated with one input. The packets are chosen such that each is destined for a different output, that is, two or more packets originating at different inputs and destined for the same output could not simultaneously be chosen; if the controller chooses a packet for output J from input I, then, in the same time slot, no other packet could be chosen for output J from any other input, and no other packet could be chosen from input I for a different output. The cells selected for concurrent delivery during a specific time slot are then routed to their correct outputs by a cell router, the job of which is to create, each time slot, a set of input-to-output paths which matches the input-to-output cell patterns chosen by the controller for that time slot. The cell router, which contains no further buffers, operates under the supervison of a cell routing controller responsible for setting up the appropriate paths as required by each time slot.

Note that the architecture of the switch appearing in Figure 3.9 has the potential to permit an applied load approaching 100% link utilization. As before, we assume input overload conditions; that is, the input rate is so great as to cause all N^2 buffers to contain cells. Under these conditions, it is possible for each time slot on every output link to be filled, if the cell router shown is *rearrangeably*

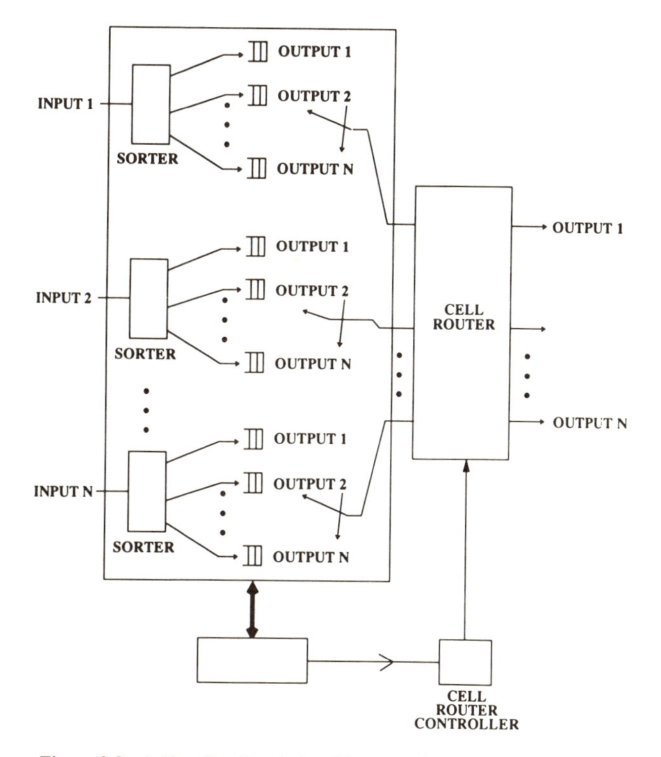

Figure 3.9. A $N \times N$ cell switch with contention resolved at the input.

nonblocking. By rearrangeably nonblocking, we mean that, as long as output port contention is avoided, the cell router is capable of delivering any set of N cells simultaneously appearing at its input to their correct destinations. Hence, if the cell router is rearrangeably nonblocking, and all N^2 input buffers contain some cells, then the controller can always select, for each input, from an appropriate queue such that in each time slot, no two cells are chosen from queues addressed to the same output. Since each of the N^2 input buffers always contains some cells, it is always possible to select a distinct output from each input.

To establish paths through the cell router, a state or setting must be found for each of the cell router's β elements. For a rearrangeably nonblocking cell router, this generally requires a collective decision: the setting of *each* β element may be dependent on all N input-to-output paths to be established through the cell router. This is to be contrasted with a strictly nonblocking cell router, for which a path can be established from any input to any (unused) output without requiring a change of state for any β elements that have already been set to provide a path from a different one of the inputs to a different one of the outputs. For example, suppose a path has been established from input 1 to output 2 and a second path is to be established from input 5 to output 6. With a strictly nonblocking router, that second path could always be established without changing the state of the β elements already used to establish the first path. To establish that same second path through a rearrangeably nonblocking cell router might require a change to the states of the β elements used to establish the first path. We can further define a cell router to be self-routing, if, to produce any set of N nonconflicting paths (each input addressed to a distinct output), the state of each β element can be found exclusively from the contents of the header fields of the cells at its two inputs. If this property does not apply, then the cell router is not self-routing. The cell router controller shown in Figure 3.9 is not needed if the cell router is self-routing. A strictly nonblocking cell router is always self-routing.

In general, a rearrangeably nonblocking cell router requires fewer β elements than a strictly nonblocking cell router. Also, a self-routing cell router requires more β elements than one which is not self-routing. We will now find the minimum number of β elements required to implement a rearrangeably nonblocking cell router.

To begin, we first ask: Within one time slot, in how many ways can a distinct output be chosen for each input? With no loss of generality, let us arbitrarily select one of N outputs (say output J_1) from input 1. Having chosen such a specific output, we see that, for input 2, we can choose an output J_2, distinct from J_1, only from one of $(N - 1)$ remaining unchosen outputs. Similarly, from input 3, we can choose an output J_3, distinct from both J_1 and J_2, only from one of $(N - 2)$ remaining unchosen outputs. Thus, if we select first from input 1, then from input 2, and so forth until we have chosen an output for input N, we see that:

Input 1 has N choices
Input 2 has $N - 1$ choices
Input 3 has $N - 2$ choices
Input N has 1 choice

The total number of ways, M_1, that a distinct output can be chosen for each input is therefore given by:

$$M_1 = N(N - 1)(N - 2) \cdots (1) = N! \qquad (3.9.1)$$

Thus, to connect each input to an arbitrarily chosen output in such a way that no input is connected to the same output at the same time, it is necessary that the cell router be capable of existing in $M_1 = N!$ states, one state for each nonconflicting input/output combination.

Let the cell router contain some number K of β elements, each of which exists in one of two selectable states. The maximum number of states, M_2, which can thereby be produced is given by:

$$M_2 = 2^K \tag{3.9.2}$$

The maximum number of states that the router can exist in, M_2, must be greater than or equal to the minimum number of states, M_1, necessary to allow each nonconflicting input/output combination:

$$2^K \geq N! \tag{3.9.3}$$

Taking the logarithm of both sides:

$$K \log_2 2 \geq \log_2 N! \tag{3.9.4}$$

$$\Rightarrow K \geq \log_2 N! \tag{3.9.5}$$

Finally, since $\log N! > N \log N - N$ (Sterling formula), we conclude that the minimum number of β elements needed to construct the nonblocking cell router is:

$$K \geq N \log_2 N - N \tag{3.9.6}$$

Equation (3.9.6) is a necessary condition for the existence of an $N \times N$ rearrangeably nonblocking cell router, but is suggestive of neither sufficiency nor an appropriate interconnection pattern among the K individual β elements. However, when N is a power of 2, it has previously been shown that a rearrangeably nonblocking cell router can be constructed from a number of β elements given by:

$$K = N \log_2 N \tag{3.9.7}$$

Furthermore, the proof of the sufficiency of (3.9.7) also produces an appropriate multistage interconnection pattern among the K individual β elements. This pattern consists of two Banyan subnetworks (to be discussed in Section 3.10) connected in tandem. Such an arrangement is known as a Benes network. Unfortunately, such a cell router is not *self*-routing, in that the state of each β element cannot be determined exclusively from the contents of the cell headers appearing on its two inputs. In general, a separate cell router controller is required to set the state of each β element in accordance with the input-to-output permutation needed for each successive time slot.

We conclude from the above that an $N \times N$ switch built in accordance with Figure 3.9, containing N sorters, N groups of buffers (each group containing N members), a cell router containing $N \log N$ binary switching elements, and an appropriate controller to nonconflictingly select an output from each input is capable of achieving a link utilization efficiency of 100%.

We note in passing that although the switch of Figure 3.9 provides a maximum link utilization efficiency of 100% (the same as output queuing), the mean delay incurred by the typical cell for all input loads less than 100% (corresponding to the cell arrival probability $p < 1$) is generally higher than that incurred with output queuing. For applied loads of less than 100%, it is not necessarily true that each of the N^2 input buffers will always contain some cells, and a situation may arise wherein, overall, some cells are present for each output and yet it may not be possible to fill each output with a cell. While head-of-queue blocking, characteristic of simple input queuing, is avoided, certain input/output combinations chosen to be served in a given time slot may preclude other combinations which would have produced a greater number of delivered cells within that time slot. Before presenting a specific example, we note that the buffer occupancy of the system shown in Figure 3.9 can be represented by an $N \times N$ matrix:

$$\mathbf{C} = \{C_{i,j}\} \tag{3.9.8}$$

where element $C_{i,j}$ represents the number of cells stored in the buffer attached to input I which contains cells bound for output J. When the applied load is less than 100%, some elements of **C** may be equal to zero. Within each time slot, the controller must select nonzero elements of **C** and cause to be delivered one cell from each chosen element in such a way that two or more cells are not bound for the same output *and* no input is selected more than once (each input can deliver, at most, one cell per time slot to the cell router). Thus, the controller must seek a set of nonzero elements from **C**, no two of which occupy the same row or the same column (selecting two from the same row would correspond to an input being chosen twice; selecting two from the same column would correspond to an output being chosen twice). A set of L such nonzero elements, no two of which occupy the same row or same column, is called a diagonal of length L. If a diagonal of length L is chosen by the controller, then, within that time slot, L output links will be filled, and $N - L$ will be empty.

Matrix **C** evolves in time as new cells arrive and existing cells are delivered to their destinations. Once per time slot, the controller selects a diagonal of matrix **C**, and causes one cell from each element of the diagonal to be delivered to its destination output. The values of the corresponding elements from **C** are then each reduced by unity, and the values of elements corresponding to new cells having arrived during that time slot (at most, one per input) are each increased by unity. At the beginning of the next time slot, the controller initiates a repetition of the above process by selecting a diagonal of the new matrix **C**.

If a diagonal of length L is chosen, then L cells will be delivered to their destination outputs within that time slot. To minimize the average delay incurred by a representative cell in passing through the switch, the controller should always seek a value of L as large as possible. In general, selection of a maximal length diagonal for large N is a very difficult procedure. When all N^2 input buffers contain some cells, it is always possible to find a diagonal of length N, but when some of the buffers are empty, a diagonal of length N is not assured, and some of the output time slots may correspondingly be unfilled, even though there are cells, somewhere within the system, which are bound for an unused output; an example of this will soon be given. Under such conditions, the rate of cell arrivals on the input links may exceed the rate of departure on the output links, and the buffers will tend to fill up with time. Eventually, all buffers will contain some cells, a diagonal of length N will be chosen, all output links will contain a cell, and the rate of departure will equal or exceed the rate of arrival. Thus, unlike simple input queuing, this switch can be loaded up to an efficiency approaching 100% (as the applied input load approaches 100%, cells will tend to occupy all buffers, a diagonal length N will be chosen, and all output time slots will be filled).

Consider now the following situation. Suppose that $N = 4$ and that, at the beginning of a particular time slot, the cell matrix $\mathbf{C}$ contains the following element values:

$$\mathbf{C} = \begin{bmatrix} 0 & 0 & 2 & 3 \\ 0 & 4 & 0 & 0 \\ 3 & 0 & 2 & 0 \\ 0 & 5 & 0 & 0 \end{bmatrix} \tag{3.9.9}$$

We note that some cells are present for each of the four outputs (cells destined for output I appear in column I, $I = 1,2,3,4$; all four columns contain at least one nonzero element). However, for the particular combination of nonzero values shown, the maximum length diagonal is of length $L = 3$. There are seven such 3-diagonals:

$$\begin{gathered} C_{1,4};\ C_{2,2};\ C_{3,3} \\ C_{1,4};\ C_{3,1};\ C_{4,2} \\ C_{1,4};\ C_{2,2};\ C_{3,1} \\ C_{1,4};\ C_{3,3};\ C_{4,2} \\ C_{1,3};\ C_{2,2};\ C_{3,3} \\ C_{1,3};\ C_{3,1};\ C_{4,2} \\ C_{1,3};\ C_{2,2};\ C_{3,1} \end{gathered}$$

For each possible diagonal which may be chosen, only these cells will be delivered in that time slot, despite the fact that cells are present for all four outputs. Consider, for example, the diagonal $C_{1,4}$; $C_{2,2}$; $C_{3,1}$. If this diagonal is chosen, then cells will be delivered from input 1 to output 4, input 2 to output 2, and input 3 to output

1; no cell will be delivered to output 3. If we attempt to deliver a cell to output 3, by taking a cell from the buffer from input 3 containing cells for output 3, ($C_{3,3}$), then a cell cannot simultaneously be taken from the buffer from input 3 containing cells for output 1 ($C_{3,1}$ is an element of the chosen diagonal) since each input can only be selected once per time slot. Similarly, for the diagonal chosen, we cannot deliver a cell to output 3 from input 1, despite the fact that element $C_{1,3}$ is nonzero. However, for output buffering (see Figure 3.6), four cells would be delivered within the time slot in question, one to each output. With output buffering, it is possible to simultaneously deliver cells which originated at one input (but arriving in sequential time slots, of course) to two or more outputs; there is nothing in the architecture of output queuing which precludes a cell contained in any buffer of column I from being delivered concurrently with a cell contained in any buffer of column J different from column I, even if the two buffers chosen, one from each column, contain cells which originated at the same input. An output queuing architecture will *always* deliver a cell to a given output if there are any cells in the system destined for that output, and cells which originated at the same input (but, again, in different time slots) may be concurrently delivered to different outputs. The architecture of the modified input queuing switch of Figure 3.9 clearly prevents any one input from concurrently delivering cells to two or more outputs, and, for all applied loads, the mean delay will be greater than that for output queuing (to date an exact computation of mean delay for the switch of Figure 3.9 has defied rigorous analysis).

However, we note that a modified input-queued switch requires only N buffers, one per input; the group of N buffers serving each input in Figure 3.9 can, in practice, be implemented with one buffer whose memory space has been divided into N sections, one for each output. We can do this because, in any one time slot, only a single cell can arrive on any input and only a single cell will be chosen for delivery from any input. After choosing an output cell from each input, the controller would then ask the memory manager (not shown) for each input buffer to select a cell from the appropriate memory location. By contrast, the output-queued switch truly requires N^2 buffers. In effect, each of N outputs is served by a group of N buffers. During a given time slot, as many as N cells may arrive for a given output, one on each input, and the group serving that output must be capable of accepting N concurrent arrivals. Furthermore, in any given time slot, multiple cells may be selected for delivery to the respective distinct outputs, all having originated at the same input over several prior time slots. Thus, N^2 buffers are needed. The modified input-queued system is therefore simpler than output queuing as measured by required number of β elements ($N \log N$ versus N^2, a minor advantage since β elements are readily fabricated in large arrays in VLSI) and number of buffers needed (N versus N^2, a major advantage). Its mean delay performance is, however, inferior, although its link utilization efficiency can approach 100%. The greatest drawback to the modified input-queued scheme is the need for a real-time controller, or scheduler, capable of selecting a maximum length diagonal in each time slot. As already mentioned, selection of a maximum length diagonal is a very difficult task requiring complicated algorithms, and is not readily amenable to real-time implementation in VLSI. Furthermore, a separate controller, synchronized with selections made by the scheduler, is needed to set the state of the cell router's β elements. More will be said of these difficulties when we revisit this basic approach in Chapter 7 in the context of lightwave networks. There, although the physical embodiment will be quite different, we shall see that, functionally, one candidate architecture for lightwaves networks will involve an approach which borrows heavily from concepts inherent in the modified input-queued switch.

3.10. The Banyan Switch

As previously mentioned, two functions which must be provided by any packet switch are (1) routing of the applied input packets to the appropriate outputs and (2) buffering of applied packets to resolve contention by statistical smoothing. We have seen that buffering may be performed at the input or output of the switch and, for the input and output buffering schemes which we have studied thus far, the packet routing is effected by means of a rearrangeably nonblocking packet routing fabric. For an N-input, N-output switch, the minimum number of β elements needed for the rearrangeably nonblocking cell router is approximately equal to $N \log N$.

The Banyan switch is a multistage self-routing architecture which uses fewer β elements than the minimum number needed to realize a rearrangeably nonblocking design. Specifically, the Banyan switch requires $(N/2) \log N$ elements. Consequently, the switch cannot be nonblocking; even if no two inputs should simultaneously seek a way to the same output, the ability to concurrently deliver the inputs to their correct outputs cannot be assured. A consequence of this blocking property is the fact that, to achieve a reasonably low lost packet rate, smoothing buffers must be deployed internally to the switch.

An 8 × 8 Banyan switch appears in Figure 3.10a. We see that the β elements are arranged in three columns, each containing four elements. The inputs to the switch are the inputs to the β elements contained in the first column; the outputs from the switch are the outputs from the β elements contained in the last column.

In general, if each switching element of a Banyan contains E inputs and E outputs ($E = 2$ for a β element), then the dimensionality of a Banyan switch, N, must be an integer power of E, i.e., $N = E^J$, J an integer. For simplicity, in our discussion, we shall consider β elements only ($E = 2$), and the Banyan switch consists of $\log_2 N$ columns, each containing $(N/2)$ switching elements. In the general case, if $E \times E$ elemental switching units are used, then the Banyan contains $\log_E N$ columns, each containing (N/E) switching elements.

The Banyan derives its name from the patterns which interconnect the

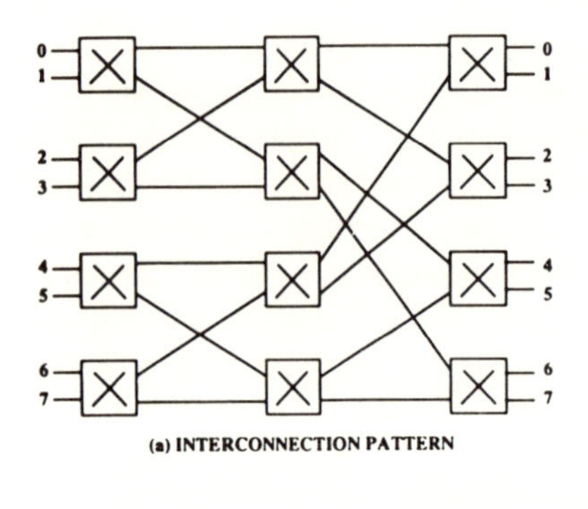

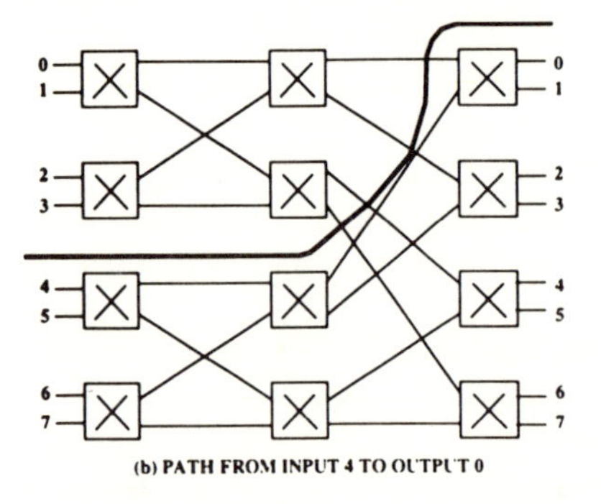

Figure 3.10. An 8 × 8 Banyan switch.

switching elements of its various stages. The basic interconnection pattern is that of a "perfect shuffle." Referring to Figure 3.10a, let us focus on the interconnection pattern from column 2 to column 3. Consider the outputs from the β elements in column 2 to be playing cards; there are eight cards in the deck. The outputs from the β element at the top of column 2 are cards 1 and 2; the outputs from the next β element of column 2 are cards 3 and 4; etc. An imaginary horizontal line drawn between the second and third β elements in column 2 will cut the deck of playing cards perfectly in half. The outputs from column 2 are connected to the inputs of column 3 in such a way as to "perfectly shuffle" the deck. If we sequentially label the inputs to the third column of β elements (first input to top β element is labeled "1," second input to the top β element is labeled "2," first input to the next β element is labeled "3," etc.), then a perfect shuffle of column 2's deck of cards results if we connect the outputs of column 2 to the inputs of column 3 as follows: (1) the first output from the first half (first output of the top β element of column 2) is connected to input 1 of column 3, and the first output from the second half of the deck (this is the first output of the third β element in column 2) is connected to the second input of column 3; (2) the second output of the first half of the deck (second output of the top β element in column 2) is connected to the third input of column 3 and the second output of the second half of the deck (second output of the third β element in column 2) is connected to the fourth input of column 3; (3) the third output from the first half of the deck (first output of the second β element in column 2) is connected to the fifth input of column 3 and the third output from the second half of the deck (first output of the fourth β element in column 2) is connected to the sixth input of column 3; and (4) the fourth output of the first half of the deck (second output of second β element in column 2) is connected to the seventh input of column 3 and the fourth output of the second half of the deck (second output of fourth β element in column 2) is connected to the eighth input of column 3.

Similarly, we can organize perfect shuffle patterns containing larger or smaller number of cards. For example, the last two stages of an $N \times N$ Banyan switch (each stage containing $N/2$ elements) are interconnected by forming a deck of cards containing the outputs from the next-to-the-last column (there are N such outputs) and dividing the deck into two halves (the upper group of $N/4$ elements from that column comprise the first half; the lower group of $N/4$ elements comprise the second half). To arrange the deck of inputs to the last column, we "perfectly shuffle" the two half-decks of the preceding column (first card from first half to first input, first card from second half to second input, etc.), as described above. Returning now to Figure 3.10a (an 8 × 8 Banyan), we again note that column 2 is connected to column 3 by means of a perfect shuffle involving eight cards. Also, we note that outputs from column 1 are connected to inputs of column 2 using two perfect shuffles, each involving four cards: the outputs of the first two elements from column 1 are perfectly shuffled to the inputs of the first two elements of column 2, and the outputs of the second two elements from column 1 are perfectly shuffled to the inputs of the second two elements of column 2.

From this observation, we see how to build large Banyan switches from modules of smaller Banyan switches. If we start with two Banyan switches, each of dimensionality $(N/2) \times (N/2)$, then these can be used to construct a Banyan switch of dimensionality $N \times N$ as follows. The last columns of the $(N/2) \times (N/2)$ switches each contain $N/4$ elements. Stack the two Banyan switches vertically, with one switch drawn above the second switch. Let the outputs of the last column of the first switch be the first deck of cards, and let the outputs of the last column of the second switch be the second deck of cards. Add a new column containing $N/2$ elements; this will be the final column of the new switch [each of the two smaller switches contains $(\log_2 N) - 1$ columns; the $N \times N$ switch contains $\log_2 N$ columns, one more than each of the two smaller switches]. Perfectly shuffle the two half-decks (remember that the last column of each of the two smaller switches is one such half-deck) to the inputs of the newly added column. The two $(N/2) \times (N/2)$ Banyans, interconnected as described above to the newly added column containing $N/2$ elements, comprise an $N \times N$ Banyan switch. This is how the 8 × 8 Banyan of Figure 3.10a was created. Note that the first two columns of the 8 ×

8 switch appearing in Figure 3.10 comprise two 4 × 4 Banyan switches. These have been augmented by, and interconnected with, a third column, containing 8/2 = 4 β elements, as described above, to construct an 8 × 8 Banyan switch. Similarly, two 8 × 8 Banyans, each containing three columns, can be augmented by, and interconnected to, a fourth column containing eight β elements, as described above, to construct a 16 × 16 Banyan switch. Two 16 × 16 Banyans (each containing four columns) can be augmented by and interconnected to a fifth column containing 16 elements to construct a 32 × 32 switch, and so forth.

Again returning to Figure 3.10a, we see that there is precisely one path connecting any switch input to any switch output. For example, to establish the path shown in Figure 3.10b to output 0 from input 4, the first β element along that path (the third element in column 1) *must* be set in the bar state, thereby routing any signal appearing on input 4 to that element's upper output. The third element in column 2 is encountered next; this *must also* be set in the bar state, thereby establishing a path from its upper input to its upper output. Finally, the first element in column 3 is encountered; this *must* be set in the cross state to establish a path from its lower input to output 0. The uniqueness of the path between any input/output pair gives rise to a simple packet self-routing algorithm which will now be described. However, this lack of alternate packet routing capability also gives rise to the blocking nature of the switch.

Note that the inputs and outputs of the Banyan are numbered sequentially from 0 to $N - 1$. Let the destination field of each cell header contain the desired output port number in binary. For an $N \times N$ switch, the routing field must contain $\log_2 N$ bits; this is equal to the number of columns in the switch. Consider a time-slotted switch, that is, all packets are of equal length and fit precisely within time slot boundaries. The information contained in the destination fields of packets arriving within a given time slot determines, for the duration of that time slot, the paths from input to output established by the switch. The switch is reconfigured each time slot in response to the next batch of newly arriving packets.

Suppose that, in each time slot, only a single input port is active (there is an arriving cell only on one input). Then, the β elements can be instructed to cell route as follows. Let each β element in the first column be programmed to set itself in the bar state if the most significant bit in the destination field of the upper input is low, and in the cross state if the most significant bit in the destination header of the lower input is low; let each β element in the second column be programmed to set itself in the bar state if the second-to-most significant bit in the destination header of the upper input is low, and in the cross state if the second-to-most significant bit in the routing header of the lower input is low; let each β element in the last column be programmed to set itself in the bar state if the least significant bit in the destination header of the upper input is low, and in the cross state if the least significant bit in the routing header of the lower input is low. We see that each of the β elements in jth column is programmed to respond to the jth bit in the destination header, $j = 1,2, \ldots, \log_2 N$. Furthermore, since we assume that only one packet is present in any time slot, no ambiguity can result; under these conditions, for each β element, there can be activity on at most one input, and the β elements are each programmed to respond to the appropriate bit on either input (the β elements will only respond to the appropriate bit of an active packet; inactive time slots are ignored). With these programming rules, a cell present on a single active input will be unambiguously routed to its output along the only path which exists between that input and output; for a given input/output combination, each β element is unambigously programmed for a particular state. Furthermore, if an error should be made at any point along the path, causing the cell to be misrouted at that stage of the switch, then it is not possible for subsequent β elements to reroute the cell to its correct destination, even if the β elements were programmed to detect cell routing errors; the uniqueness of the path between each input/output pair prevents a cell which is misrouted at any stage from subsequently arriving at its correct destination.

We now see how the routing rules and rigidity can cause contention to arise within the switch. Suppose that a cell bound for output 1 arrives on input 0 during the same time slot that a cell bound for output 0 arrives on input 2, and further suppose that during this particular time slot, all other inputs are idle. Since the two outputs are distinct, a nonblocking switch would concurrently deliver each arriving packet to its correct destination. The first β element in column 1 must be set in the bar state to route the cell arriving on input 0 to output 1; the first β element in column 2 will be the next β element encountered. The second β element in column 1 must be set in the bar state to route the packet arriving on input 2 to output 0; the first β element in column 2 will be the next β element encountered. We now see the conflict which develops at the first β element in column 2. Routing a cell from input 0 to output 1 requires that this β element be set in the bar state; simultaneously routing a cell from input 2 to output 0 requires the same β element to be set in the cross state. Hence, both of the required paths cannot concurrently be established; the establishment of either path will inhibit, or block, the establishment of the other. This is an example of the type of internal congestion which can arise within the Banyan switch.

To resolve this type of conflict, either or both of the following remedies might be attempted: (1) provide buffers within each β element so that cells which cannot be immediately delivered because of a path conflict are stored and delivered on a deferred basis, or (2) run the internal links which interconnect the columns of β elements at a rate which is an integer multiple of the input/output line speed so that several paths can be sequentially established with one time slot. These techniques improve the lost packet rate of the Banyan switch, at the expense of additional complexity in each β element. A β element which mitigates path blockage by buffering contending packets is functionally shown in Figure 3.11a. Note that the complexity is significantly greater than that of a simple β element, thereby adversely impacting the number of such elements which can be synthesized on a single VLSI chip. The β element becomes, in essence, a 2 × 2 self-routing switch with output buffering. Rather than having a single 2 × 2 router which is set in either the bar or cross state in response to the routing headers appearing within the arriving packets, the enhanced β element requires two routing elements, since each input must be independently steerable to either output. Also, four buffers are needed, two for each output, each of which accepts packets from one of the inputs. The depth of the buffers (i.e., the number of packets which can be stored in each) will affect the lost packet rate of the resulting switch.

To run the internal links at a multiple of the input/output line speed, each β element might be "clocked" at the corresponding multiple, in which case the clock speed may become quite substantial. An alternative approach is shown in Figure 3.11b which depicts an enhanced β element for which the internal operations are effectively run at twice the speed of that shown in Figure 3.11a. The speedup has been effected by duplicating each link and function, and running the dual sets of apparatus in parallel. At the very input of a Banyan switch implemented with such β elements, the bits of each arriving cell would be converted into two-bit parallel words and would subsequently be routed and stored as two-bit word parallel cells; the "word rate" of the parallel β elements would equal the bit serial data rate of the arriving cells. At the very output of the switch, two-bit parallel to bit serial converters would be deployed to reconstruct the bit serial cells. Since each internal β element processes all cells in a two-bit word parallel format, and since each input must be independently steerable to either output, a total of four cell routers and eight buffers are needed. By extension, such an approach can effectively process packets at a multiple of M times the speed of the switch input/output links if arriving cells are converted into M-bit parallel words and M sets of apparatus are run in parallel.

Throughput for a large buffered Banyan network (1024 × 1024) is shown in

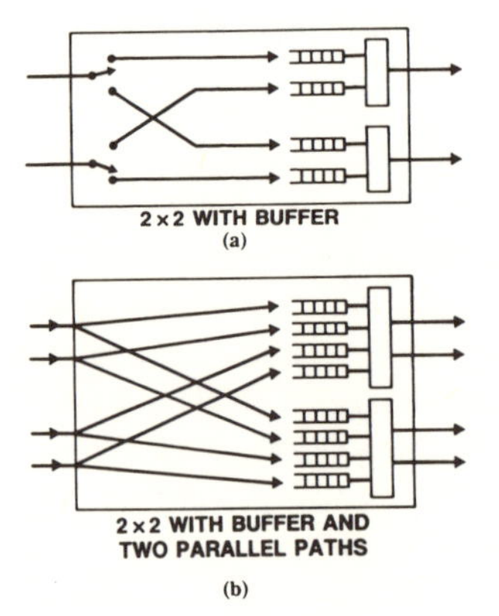

Figure 3.11. Possible β-element enhancements to mitigate blocking behavior of the Banyan switch.

Figure 3.12. These curves were obtained by means of extensive numerical simulation. Shown is the carried load as a function of offered load. The source model is that appropriate for a time slotted system wherein, for each switch input, a cell is present in a particular time slot with probability p (the offered load); packet occurrences are independent in time, and arrival events on different inputs are uncorrelated. If present, a packet is equally likely to be destined for any output. The difference between offered load and carried load represents those cells lost by the switch. Cell loss occurs via overflow of one of the β element buffers. The curves are distinguished by two parameters: the number of parallel paths internal to each β element (either one or two) and the depth of each buffer appearing in the corresponding one of Figure 3.11a or b, measured in units of fixed-length cells. The ideal is a straight 45° line, corresponding to carried load–applied load (no packet loss). We see that for a single path with a buffer depth of one, the carried load saturates at about 25%; for large offered load (approaching 100%), as much as 75% of the applied traffic would be lost within such a switch. If the buffer depth is doubled, the saturation value rises to about 50%. With a speedup of 2 (two paths), the carried load improves, and the saturation value slowly increases with buffer depth. However, saturation carried load is usually not the most significant measure of performance; lost packet probability is usually much more important. The curves of Figure 3.12 strongly suggest that to achieve a lost packet rate of one part in 10^6 or better (only one out of every million applied packets is not delivered), the offered load would need to be reduced to a tiny fraction, i.e., for large

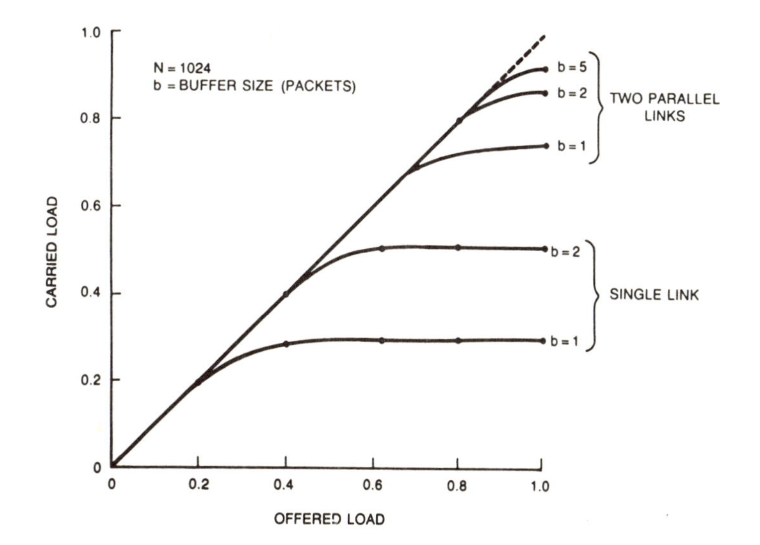

Figure 3.12. Carried load versus offered load for a 1024 × 1024 Banyan switch with various types of enhanced β elements.

switch dimensionality the traffic-bearing capability of a buffered Banyan switch is quite poor.

Problems arise with the Banyan switch at least in part because the switch is not nonblocking (congestion can arise internally even if all cells simultaneously presented to the inputs are intended for distinct outputs). However, as mentioned earlier, Banyans can be connected in tandem to produce a rearrangeably nonblocking design known as a Benes switch. To achieve the desired routing, the state of each switching element in a Benes switch is, in general, a complex function of the destination fields of all cells arriving within a common time slot. Hence, although it is rearrangeably nonblocking, the Benes switch is not self-routing (that is, each β element cannot determine its correct state only from information contained in the routing fields appearing on its two inputs), and the Benes network will not be further discussed. Rather, a different example of a multistage nonblocking switch which is self-routing (the Batcher–Banyan) will be discussed in Section 3.13.

3.11. The Knockout Switch

By definition, the output-buffered switch shown in Figure 3.6 is strictly nonblocking; there is a distinct path from each input to each output. Such a switch is said to be fully connected and internal congestion (path blocking) cannot occur because each input has its own dedicated path to every output which is not shared with, or contended for, by any other input. The only type of congestion which can occur is that caused by multiple inputs concurrently seeking access to the same output. In such an event, all but one of the contending packets is stored in the appropriate output buffer for subsequent delivery on a deferred basis. New arrivals destined for that ouptut will queue up while the prior backlog is being cleared.

As already mentioned, a major implementation issue without output buffering involves not the fully connected architecture (N^2 β elements are needed) but, rather, the fact that N^2 distinct buffers are needed. For large N, this becomes a significant issue. If we measure switch complexity by the required number of distinct buffers, then the complexity of a fully connected output-buffered switch grows quadratically with the switch dimensionality. By contrast, the complexity of the simple input-queued switch of Figure 3.5 grows only linearly with N, but the carried load saturates at 58%. The saturation throughput of an output-buffered switch is, however, 100%; if the buffers are sufficiently deep, then the carried load will always equal the offered load, even for an input/output link utilization efficiency approaching 100%.

The Knockout switch is a fully connected architecture which attempts to provide the implementation simplicity of input queuing (complexity grows linearly with N) with the throughput performance of output queuing (saturation throughput approaches 100%). The architecture of the Knockout switch is based on the observation that in any packet switch, some (hopefully, low) level of packet loss is unavoidable; since originating source transmissions are unscheduled (remember, unlike circuit switching, path and output port reservation is not performed), congestion will inevitably, eventually, and unavoidably occur. In a packet switch, the resources needed to guarantee delivery of each and every applied packet are not reserved but, rather, the resources are statistically shared among many packets; buffers are used to temporarily store packets that simultaneously contend for a common resource, with contending packets delivered on a deferred basis as resources become available. In effect, the buffers "smooth" the statistical variability (burstiness) of the arrivals, producing a smoothed utilization of switching and transmission resources. Even with output queuing, for which the internal fully connected links are congestion free, contention arises for access to each output link and, even for an arbitrarily low (but not zero) applied load, there is a nonzero probability that the output buffer behind each output link will overflow, thereby causing some packets to be lost. The degree of packet loss can be controlled (but packet loss cannot be totally avoided) by properly sizing the depth of the buffers and by controlling the average load applied to the switch.

We note in passing that, in addition to buffer overflow, another source of packet loss in any real system is random channel errors; packet data are typically protected by error detection codes, and a packet which suffers a random transmission error will generally be discarded if and when that error is detected. Although this source of packet loss is not congestion-related, some of the information applied to the network will, nevertheless, not be undelivered. The rate of packet loss from this mechanism can be controlled by improving the quality of the transmission links and/or by using error correction codes in addition to error detection codes. The subject of channel coding is well developed, and the interested reader is referred to any of many excellent texts on the subject.

The Knockout switch recognizes the unavoidability of packet loss within an unscheduled packet switch and intentionally introduces yet one additional loss mechanism, the rate of loss from which can be readily controlled, to greatly reduce the complexity of an output-buffered switch. This new loss mechanism is known as buffer blocking.

The overall architecture of the Knockout switch appears in Figure 3.13. The system is time slotted, and each time-aligned, fixed-length packet (cell) arriving at one of N input ports is placed onto a broadcast bus from which it is passively fanned out to each of N output modules (multicast and broadcast services are readily supported by the Knockout switch). Each output is served by an output module which interrogates the header of each cell arriving on each input, accepting those cells intended for that output and buffering those cells which cannot immediately be placed onto the output link because of contention (an output

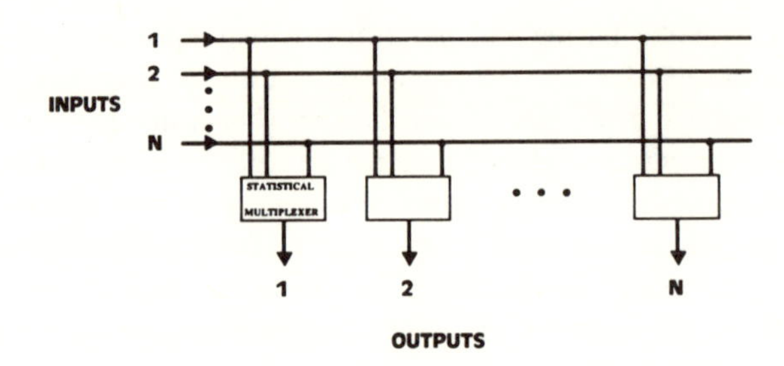

Figure 3.13. Overall architecture of the Knockout switch.

module statistically multiplexes those cells intended for a given output onto that output link).

A block diagram of a Knockout switch output module is shown in Figure 3.14. Each of the N inputs to an output module contains the time-aligned, fixed-length cells broadcasted from the corresponding input bus. Each packet filter is 2 × 2 β element, only one input of which is used. The packet filter operates on the cell header and, for each cell, a binary decision is made: either the cell header identifies the cell as belonging to that output, in which case the cell is delivered to the concentrator, or the cell is ignored. Thus, one output of each packet filter (β element) is connected to the concentrator, and the other is undetermined; only

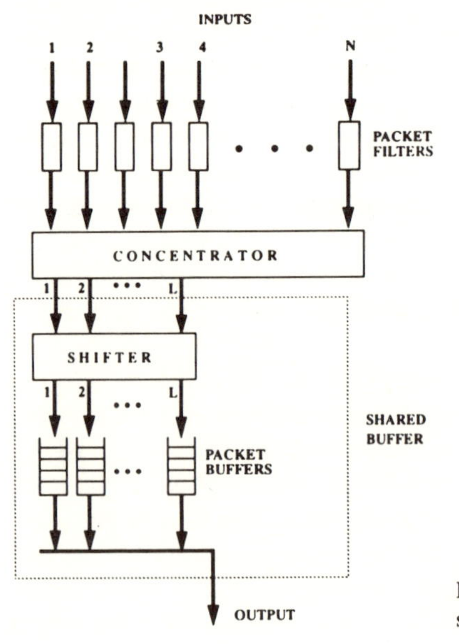

Figure 3.14. The output module of the Knockout switch.

the relevant outputs from the packet filters are shown in Figure 3.14. As already mentioned, the routing header of a given cell can identify that cell as belonging to a specific one of N outputs, or it can identify the cell as being a broadcast or multicast cell, in which case the cell will be accepted by the appropriate group of outputs; at call setup time, each output module of a multicast connection will have been instructed to accept cells which contain in their headers the unique multicast connection number associated with that connection.

Cells which appear at the outputs of the packet filters shown in Figure 3.14 are known to belong to the switch output being served by that output module. In any given time slot, some or all of the inputs to the concentrator may contain active cells (on a long-term statistical basis, we would expect most of the time slots appearing at the inputs to the concentrator to be empty since cells can leave a given output at a rate of, at most, one cell per time slot, and cannot, therefore, arrive at a greater rate on a long-term basis). The role of the concentrator is to identify, on a time slot-by-time basis, those inputs which contain active cells, and to connect all active cells so found to the left-most concentrator outputs, one per concentrator output line. We shall soon show the functional block diagram for such a concentrator. For now, we note that the concentrator contains N inputs, but only $L < N$ outputs. In any time slot, the concentrator will seek all active cells appearing at its inputs and, if L or fewer are found, these will be placed onto the concentrator outputs with the left-most outputs being filled first. However, if $L+1$ or more cells should simultaneously arrive, only L such cells will be processed through the concentrator; the concentrator does not contain any cell storage capability and all but L concurrently arriving cells will be lost. This is the additional loss mechanism introduced by the Knockout switch: if more than L packets should simultaneously appear among the N inputs, all but L will be lost from the system. By properly choosing L, the rate of loss from this mechanism can readily be controlled and maintained at an acceptably low level.

Furthermore, as we shall see, as N increases, the value of L needed to maintain the objective lost packet rate becomes independent of N; this is the key to realizing an overall Knockout switch implementation complexity which scales linearly, rather than quadratically, with N. Looking ahead, we see that subsequent to the concentrator is an $L \times L$ unit known as a shifter, and a bank of L packet buffers. Note that the number of packet buffers is L, independent of N. If we measure implementation complexity by the required number of buffers, we conclude that an $N \times N$ Knockout switch requires N output modules, each with L packet buffers, and its implementation complexity is therefore proportional to $L \times N$, which is linear in N.

The function of the shifter appears in Figure 3.15, drawn for $L = 8$. If the L packet buffers of Figure 3.14 were to be loaded directly from the concentrator, then the left-most buffers would tend to fill up, and might even overflow, despite the fact that buffers farther to the right might be empty. To prevent the left-most

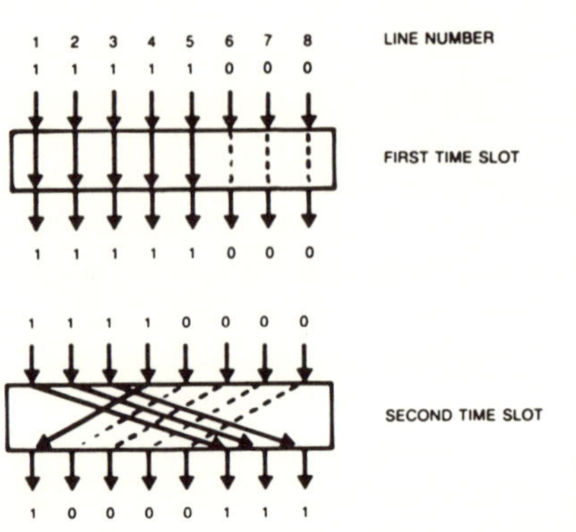

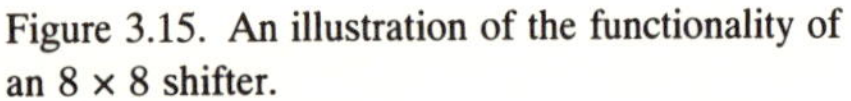
Figure 3.15. An illustration of the functionality of an 8 × 8 shifter.

packet buffers from overflowing while the buffers farther to the right are unfilled, a shifter is used to ensure that the L buffers are cyclically filled. The shifter is also needed to maintain a first-in first-out service for cells arriving on the same input. Operation is as follows. Suppose that during a first time slot, L_1 (less than or equal to L) packets appear at the L_1 left-most outputs of the concentrator. One of these will be loaded into each of the left-most L_1 packet buffers, and the shifter "remembers" that buffer number L_1 was the last to receive a packet. During the next time slot, L_2 (less than or equal to L) packets appear at the L_2 left-most outputs of the concentrator. The shifter will then begin to fill the output buffers beginning with that buffer just to the right of the last one filled (i.e., buffer $L_1 + 1$, unless L_1 happens to equal L; see below). Assuming that $L_1 + L_2 > L$, the shifter will fill buffers $L_1 + 1, L_1 + 2, \ldots, L_1 + L_2$ with the new batch of arrivals; during the next time slot, the shifter will fill the buffers, beginning with buffer $L_1 + L_2 + 1$, with any new arrivals. If $L_1 + L_2 > L$, then the shifter will place the new arrivals during time slot 2 in buffers $L_1 + 1, L_1 + 2, \ldots, L, 1, 2, \ldots, (L_1 + L_2) \bmod L$. In effect, the shifter performs a (mod L) addition of the number of packets concurrently arriving in consecutive time slots. If the cumulative number of packets having arrived to a given output after k time slots is A_k, then the state of the shifter, S_k, is simply

$$S_k = (A_k) \bmod L \tag{3.11.1}$$

If i_{k+1} new cells arrive during the (k+1)st time slot, then

$$S_{k+1} = (S_k + i_{k+1}) \bmod L \tag{3.11.2}$$

and the shifter will fill the output buffers beginning with buffer number $(S_{k+1} + 1) \bmod L$.

For the example shown in Figure 3.15, five cells arrived during the first time slot and were passed through to the first five shifter outputs. During the second time slot, four cells arrived and were sequentially placed onto shifter outputs 6, 7, 8, and 1.

Overall, we see that inclusion of a shifter causes the L packet buffers for a given output to be filled in a cyclical fashion with newly arriving cells intended for that output. Physically, the shifter can be implemented with an $L \times L$ Banyan network. Since the buffers are being filled cyclically, they can also be emptied cyclically. The output line will first fetch a cell from output buffer 1. In the next time slot, the output will fetch a cell from buffer 2, and in the next from buffer 3, and so forth. After a cell has been fetched from buffer L, the next cell is fetched from buffer 1. Although buffers can be filled in parallel (multiple cells may arrive in any given time slot for the output in question), the buffers must be emptied sequentially (only one cell can be placed on the output link per time slot). While sequentially emptying the buffers, if the output should encounter an empty buffer, it pauses at the location of that buffer; because of the sequential nature of the buffer filling and emptying processes, if an empty buffer should ever be encountered by the output link, then it is certain that (1) all buffers are empty at that time and (2) one of the cells arriving in the next nonempty time slot will be directed to that buffer where the output has paused. The process of sequentially emptying the buffers can then restart from that point.

Sequential filling and emptying of the L buffers associated with a given output effectively results in an equitable sharing of the buffers. It is no longer possible for one buffer to be completely filled while others are empty. Moreover, at any point in time, the difference between the number cells stored in any two buffers is, at most, one.

A block diagram of the Knockout concentrator is shown in Figure 3.16 (for simplicity, an $N = 8$ input, $L= 4$ output concentrator is illustrated). The concentrator inputs consist of arriving cells which have already been passed by the packet filters and are known to be intended for the switch output port served by the concentrator. There are $L = 4$ stages associated with the Knockout concentrator shown. Confining our attention to the first stage on the left side of Figure 3.16 we see that this is designed and intended to operate like a round-robin elimination tournament. Each of the small square elements is a simple β element which is set in the bar state if a packet is present on its left input (activity bit = 1) and is set in the cross state if the left input consists of an empty time slot (activity bit = 0). The first row of four β elements is the first level of round-robin competition; the second row of two β elements is the second level, and the third row containing a single β element is the third and final level.

Consider now the β element appearing in the upper left-hand corner of the first stage of the Knockout concentrator. If a cell is present on the left input, the β element is set in the bar state, and that cell will be passed from the first to the second level of competition; if a cell is also present on the right input, it loses the

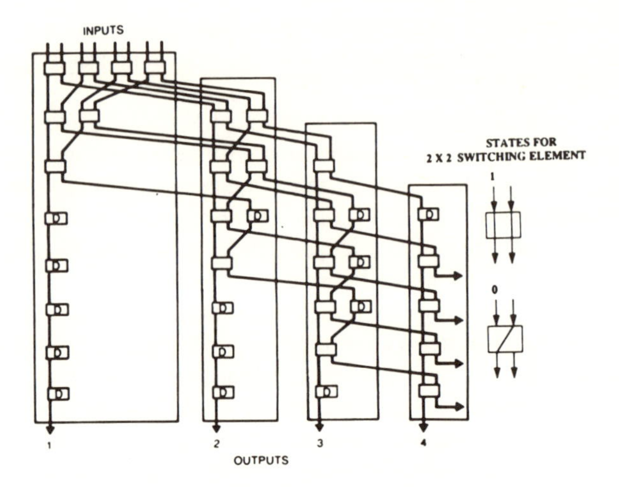

Figure 3.16. An 8-input, 4-output Knockout concentrator.

competition at the first level and is "knocked out" of the first Knockout concentrator stage, leaving to the right along the path shown. If no cell is present on the left input, then the β element is set in the cross state, and any cell present on the right input will be passed from the first to the second level of competition. Similarly, a pairwise competition occurs among the inputs to each of the three other β elements appearing in the first row. For each of those elements, if a cell is present on only one input (it does not matter which one), that cell will pass from the first to the second level of competition; if a cell is present on both inputs, then the cell on the right input is "knocked out" of the competition.

In the second level of competition, the winners from the first level (if any) again compete pairwise. Referring now to the second row of β elements in the first Knockout concentrator stage, if a packet is present on the left input, then the β element is set in the bar state and that packet will win the second level of competition and be passed to the final level; if a packet was also present on the right input, it is "knocked out" of the competition and leaves to the right along the path shown. If no packet is present on the left input, then the β element is set in the cross state, and the packet on the right input (if any) will win the second level of competition and be passed to the third level. At the same time; a similar competition takes place among the packets on the inputs to the second β element on row 2, if any. At the third level of competition, the winner of the level two competition, if any, compete for first place. If only one "finalist" was present, it will be declared the winner. If two finalists were present, then the one on the right input to this level three β element will be "knocked out."

As a result of this first stage of round-robin competition, a single packet present on any of the eight inputs to the Knockout concentrator will be passed to the output of the first stage. If more than one packet was present on the inputs, then all but one are "knocked out."

We now allow those packets which were "knocked out" of the first stage (any packets which left the first stage along the paths to the right leading to the second stage) to compete pairwise in the second stage of competition. In the first level of competition at the second stage of the Knockout concentrator, as many as four packets which were "knocked out" of stage 1 at the first level of competition compete pairwise in the two β elements shown. These β elements were drawn on the same horizontal line as those of the second level of competition for stage 1 because the first level of competition in stage 2 occurs at the same time as the second level of competition in stage 1, i.e., immediately after the headers of the packets were processed in the first level of competition of stage 1 and the outcomes determined. Looking ahead just a bit, we see that some of the β elements in stage 2 and subsequent stages contain only one input and one output and are labeled with the letter D; these are used only to equalize path delays so that, at each level of competition, the competing headers have (1) all encountered the same number of β elements, (2) all incurred an equal number of header processing time delays, and (3) are all time-aligned. As before, for each of the β elements in the first level of stage 2 competition, if a packet is present on only one input, then the state of the β element will be set such that the packet is passed to level two of the competition. If packets were present on both inputs, then one of the packets is "knocked out" of the stage 2 competition and leaves to the right. The winners (if any) of level 1 compete in the first β element at the left side of row 2, at the same time as the packets which were "knocked out" of stage 1 at level 2 compete at level 2 of the second stage using the second β element to the right of row 2. The winners (if any) of level 2 compete in level 3; at the same time, the packet which was "knocked out" at level 3 of stage 1 (if any) is delayed by one header processing time interval. Finally, up to two packets compete at level 4. If only one packet was present, it is passed to the output of stage 2 of the Knockout concentrator. If two packets were present, one is passed to the output of stage 2 and the other is "knocked out" of stage 2.

If two and only two packets were present among the inputs to the Knockout concentrator, then one of these two packets will appear at the output of stage 1 and the other at the output of stage 2. If more than two packets were present, the excess beyond two appear as inputs to the third stage of the knockout concentrator.

Proceeding in like fashion through the third and fourth stages of the Knockout concentrator, we conclude that if packets are present on any three and only three inputs to the Knockout concentrator, these will appear at the outputs of stages 1, 2, and 3, respectively. If packets were present on any four and only four inputs, then these will appear at the outputs of stages 1, 2, 3, and 4, respectively. If packets were present on more than four inputs, then any number in excess of four are "knocked out" of the fourth stage; since there is no fifth stage, these excess packets are lost.

The principles of the Knockout concentrator can be applied to find up to L active packets among N inputs, and to route these to the L outputs. If L_1 active packets are present, these will appear on the first L_1 outputs, L_1 between 1 and L.

Now, how large should we make L? Let there be N inputs to the Knockout switch. The probability that an input time slot is filled is equal to p and if a cell is present, it is equally likely to be destined for any output. Cell arrivals are uncorrelated among the inputs and are uncorrelated in time for any one input.

An analysis of the rate of cell loss from the Knockout mechanism is given in the following section. Results of that analysis are plotted in Figure 3.17. Shown there is the lost cell rate (or lost cell probability) as a function of the number of concentrator outputs, L, for $p = 0.9$ and various values of N. We note that the large switch asymptote regime (N approaches infinity) is approached rapidly, implying that the value of L needed to produce some desired lost cell rate is independent of N for N large. Furthermore, since the cell loss probability diminishes as N gets smaller, use of the N = infinity curve produces a conservative design for any finite value of N, along with a value of L which is totally independent of N; the value of L needed to produce a desired lost cell rate depends only on the offered load p. Even here, we note from Figure 3.18, which shows the lost cell probability as a function of L for a large switch and various offered loads, that the lost cell probability arising from the Knockout concentrator is an increasing function of offered load but remains small even for offered loads of 100% (every input time slot filled). This is certainly not true for the lost cell probability arising from buffer overflow, which approaches unity as the load approaches 100%.

Use of the $p = 100\%$, N=infinity curve again produces a conservative design. We note from Figure 3.18, for example, that a lost cell probability of 10^{-6} can

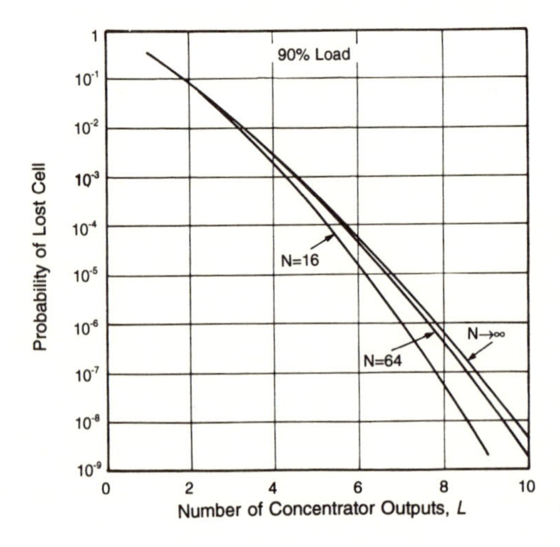

Figure 3.17. Rate of cell loss from the Knockout mechanism for an offered load equal to 90% and various size switches.

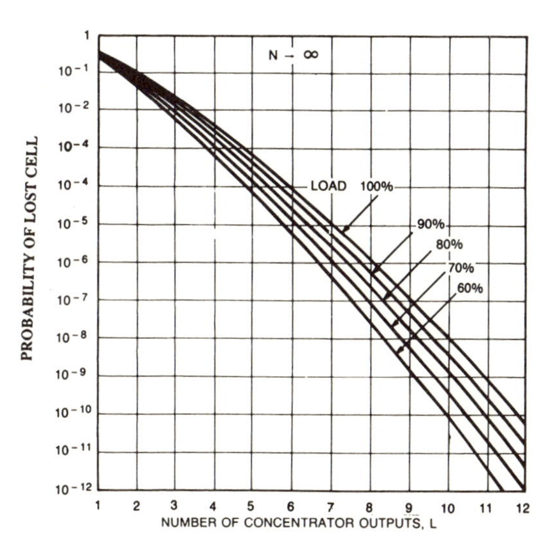

Figure 3.18. Rate of cell loss from the Knockout mechanism for a very large switch and various values of applied load.

always be guaranteed by choosing L = 8. Furthermore, the lost cell probability diminishes approximately by an order of magnitude every time L increases by one. Thus, to produce a lost cell probability of 10^{-7} requires L = 9; 10^{-9} requires L = 11. An important observation is that the value of L needed to guarantee a very low lost cell rate is modest *and* is independent of N. Thus, the Knockout switch achieves its objective: if we measure switch complexity by the number of buffers needed, then the complexity of the Knockout switch grows linearly with N (some relatively small number of buffers, L, for each of N outputs); the offered load, however, may approach 100%, that is, the saturation throughput is 100%, implying up to 100% link utilization efficiency.

The Knockout principle is a high-tech example of the law of diminishing returns. If we insist on a switch which exhibits no buffer blocking, and hence no cell loss from buffer blocking mechanisms, then N buffers are needed for each of N outputs, and the switch complexity grows quadratically with N. However, if we are willing to accept buffer blocking with an accompanying small cell loss rate, then the switch complexity is dramatically reduced and scales only linearly with N. The vast majority of the cells are captured with very modest complexity (small L); the capturing of *every* cell requires much greater complexity ($L = N$). The Knockout principle, in various contexts, appears in several aspects of broadband networks. We will encounter one more example of the Knockout principle, also in the context of self-routing packet switches, in Section 3.14.

To complete the design of the Knockout switch, we must determine the depth needed of the output module buffers. Because a shifter is used, the L buffers belonging to each output are totally shared, and it shall suffice to determine the total storage required, in units of packets, to achieve the target lost cell rate. Again, the homogeneous cell arrival model will be used (on each input, a cell is present with probability p; all cells are equally likely to be destined for any output, all packet occurrences are independent events). Under these assumptions, the number of cells arriving in any time slot and bound for one particular output (say, output J) is binomially distributed [see (3.12.1)]. For large N, the binomial distribution approaches that of the Poisson, and we shall use this large N approximation. The probability that the number K of arrivals in any time slot, bound for output J, is equal to k is therefore

$$P(K = k) = e^{-p}p^k/k! \tag{3.11.3}$$

Since, by design, buffer blocking will contribute negligibly to the lost cell probability, we will assume that no cells are lost because of this mechanism and thereby produce a conservative design.

Since the output port can accept at most one packet per time slot, we can model the input–output process for output module J as an M/D/1 queue. For this system, which was studied in Section 2.11, one can find the steady-state probability distribution for the number of cells in the system, assuming an infinite queue. From this, we can find the probability that the queue size exceeds some value M. This probability is plotted in Figure 3.19 as a function of M for various offered loads p. As expected, we see that the queue size is strongly dependent on the offered load, and grows without bound as p approaches unity.

Armed with Figures 3.18 and 3.19, we can now design the buffer for a Knockout switch output module. The procedure will be illustrated by example. Suppose the system is to be designed for an offered load of 85% and a lost cell probability of 1×10^{-7}. From Figure 3.19, we see that for an offered load of 85%, the probability that the number of cells in an infinite buffer is greater than 50 is about 1×10^{-7}. Hence, if we design a buffer that can store 50 cells total, the rate of cell loss from buffer overflow will be 1×10^{-7}. We would like the loss rate from buffer blocking to be small compared with that resulting from buffer overflow (at least an order of magnitude smaller, or 1×10^{-8}). Using 100% load as a worst-case bound, we see from Figure 3.18 that an $L = 10$ output Knockout concentrator will produce the desired loss rate attributable to buffer blocking. Thus, a total of 10 buffers is needed, along with a total storage capacity of 50 cells, and each of the 10 buffers must be capable of storing 50/10 = 5 cells.

Some final comments concerning the Knockout switch. First, we note that the assumption of arriving packets being equally likely destined for any output is a worst-case approximation and results in a conservatively designed switch. This observation follows from Figure 3.17 which shows that smaller switches always outperform larger switches. Now, consider a very large switch, but one which is serving a user population with strong communities of interest. The user population

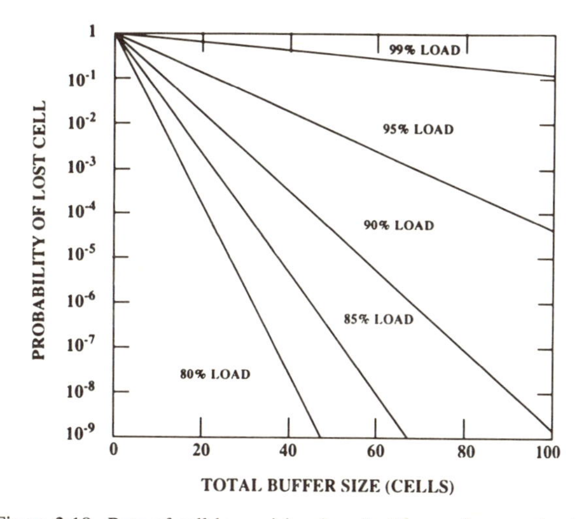

Figure 3.19. Rate of cell loss arising from buffer overflow mechanism.

is divided into distinct groups, each containing, say, 16 users, and groups have the property that each user within a given group communicates only to another user within that group. Let each group be served by a corresponding group of 16 inputs and outputs of the Knockout switch. Since cells from the first group of 16 inputs only seek access to the first group of 16 outputs, and so forth, the large Knockout switch experiences the loading characteristics of a bank of small 16 × 16 switches, operating in parallel, each serving one group. Since, from Figure 3.17, small switches outperform large switches, we conclude that strong communities of interest will reduce the lost cell probability, and the assumption of no communities of interest (packet equally likely destined for any output) is indeed worst case.

Second, we note that Knockout concentrators can be cascaded as shown in Figure 3.20. Suppose an LM-input, L-output Knockout concentrator is synthesized in VLSI, where M is an integer greater than or equal to two. Then, replicas of the same chip can be used to implement an LM^J-input to L-output concentrator, where J is an integer greater than or equal to two. In Figure 3.20, $L = 8$, $M = 4$, $J = 2$. Each of the 32-to-8 concentrators in the upper row of four concentrators finds up to eight active cells among the corresponding 32 input lines. In total, up to 32 active cells may be found, eight from each of the four concentrators. The lower 32-to-8 concentrator now selects up to eight active cells from the up to 32 active cells offered. Overall, if eight or fewer active cells are present among the 128 inputs, each of these will appear on one of the outputs from the lower concentrator. A 512-to-8 Knockout concentrator ($L = 8$, $M = 4$, $J = 3$) could be made by dividing each of the 32 inputs to each of the upper row of concentrators in Figure 3.20 into groups of eight, with each such group fed by one 32-to-8 concentrator chip.

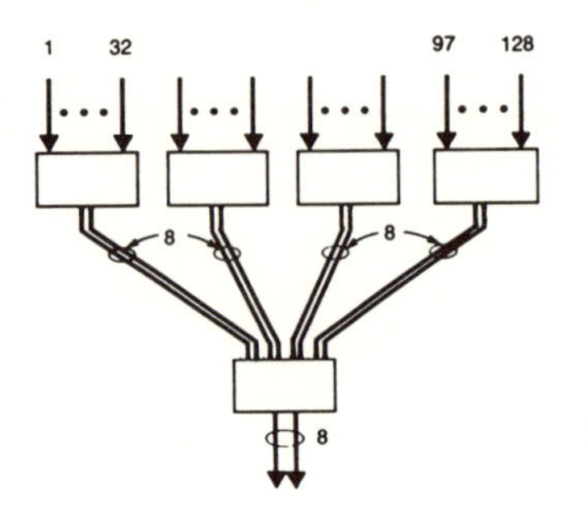

Figure 3.20. Illustrative cascading of Knockout concentrators.

Third, we note that the dimensionality of a Knockout switch can be modularly grown from $N \times N$ to $JN \times JN$, where J is an integer. Referring to Figure 3.21, we note that this is done through the use of an $(N+L)$-to-(L) Knockout concentrator. Each concentrator shown selects active cells from one group of N inputs and, within a given column of concentrators serving one output, the L cells selected from the upper concentrator are fed to a corresponding set of L inputs to the lower concentrator. In this fashion, if up to L packets destined for a given output are present anywhere among the JN inputs, then these will appear at the outputs of the concentrator at the bottom of the column serving that switch output;

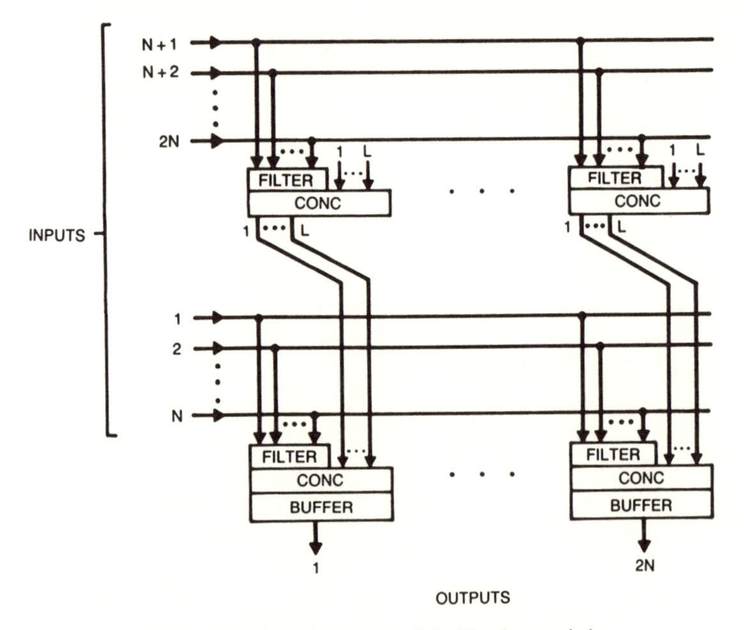

Figure 3.21. Modular growth of the Knockout switch.

the L-input, L-output shifter and the L shared buffers need to be attached only to the bottom concentrator in each column.

Finally, the Knockout switch is fault tolerant and easily maintainable. Referring back to Figure 3.13, we note that all of the active circuitry is divided among identical modules, one serving each output. If a module should fail, only the output served by that module is affected; all other modules operate as usual. Furthermore, because of the broadcast nature of the input buses, and the fact that all modules are identical, spare output modules, each passively connected to the input buses, can be provided to serve as "hot standbys." If an output module should fail, one of the "hot standbys" can be instructed to take over (i.e., its cell filters will be instructed to accept those packets bound for the output served by the failed module), and the output of that spare module would then be mechanically switched to feed the corresponding switch output link.

3.12. Analysis of Cell Loss for the Knockout Switch*

Consider any arbitrarily chosen output from the Knockout switch, say, output J. Let K be the number of cells arriving concurrently in one time slot, bound for output J. For the traffic model described in the prior section (input fill factor $= p$, equally likely bound for any output, independent arrivals), the number K is binomially distributed random variable: for each input, the probability that a cell is present and is bound for output J is equal to p/N; up to N such events may independently occur within any time slot. Thus, for any integer k between 0 and N, the probability that the number of arrivals, K, is equal to k is given by:

$$P(N = k) = \binom{N}{k}(p/N)^k (1 - p/N)^{N-k} \tag{3.12.1}$$

Up to L cells can arrive within any time slot without incurring any cell loss. If $L + 1$ cells arrive, then 1 cell is lost; if $L + 2$ cells arrive, then 2 cells are lost, etc. The expected number of lost cells per time slot, bound for output J, is therefore given by:

$$\begin{aligned} E\{\text{lost}\} &= \sum_{k=L+1}^{N} (k - L)\, P(K = k) \qquad (3.12.2) \\ &= \sum_{k=L+1}^{N} (k - L)\binom{N}{k}(p/N)^k (1 - p/N)^{N-k} \end{aligned}$$

There are, on average, pN cell arrivals per time slot. Since each arriving cell is equally likely to be bound for any of N outputs, the average number of arrivals per time slot bound for output J is simply $(p/N)/N = p$. The lost cell probability, P_{loss}, is simply the expected number of lost cells per time slot at any given output, divided by the average number of cell arrivals per time slot bound for that output.

Thus,

$$P_{\text{loss}} = 1/p \sum_{k=L+1}^{N} (k - L)\binom{N}{k}(p/N)^k (1 - p/N)^{N-k} \tag{3.12.3}$$

The cell loss rate (3.12.2) arising from the Knockout mechanism is plotted in Figures 3.17 and 3.18.

The cell loss rate from buffer overflow can be found by extending the analysis of Section 2.11. For any nonnegative discrete random variable b, the generating function given by (2.11.8) can be used to find the probability that $b = i$, $i = 0,1, \ldots$. From (2.11.15) the generating function for the number of cells in a queue can be found if the generating function of the number of arrivals within a service interval is known. Although the generating function for binomial arrivals is easily found, we assume for Figure 3.19 that the switch is large and that the cell arrival process for any given output may be approximated as Poisson. From (2.11.19) one can easily find the generating function for Poisson arrivals; hence, the generating function for the number of cells in the queue for any particular output is known. Finally, we note from (2.11.8) that

$$\left.\frac{d^k \psi_n}{dx^n}\right|_{x=0} = k!\, p(n = k) \tag{3.12.4}$$

Thus, from (3.12.4), we can find the probability that there are exactly k cells in a given output buffer [see also (4.3.2)–(4.3.6)] From this we approximate the overflow probability P_{L} from a buffer having B cell storage elements by:

$$P_L = \sum_{k=B+1}^{\infty} p(n = k) \tag{3.12.5}$$

This is the result plotted in Figure 3.19.

3.13. The Batcher–Banyan Switch

We saw earlier that the multistage Banyan self-routing packet switch represents a realization in which the number of binary switching elements is minimized, but suffers from internal congestion even when each input is addressed to a distinct output. It is possible to augment the Banyan switch by additional switching stages, thereby producing a nonblocking architecture; the Benes switch is one such example. However, although the Benes switch is nonblocking, it is not a self-routing architecture: the setting of each binary switching element is a complex function of the distinct outputs intended for all inputs, and it is not possible for any β element to determine its correct state from only the destination fields of the cells appearing on its two inputs.

The Batcher–Banyan is a switch architecture which is both nonblocking and self-routing. For this architecture, the number of inputs and outputs must be a power of two, and the switch operates on time-aligned, fixed-length packets.

A high-level block diagram of an 8 × 8 Batcher–Banyan is shown in Figure 3.22. There are two functional blocks. The first of these is a Batcher sorting network which has the following property. Assume that all active inputs are addressed to distinct outputs. Then, if we regard the destination fields of the cells applied at the inputs to a Batcher sorting network to be an unordered list, the network will produce an ordered list (or sort) the output fields in ascending order: the input addressed to the lowest-numbered output will be routed to the bottom (lowest output) of the sorted list, and the input addressed to the next higher-numbered output will be routed to the next-to-the-bottom output of the sorted list, etc. The outputs of the sorting network are filled sequentially, with the higher-order outputs remaining unfilled if not all inputs are active. Thus, referring to Figure 3.22, we see that, for this example, there are six active inputs. The first bit in each cell is the activity bit. This is followed by the destination field, where outputs are numbered starting with zero. For the 8 × 8 example of Figure 3.22, the

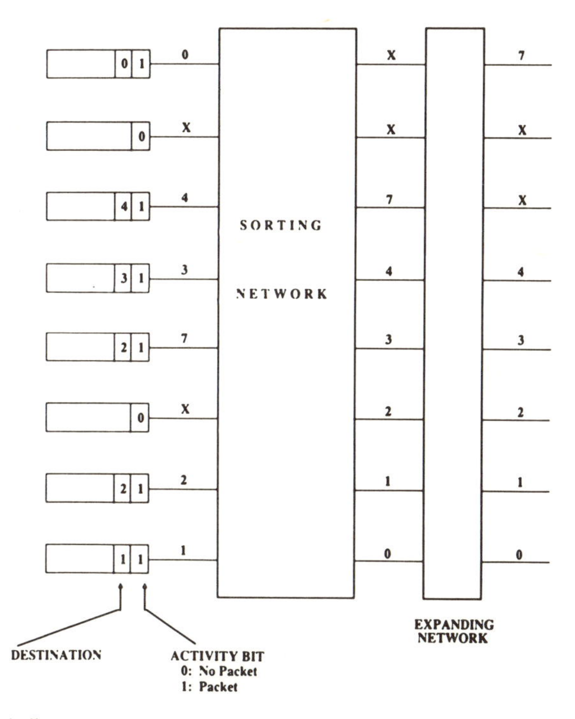

Figure 3.22. Block diagram showing the operation of a Batcher–Banyan switch, drawn for $N = 8$ and no output port contention.

outputs are correspondingly labeled from zero to seven. We note that all outputs shown are distinct; conflicts which will certainly arise in unscheduled packet switching will be discussed later. The input appearing at the top of the input list is addressed to output zero. The next input is idle, as denoted by an "X." Continuing down the input list, we see that the next three inputs are addressed to outputs four, three, and two, respectively. These are followed by another idle input. The last two inputs are addressed to outputs two and one, as shown. For this particular choice of distinct outputs, the sorting network will produce the ordered list shown at its output; an unordered list has been converted to an ordered list, but one which is not sequentially numbered, since outputs five and six are idle. A second module known as an expander network is therefore needed, since it is not enough for the switch to merely produce an ordered list; it is also required that the inputs be routed to their correct outputs. (We shall see that if an ordered but nonsequential list is presented to the input of a Banyan switch, then each input will, in fact, be routed to its correct destination as shown in Figure 3.22.)

An 8×8 sorting network appears in Figure 3.23 and consists of a multistage arrangement of self-routing β elements, each programmed to respond in a specific fashion to the activity and destination fields of its inputs. Each column contains $4(=8/2)$ β elements. For the 8×8 example shown, there are three types of groupings of β elements. The first grouping type consists of a single β element;

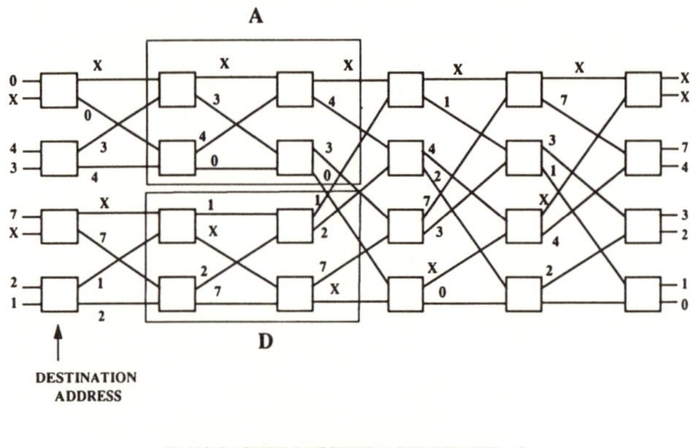

X: NO PACKET PRESENT (ACTIVITY BIT = 0)
A: ROUTE IN ASCENDING ORDER
D: ROUTE IN DESCENDING ORDER

ASSUMPTION: NO TWO DESTINATION ADDRESSES ARE THE SAME
(CONTENTION TREATED LATER)

Figure 3.23. Illustration of the operation of an 8×8 Batcher sorting network.

the four β elements in the first stage each fall into such a grouping. The next grouping type consists of four β elements; there are two such groupings as shown in stages 2 and 3. The final grouping contains the 12 β elements appearing in stages 4 through 6. With the exception of those groupings which contain only a single β element, the β elements of each grouping are interconnected by perfect shuffle patterns. Furthermore, the first stage of a particular grouping is interconnected with the last stage of the next-smaller grouping using a perfect shuffle pattern. For example, in Figure 3.23, the four β elements in the grouping marked "A" are arranged in two columns which are interconnected by a 4-input, 4-output perfect shuffle; the first column in this grouping is also interconnected with two β elements of the prior column by means of a 4-input, 4-output perfect shuffle. Similarly, the 12 elements of the largest grouping are arranged in three columns, each with four β elements. Column 1 is connected to column 2 by an 8-input, 8-output perfect shuffle, and column 2 is connected with column 3 by another 8-input, 8-output perfect shuffle. The first column of this grouping is connected with the prior column (containing β elements from smaller groupings) by yet another 8-input, 8-output perfect shuffle.

In general, for an $N \times N$ Batcher sorting network, where N is a power of 2, each column contains $N/2$ self-routing β elements, and there are $\log N$ types of groupings. The first grouping type contains a single element and occupies a single stage in the switch; there are $(N/2)$ such groupings. The second type contains four elements and occupies two stages in the switch; there are $N/4$ such groupings, etc. The last type contains $(N/2) \log N$ elements and occupies $\log N$ stages in the switch; there is one such grouping. We see that the total number of columns in the Batcher sorting network is given by $1 + 2 + 3 + \ldots + \log N = \frac{1}{2} \log N(1 + \log N)$, and each column contains $N/2$ self-routing β elements. The total number of β elements is therefore equal to $(N/4) \log N(1 + \log N)$. The columns of a grouping which overlaps k columns are interconnected by 2^k input, 2^k output perfect shuffles, and the first column in such a grouping is interconnected with β elements in the last column of the next-smaller groupings with another 2^k input, 2^k output perfect shuffle.

Referring again to Figure 3.23, β elements within a given grouping are all programmed to sort in either ascending or descending order. Each β element within the first column is contained in a single-element group; the β element at the top of the column sorts in ascending order, with the sorting rule alternating between ascending and descending order as we proceed down the column. The β elements in the upper group of four (within the second and third columns) all sort in ascending order, and those in the lower group of four all sort in descending order; for a larger switch, the sorting rule would again alternate as we proceed down these two columns containing groups with four β elements each. Each β element in the final group of 12 is programmed to sort in ascending order. In general, for a switch of larger dimensionality, the sorting rule of the groupings containing the same number of β elements would alternate as we proceed down the columns comprising those groupings.

When sorting in ascending (descending) order, a β element inspects the activity and destination fields of its inputs and, if both inputs are active, sets its state such that the packet containing the higher (lower) numbered destination appears on its upper output line. If either input is idle, the β element obeys its assigned sort rule, but always assumes that the idle input is addressed to the higher-numbered destination. Thus, for example, if we look at the β element at the top of column 2, we see that its upper input is idle; since that β element is programmed to sort in ascending order, it sets itself in the bar state (it assumes that the idle input is addressed to a higher-numbered output than is the active lower output). Similarly, if we look at the third β element in column 2, we see that its upper input is also idle; since that β element is programmed to sort in descending order, it must set itself in the cross state. If both inputs to a β element are idle, then that element can choose its state arbitrarily (no information is transported through that β element during that time slot).

The destinations of the inputs in Figure 3.23 are identical to those of Figure 3.22. The sort rules are applied for each of the β elements; we see that the desired ordered list is produced at the output. It remains now to "expand" this ordered list to the desired outputs. This is accomplished by means of the 8 × 8 network shown in Figure 3.24. This network contains $(N/2) \log N$ switching elements, and there is exactly one path from any input to any output. The switch exhibits the same blocking characteristics as the Banyan switch of Figure 3.10. Because of the similarities, we shall refer to the network of Figure 3.24 as a Banyan, despite the fact that an interstage interconnection pattern different from that of figure 3.10 is used. In general, an $N \times N$ Banyan with the interconnection pattern of the 8 × 8 example of Figure 3.24 would be needed for an $N \times N$ Batcher–Banyan switch. The first stage of the Banyan network is connected with the last stage of the Batcher network by means of an 8-input, 8-output perfect shuffle, and consecutive columns of the Banyan are similarly connected by 8-input, 8-output perfect shuffles. Since $N = 8$, there are three bits in the destination address of each packet, and there are three stages in the Banyan switch. Each β element contained in any given column is programmed to set its state in response to the activity bit and the *one* particular bit of the destination field corresponding to that of the column in question; the first column is reactive to the first (most significant) bit of the destination address, and so forth. In general, there are $\log N$ stages in the expander network and a one-to-one correspondence with the $\log N$ bits in the destination field.

The fact that the packets appearing at the inputs to the expander network were presorted by the Batcher sorting network precludes the controlling bit in the destination fields of two active packets from being the same on both inputs of any β element in the first stage of the expanding network. This, coupled with β element

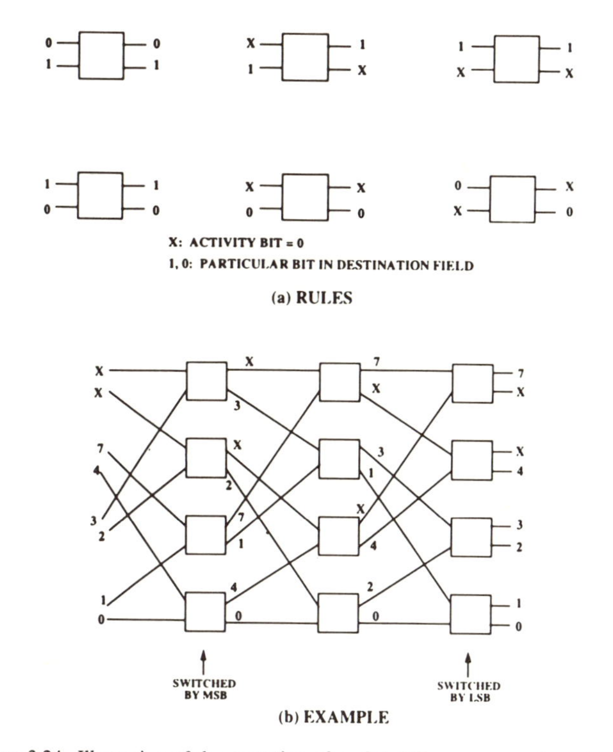

Figure 3.24. Illustration of the operation of an 8 × 8 Banyan expander network.

routing in accordance with the rules appearing in Figure 3.24, similarly preclude the controlling bit in the destination fields of two active packets from being the same on both inputs of any subsequent β element. Thus, for each β element, there are only six possible events, each of which must produce either the bar or cross state. If both inputs contain active packets (corresponding to the control bit on the upper input being a "1" and that on the lower input being a "0," or vice versa; the control bits cannot be the same on both inputs per the above), then the β element is set in the bar state. If the upper input is idle and the control bit on the lower input is "1," then the β element is set in the cross state; if the upper input is idle and the control bit on the lower input is "0," then the β element is set in the bar state. If the lower input is idle and the control bit on the upper input is "1," then the β element is set in the bar state; if the lower input is idle and the if the control bit on the upper input is "0," then the β element is set in the cross state. If these rules are applied to the sorted list appearing at the input to the expander network of Figure 3.24 (which corresponds to the output of the Batcher network for Figure

3.23's example), then we see that each active packet will be routed to its correct destination port.

Thus, we see that if each active input is addressed to a distinct output port (i.e., if there is no output port contention), then the tandem combination of a Batcher sorting network containing $(N/4) \log N(1 + \log N)$ self-routing β elements, and a Banyan expanding network containing $(N/2) \log N$ self-routing β elements, produces a self-routing packet switch in which the state of each β element of the multistage architecture is uniquely determined exclusively on the basis of information contained in the activity and destination fields of the two cells appearing at its input. A Batcher–Banyan switch might therefore be used as the cell router for the modified input queuing system depicted in Figure 3.9; in such an arrangement, the switch controller selects cells from the input queues such that no two cells appearing at the inputs to the cell router are destined for a common output. Because the Batcher–Banyan is self-routing, it might represent a better choice for the router than would a Benes network in which the state of each β element is a complex function of the input and destination ports of all arriving cells. (Recall that the Benes network is *not* self-routing; for each of the $N!$ possible ways that N applied inputs could be routed to a distinct output, each β element would need to be told its correct state.)

Can the Batcher–Banyan be used when the simultaneously applied input cells are not destined for distinct outputs? The answer to this question is yes, provided that some modifications are made. As presented thus far, the Batcher–Banyan provides the required routing function of a packet switch in which the inputs are scheduled to avoid output contention. The smoothing function, which temporarily stores and delivers contending packets on a deferred basis, must yet be provided. We have already seen one approach to such smoothing, namely, the modified input-queued switch of Figure 3.9 in which buffering is done at the input and the Batcher–Banyan is used as the cell router. Another approach, which is more in the spirit of output queuing and, more importantly, totally avoids the need for scheduling at the input to the router, appears functionally as Figure 3.25.

The N-input, N-output switch depicted in Figure 3.25 contains a Batcher sorting network of dimensionality greater than N, followed by an $N \times N$ expander network. As before, the cells simultaneously incident upon the expander network within any time slot are, under normal conditions, sorted in ascending order by destination and, as before, the expander routes each cell from this ordered list to the correct switch output. However, the cells simultaneously applied to the N inputs of the Batcher network need *not* be prescheduled so as to be destined for distinct outputs. Rather, the β elements of the enlarged Batcher sorting network are programmed such that if two or more cells with a common destination field are simultaneously present, then, with very high probability, all but one of these cells will be deflected to one of the output ports of the sorting network which does not interface to the expander; the depth of the sorting network (and, therefore, its

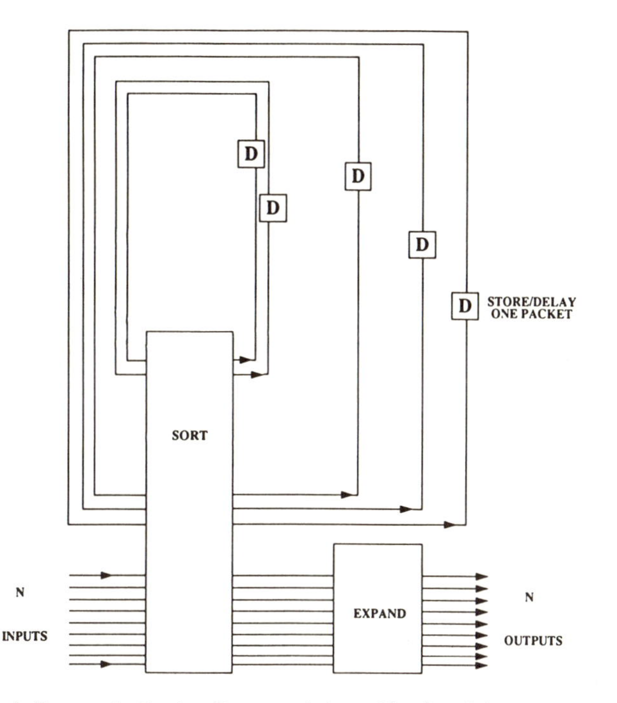

Figure 3.25. Block diagram of a Batcher–Banyan switch capable of resolving output port contention.

overall dimensionality) is chosen such that the probability that two or more cells with the same destination simultaneously arrive at the expander network is very small. By increasing the depth of the sorting network, we increase the likelihood that contending cells will be processed by a common β element, which can then detect the contention and route one of the contending cells toward an output different from those leading directly to the expander. Essentially, upon detection of contention, one of the two contending cells can be "tagged" such that, within subsequent stages of the sorting network, the "tagged" cells are regarded as having a destination address sufficiently high as to cause their deflection away from the expander network.

What becomes of the "tagged" cells? We see from Figure 3.25 that these are deflected toward outputs of the sorting network which are connected to storage elements, each of which can store one single fixed-length cell. These storage elements essentially provide one time slot of delay, prior to reintroducing the "tagged" cells to the inputs of the enlarged sorting network. In the next time slot, these "recirculated" cells contend (using their original destination fields), both among themselves and with any new arrivals, for their place in the sorted list appearing at the input to the expander. All but one of the contending cells intended for a common destination are again routed by the sorting network toward the one-cell storage elements where, after a delay of one time slot, they are again reintroduced at the input to the sorting network to contend among themselves and with any new arrivals for their place on the sorted list appearing at the input to the expander.

We see that, collectively, the one-cell storage elements essentially comprise one large output queue containing those cells which, because of contention, could not be immediately delivered to their correct destinations. Those cells which could not be immediately delivered are recirculated and try again in the subsequent time slot. Except under unusual circumstances, to be subsequently described, the expander is always presented with a "list" of cells, sorted by destination in ascending order, each of which can then be immediately delivered to its correct output.

We note that the recirculating buffers represent a "closed" queuing system: once a cell is deflected toward a recirculating buffer, it cannot cause the buffer to overflow (each buffer can store one cell which it redelivers to the sorting network in the next time slot, and during a given time slot, at most one cell can be delivered to any buffer). How does such a self-routing packet switch, containing an enlarged routing network, an expander network, and a set of recirculating buffers, lose packets? Recall that in any packet switch with finite buffer space, some packet loss caused by buffer overflow is unavoidable. In the Batcher–Banyan switch with recirculating buffers, such "buffer overflow" would occur within some given time slot if there are an insufficient number of recirculating ports to store all of the cells which could not be delivered to their correct outputs within the time slot. Suppose, for example, that there are M recirculating buffers and that, because of their prior cell arrival patterns, each recirculating buffer contains a cell intended for output No. 1. Then, during that time slot, there are $M+2$ cells contending for output No. 1 (two new arrivals plus M recirculated cells). Only one of these $M+2$ cells can be successfully delivered to output No. 1 during that time slot, implying that $M+1$ cells must be stored for deferred delivery. However, there are only M storage buffers, one per recirculating port. The result is that the Batcher sorting network will cause one of the $M+1$ cells which cannot be immediately delivered to be misdirected toward the expander network. The expander network operates properly only when it is presented with an ordered list, with each destination appearing at most once. The end result is that, during this time slot, one cell will be correctly delivered to output No. 1, any cells appearing at the input to the expander addressed to distinct outputs different from output No. 1 will be misdelivered to the wrong output from the expander. One final address filter connected to each output (not shown in Figure 3.25) is then responsible for checking the destination field of every cell passing through that port against that port number; if a cell should be discovered which does not belong on that output, then the channel filter will drop the cell (the channel filter is, in reality, a 1-input, 2-output β element which is normally set in the bar state, connecting its input to the corresponding output line, but which, when set in the cross state, deflects its input to its open output which is not connected to the output line). This final act of dropping cells by checking their destination fields against physical output ports is the buffer overflow mechanism of the buffered Batcher–Banyan switch.

Because the recirculating buffers effectively comprise a single output queue shared among all switch outputs, calculation of buffer overflow probability is extremely difficult, even for simple cell arrival statistics. Extensive simulations performed under the assumption of cell arrivals which are uncorrelated in time, independent among inputs, and equally likely to be bound for any output, having yielded the following performance benchmark: if, for each input, the arrival rate is 85% (i.e., the probability that, during a given time slot, a cell is present on a given input is equal to 85%), then a lost cell rate of 1×10^{-6} can be achieved if there are approximately five recirculating lines for every one active line. Thus, an $N \times N$ switch requires a sorting network of dimensionality $6N \times 6N$. However, since the number of inputs or outputs must be a factor of 2, and since N must also be a factor of 2, we conclude that the sorting network must be made slightly larger, namely $8N \times 8N$. The required number of β elements is therefore $(8N/4) \log 8N (1+\log 8N)$. Fortunately, the required number of β elements is not a good measure of switch complexity; a better measure is the number of distinct buffers needed. For the Batcher–Banyan with recirculating buffers, the required number of buffers is $8N$. We note that, as with the Knockout switch, the Batcher–Banyan with recirculating buffers achieves the performance of output queuing (no head-of-queue blocking) with the simplicity of input queuing (number of buffers needed grows linearly with N). Further, we note that for the same offered load, a lost packet rate of 1×10^{-6} is achieved by the Knockout switch if there are eight buffers per output, for a total buffer count of $8N$, each of depth five. Although each of the Batcher–Banyan buffers is of depth one while, for the same offered load and lost cell rate, each Knockout buffer must be depth five, it is the total number of distinct buffers needed, and not the required depth of each buffer, that determines switch complexity and, by this measure, both switches are of comparable complexity.

Finally, we note that the use of recirculating buffers with a Batcher–Banyan switch can cause cells to be delivered out of sequence; an earlier, recirculating packet might be delivered later in time than a newly arriving packet. To minimize such desequencing, an age field can be added to each cell header by the input module to the switch; this field would be detected by the output modules of the switch since the age field has no significance outside the switch. Newly arriving cells would be assigned an age of zero. Each time a cell is deflected toward a recirculating buffer, the value in its age field would be increased by unity. Each β element would then be programmed to give priority among cells contending for access to the expander network to that cell with the higher age. If the two inputs to a given β element have different destinations, the age field would be ignored since these cells are not contending for the same output. If both inputs to a given β element have the same destination and the same age, then they must have originated from different inputs and, again, the age field can be ignored without causing cells which originated at the same input from arriving at their common destination out of sequence. By means of the age field, the time sequence of cells delivered to each output from the same input is effectively maintained to be the same as the sequence of arrivals to the input. However, since the age field contains a finite number of bits (say, k), then the highest age possible is $2^k - 1$; if two or more cells from the same input should remain in the system sufficiently long that more than one such cell has achieved this maximum age, then desequencing might occur since age priority will have become meaningless. The likelihood of such an occurrence can be reduced by increasing the number of bits in the age field or by reducing the intensity of the offered load.

3.14. The Tandem Banyan Switch

An interesting approach which combines the routing simplicity of the Banyan switch with the output buffering simplicity of the Knockout switch is shown in Figure 3.26. Known as the Tandem Banyan switch, the approach, like the Knockout switch, allows a maximum of L packets arriving among N inputs to be accepted by any given output in each time slot. As shown, each output module consists of L channel filters, each of which checks the destination address of each

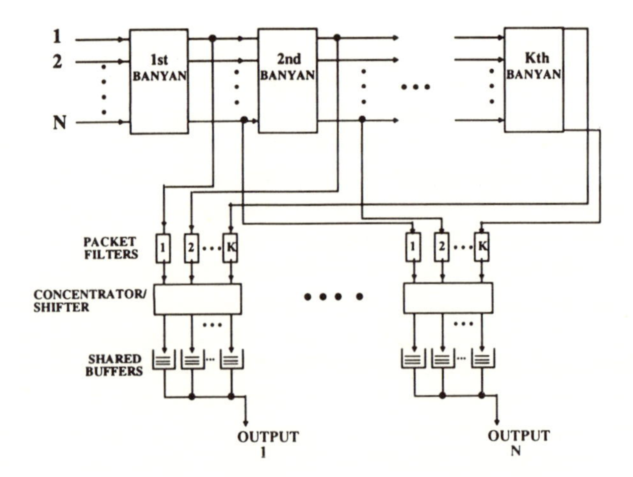

Figure 3.26. Block diagram of the Tandem Banyan switch.

cell appearing at its input and passes that cell to the concentrator/shifter only if the destination is that of the switch output port being served by that filter. There are L buffers per output which are cyclically filled by the concentrator/shifter as per the Knockout switch. Again, just like the Knockout switch, if we measure switch complexity by the number of buffers needed, then, since each output requires L buffers (where L is independent of N), the complexity of the Tandem Banyan switch grows linearly with N. And, like the Knockout switch, the Tandem Banyan can lose cells from two mechanisms: buffer overflow (the rate of loss from which is determined by the depth of each buffer and the magnitude of the applied load) and buffer blocking (the number of cells which can be accepted per time slot by each output is at most $L < N$). The Tandem Banyan does not use a Knockout concentrator to find the simultaneously arriving cells which are bound for the same output. Rather, cells bound for the same output are found by a series of cascaded memoryless $N \times N$ Banyan switches. As we discussed earlier, the Banyan switch provides exactly one path from each input to each output, but the switch is not nonblocking, that is, even if each input is addressed to a distinct output, conflicts may arise on one or more of the internal links. Since, within any time slot, each arriving cell may be bound for any output, contention will arise if two or more cells are destined for the same output or if two cells bound for different outputs both need to use a common internal link. In either case, the β element where the conflict appears will correctly route one cell and misroute the other. Since there is only one path from any input to any output, a misrouted cell cannot possibly find its way to its intended output from that Banyan, and will therefore emerge from that Banyan on the wrong output line. Because misrouted cells cannot possibly find their way to their correct outputs, they are given lower priority at successive stages within that Banyan (this can be done by adding priority bits to their headers which are only used internally to the switch and which are therefore removed at the output of the switch). Thus, if the two arrivals to a given β element should compete for the same output, and if one of the arrivals is proceeding along its correct path while the other has already been misrouted one or more times, the β element will set itself up such as to route that cell, which has not yet been misrouted, to its desired link. In this way, each Banyan seeks to maximize the likelihood that cells reach their intended outputs. As a result of misrouting, each cell appearing at the output of each Banyan may or may not be on the output line which corresponds to its intended output line from the overall Tandem Banyan switch.

After the first of L cascaded Banyan switches, the cell (if any) appearing on each Banyan output is interrogated by packet filters serving the corresponding output for the overall Tandem Banyan switch. Thus, the cell (if any) appearing on the first output of the first Banyan switch is interrogated by the first of L packet filters serving the first output from the overall switch; the cell (if any) appearing on the jth output of the first Banyan switch is interrogated by the first of L packet filters serving the jth output from the overall switch, j between 1 and N. After the first Banyan, each cell having appeared at its correct output is passed by the corresponding packet filter to the corresponding concentrator/shifter, and is *not* passed to the second Banyan switch. Thus, only cells which were misrouted by the first Banyan switch appear at the inputs to the second Banyan switch, and since some cells (those appearing at their correct output) may have been removed by the first Banyan, the load on the second Banyan is reduced, thereby lowering the likelihood of misrouting within the second Banyan switch.

After the second Banyan switch, each cell appearing at its correct output is removed by the packet filter serving that output and is passed to the corresponding concentrator/shifter. Thus, only cells having been misrouted by the second Banyan appear at the inputs to the third Banyan. This process continues sequentially through the various Banyans, with each cell having arrived at its correct output at any given stage being removed and delivered to the corresponding concentrator/shifter. After L consecutive attempts, all cells not yet having arrived at their correct outputs are lost.

The block diagram of Figure 3.26 assumes that the processing delay of each Banyan switch is negligible, and that cells arriving within the same time slot appear at the inputs to the shared buffers in perfect synchronism. In practice, a small amount of delay would need to be added after each packet filter to equalize the delay of subsequent Banyan switches. The first packet filter serving each output would need the longest delay since the delay of $L - 1$ subsequent Banyans must be equalized. The second filter serving each output would need a slightly smaller delay, since the delay of only $L - 2$ subsequent Banyans must be equalized. The last packet filter serving each output would need no delay. In all cases, the amount of equalizing delay actually needed is small, since the β elements operate only on cell headers, and subsequent Banyans can process the cell header before the entire cell has emerged from the preceding Banyan. Thus, the delay is measured in units of cell *headers*, not total cells.

The lost packet performance caused by Tandem Banyan buffer blocking is shown in Figure 3.27. Plotted is the lost cell probability as a function of the number of cascaded Banyans, L, under conditions of 100% loading (each input time slot contains an active cell, equally likely to be bound for any output). These curves were obtained by Monte-Carlo simulation; an exact analysis is intractable because of the blocking nature of the Banyan switch and the resulting dependence of cells lost per time slot on the specific pattern of destinations for cells arriving in each time slot. We see that, unlike the Knockout switch, the number of buffers (L) needed per output to provide a given lost cell probability is not independent of the switch dimensionality (N). For example, a small 32×32 switch needs nine Tandem Banyans to maintain a 1×10^{-6} lost cell probability; a large 1024×1024 switch needs 16 Tandem Banyans to maintain the same lost cell probability. By contrast, with 100% applied load, a very large Knockout switch with dimensionality approaching infinity requires $L = 8$ buffers to maintain a 1×10^{-6} lost cell

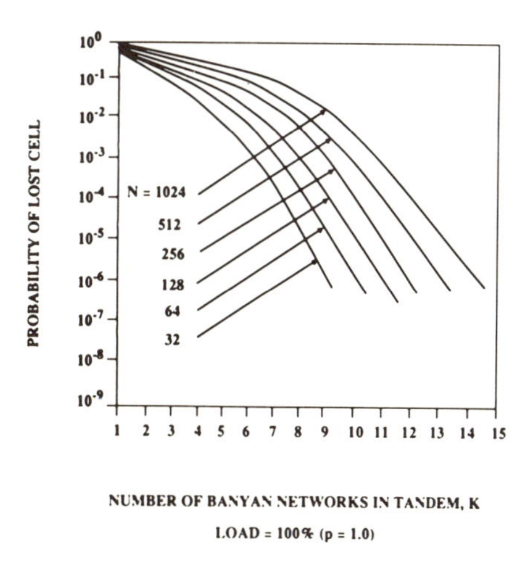

Figure 3.27. Lost cell performance of the Tandem Banyan switch.

probability caused by buffer blocking. The total amount of buffering needed per output (number of buffers × depth of each buffer) to produce some given probability of cell loss arising from buffer overflow is the same for both switches, since the same mechanisms and formulas apply for both switches (provided, of course, that the switches have been designed such that the probability of buffer blocking is negligible compared with the probability of buffer overflow and that the effect of buffer blocking can be ignored when calculating the probability of buffer overflow).

The number of β elements needed to implement a Tandem Banyan switch is substantially lower than that needed for the Knockout switch. For the Knockout switch, N packet filters are needed per output, each requiring one β element, yielding a total of N^2 self-routing β elements needed for packet filtering. For the Tandem Banyan, L packet filters are needed per output, yielding a total of $LN < N^2$ self-routing β elements needed for packet filtering. To select up to L cells per output, each Knockout concentrator requires on the order of LN self-routing β elements, and N Knockout concentrators are needed. The total number of β elements needed for the Knockout concentrators is therefore on the order of LN^2. The number of β elements needed to implement one Banyan switch is $(N/2)\log N$, and L such Banyans are needed for the Tandem Banyan to permit the selection of up to L cells per output. The number of β elements needed for selection is therefore on the order of $(LN/2)\log N < LN^2$ as needed for the Knockout switch. The shifter is the same for both switches and requires a number of β elements which is independent of N. For the Knockout switch, the required L is independent of N; for the Tandem Banyan, there is a dependence which, as shown in Figure 3.27, becomes weaker for large N. Thus, ignoring this dependence for large N, we conclude that the number of β elements needed for the Tandem Banyan is dominated by the needs of the Banyan switch and is on the order of $N \log N$. For the Knockout switch, the packet filters and Knockout concentrators require on the order of N^2 self-routing β elements, and we conclude that, of the two, the Tandem Banyan requires a substantially smaller number of β elements.

To summarize, both the Tandem Banyan and the Knockout switches achieve the simplicity of input queuing with the performance of output queuing. The number of β elements needed by the Tandem Banyan is substantially lower than that needed for the Knockout switch but for the Tandem Banyan switch, the total number of distinct buffers needed per output grows slowly as N increases, whereas for the Knockout switch the number of distinct buffers needed per output is independent of N.

3.15. The Shared Memory Switch

The final switch architecture which we shall discuss is that of the Shared Memory switch, shown in Figure 3.28. A distinctive feature of an $N \times N$ Shared

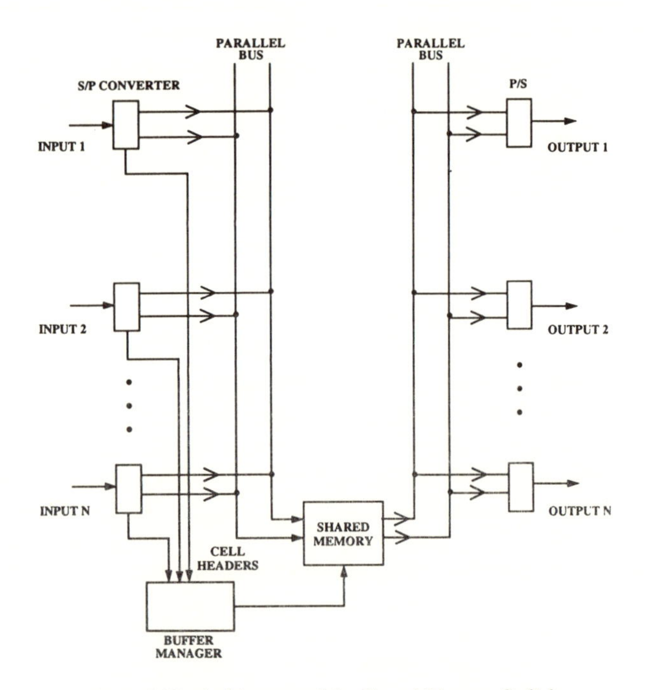

Figure 3.28. Architecture of the Shared Memory Switch.

Memory switch is its use of a high-speed internal bus whose bit rate is a factor of N times as large as the bit rate used on each individual input/output line. For example, if each input/output line operates at 155 Mbit/sec and $N = 16$, then the internal bus operates at 2.48 Gbit/sec.

The Shared Memory switch operates on time-aligned, fixed-length packets or cells. Let F be the time duration of each fixed-length packet applied to the switch. Then, since the internal bus operates at a speed N times as great as the speed of an individual input, the bus can accept a total of N cells per F seconds, or 1 cell per input per time slot F. Time on the high-speed bus can therefore be subdivided into N minislots every F seconds, each of duration F/N, and each minislot can accept one full-length cell since the bus data rate is N times as great as the input data rate. Thus, as shown in Figure 3.29, each input can be allocated a given minislot time slot in each Time Multiplexed frame (over frame length = F seconds).

Once per frame, any input line having a newly arrived cell will place that cell onto that input's assigned minislot. If, for a given frame, no cell should arrive from any particular input, then that input's assigned minislot remains unfilled. Also, since cells arrive on each input at a maximum rate of one cell every F seconds, and each input can discharge a cell onto the internal high-speed bus at a rate of 1 cell every F seconds, no congestion can develop at the input to the switch.

Referring again to Figure 3.28, we note that the cells arriving at each input are converted by a serial-to-parallel converter from bit serial to word parallel form. This avoids the need to operate the internal bus at a bit serial clock speed which is N times as great as the speed of each individual I/O line, while still permitting the data rate of the word-parallel internal bus to be N times as great as that of each individual I/O line, as needed for correct operation of the Shared Memory Switch. Using the earlier example, suppose $N = 16$, the I/O data rate is 155 Mbit/sec per line, and each word is 8 bits wide. Then, the data rate of the internal bus is again 2.48 Gbit/sec but the clock speed is "only" 2.48 Gbit/sec/8 = 310 MHz.

Each frame, the header of each arriving cell is read by a buffer manager, and the contents of the minislots are read into a shared memory under the control of the buffer manager. The shared memory is partitioned into a separate space for each output and, since the buffer manager has read the header of each arriving cell, it can load each cell into the appropriate space in the shared memory.

Finally, to complete its operation, frames of length F are created on the

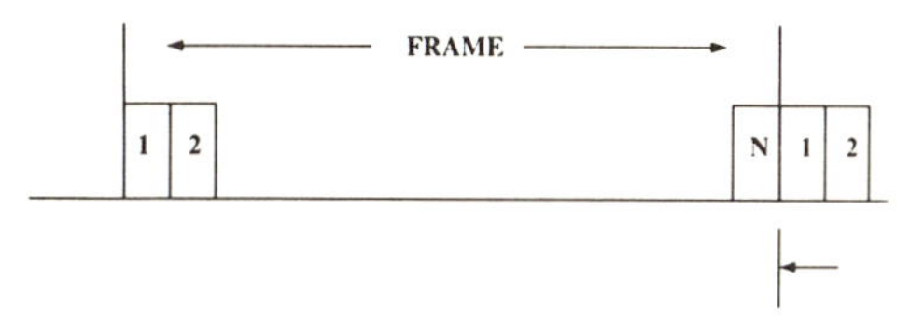

Figure 3.29. Frame and minislot structure for the Shared Memory switch.

high-speed bus leading to the switch output ports. Each frame is divided into N minislots, each of duration F/N, with one such minislot assigned to each output port. In this way, every F seconds, one cell can be read from the shared memory for each output; the buffer manager simply fetches cells from the appropriate memory spaces to fill the outbound minislots. If there are no cells awaiting delivery to a particular output, then the minislot assigned to that output remains unfilled.

Once per frame, the cell (if any) intended for a particular output is converted by that output's parallel-to-serial converter for delivery on the output line in bit serial format. Once converted, the cell again occupies F seconds, as required. Since cells cannot be delivered to any given output port at a rate greater than 1 cell every F seconds (each output is assigned only one minislot per frame F), and the line leading from each output port can deliver one cell every F seconds, no congestion can arise at the output ports.

With the Shared Memory switch, output port contention is resolved by the shared memory, under control of the buffer manager. If, during an input time slot of duration F, cells should arrive on several input lines all bound for the same output, the buffer manager will read these into the buffer space previously set aside for that output (as previously explained, this operation takes only F seconds since the internal bus operates at a data rate N times faster than the data rate of any input, and each input has one access to the shared memory every F seconds). These cells are then sequentially read to their common output port at a rate of one cell every F seconds (again, each output port can receive from the shared memory at a rate of one cell every F seconds).

The shared memory can be operated with rigid partitioning between the spaces set aside for the various outputs, in which case the delay and lost cell performances are indistinguishable from those of ordinary queuing (cells arriving for a given output will be lost whenever the space assigned to that output is already filled by cells having previously arrived). Alternatively, the buffer manager may dynamically repartition the total buffer space in response to the actual distribution of cells stored for each output. For example, if there are no cells present for output J_1 but many cells present for output J_2, some space previously set aside for output J_1 may be reassigned to J_2. In this way, for a fixed total amount of memory, the lost cell performance of rigid partitioning may be improved upon.

Because the data rate of the internal bus grows linearly with the number N of I/O ports, the shared memory is not modular, and the number of I/O ports is generally much smaller than that possible with the Knockout, Batcher–Banyan, and Tandem Banyan switches. However, a large switch can be made by using multiple shared-memory switches in an appropriate multistage arrangement. Because bus speed increases linearly with N, the number of buffers which must be operated in parallel to synthesize the shared buffer also scales linearly with N since the access rate of a buffer (number of I/Os per second) is generally limited.

3.16. Concluding Thoughts on Packet Switch Interconnecting Fabrics

We see that, for any packet switch, traffic congestion always gives rise to some unavoidable packet loss. The lost packet rate from such a switch is affected by

1. The switch architecture (switches without smoothing buffers and switches such as the Banyan which are not nonblocking generally exhibit poor lost packet performance)
2. Placement of the smoothing buffers (for fixed-size buffers, input queuing gives rise to a higher rate of packet loss than does output queuing)
3. The depth of the smoothing buffers (larger buffers will overflow less frequently than smaller buffers)
4. The intensity of the applied traffic (for a given switch, the target lost packet rate can be maintained by appropriately limiting the intensity of applied traffic by means such as admission control)

Rate of loss from the first three factors can be controlled by proper switch design. For a given switch design, rate of loss from the fourth factor can be controlled by proper traffic management.

Switch architectures can be distinguished by their complexity (a good measure of complexity is the number of buffers needed) and their traffic-handling capability. We have seen that several switch architectures (Knockout, Batcher–Banyan, Tandem Banyan, and Shared Memory) are capable of operating at traffic loads approaching 100% (a characteristic of output queuing) with a complexity which scales only linearly with N (a characteristic of input queuing). Of course, as the applied traffic load approaches 100%, both the queuing delay and the lost packet probability grow large.

3.17. Problems

P3.1. Consider an N-input, N-output, time-slotted memoryless packet switch. On each input, the probability of an arrival during any time slot is p; arrivals are uncorrelated in time and are independent among inputs. Each arriving packet is equally destined for any output. For p sufficiently small, show the packet loss probability scales linearly with p.

P3.2. For a time-slotted memoryless packet switch, suppose that the number of packets arriving per time slot for a particular output is a Poisson-distributed random variable with mean value n. Find the packet loss probability. Compare your answer with that for Problem 3.1 when the loss probability is small and the number of switch ports is large.

P3.3. A particular data communication system uses fixed-length packets, each containing 1000 bits. The data rate on all transmission links is 10 Mbit/sec. Service is connection-oriented, and each source involved in a connection independently generates a packet in each time slot at an average rate of 100 packets/sec. Sources are interconnected by an N-input, N-output memoryless time-slotted packet switch, and a given connection is equally likely to be destined for any output. (a) Find an expression for the packet loss probability as a function of the number of active connections bound for a given output. (b) What is the maximum number of connections which can be made to any output such that the loss probability is less than 10^{-6}?

P3.4. Consider an $N \times N$ memoryless time-slotted packet switch. Arrivals are independent in time and uncorrelated among inputs, and the activity factor per input is p. The switch contains two stages in cascade. Packets that are lost from the first stage are given a second chance in the second stage; in effect, the composite switch can accept up to two packets per time slot bound for a given output. (a) Find the overall packet loss probability and plot as a function of p for $N = 2, 3, 8$. (b) Find and plot the asymptotic result for large N and compare against part a.

P3.5. Consider an $N \times N$ memoryless time-slotted packet switch. Arrivals are independent in time and uncorrelated among inputs, and the activity factor per input is p. Each time slot, any packet which arrives on input 1 is given priority access to its intended output. (a) Find an expression for the loss probability for packets which arrive on inputs other than input 1. (b) Plot this expression as a function of p for $N = 2, 3, 8$. (c) Find and plot the asymptotic result for large N.

P3.6. Consider an $N \times N$ time-slotted packet switch with input queuing where N is an integer multiple of 4. Arrivals are correlated: arrivals on the first group of 4 inputs are equally destined only for any of the first 4 outputs; arrivals on the next group of 4 inputs are equally destined only for the second group of 4 outputs; and so forth. Each time slot, a packet is chosen at random from each queue containing a nonzero number of packets and applied to the switch fabric. Find an expression for the saturation throughput and compare against the case where arrivals are uncorrelated among inputs.

P3.7. For an $N \times N$ time-slotted packet switch with input queuing, arrivals are independent in time, uncorrelated among inputs, and equally likely to be destined for any output. The input activity factor is p, and the slot duration is T. A packet is chosen each time slot from each queue having a nonzero number of packets, and applied to the switch fabric. In choosing packets from each queue, priority is given to those intended for output 1, and, among these, one (if any) is chosen at random and delivered to output 1. Find an expression for the average delay (including service time) for packets bound for output 1.

P3.8. Arrivals to an $N \times N$ input-queued time-slotted packet switch are independent in time, uncorrelated among inputs, and equally likely to be destined for any output. The activity factor for each input is p, and the slot duration is T. Each time slot, a given input is assigned to a randomly chosen output such that no two inputs are simultaneously assigned to that output. If an output queue contains a packet bound for the output assigned to that input, the packet is transferred to its output; if not, then that output's time slot remains unoccupied. Find an expression for the average packet delay, including the service time.

P3.9. Consider an output-queued time-slotted packet switch. Arrivals are independent in time, uncorrelated among inputs, and equally likely to be destined for any output. The activity factor for each input is p, and the slot duration is T. The switch is preceded by a processor which accepts only up to L out of N possible arrivals per time slot bound for any given output, with other packets discarded. Accepted packets are delivered to their respective output buffers, each of which can store an arbitrarily large number of packets. (a) Find an expression for the probability of packet loss. (b) For those packets which are accepted, find an expression for the average delay, including service time.

P3.10. Parameters of an output-queued time-slotted packet switch are as follows:

Packet size = 1000 bits
Link speed = 10 Mbit/sec
Activity factor = 0.7
Number of switch ports = 8

Arrivals are independent in time, uncorrelated among inputs, and equally likely to be destined for any output. Find the average packet delay, including service time.

P3.11. Consider an $N \times N$ output-queued time-slotted packet switch of slot duration T. Arrivals are independent in time and uncorrelated among inputs. Arrivals not intended for output 1 are equally bound for any other output; each arrival is 10% more likely to be bound for output 1 than for any other specific output. Assuming the existence of buffers that can each store an arbitrarily large number of packets, find the maximum permitted input activity factor such that the average delay experienced by packets bound for output 1 remains finite.

P3.12. For the same conditions as apply for Problem 3.11, find and plot the average delay of packets bound for output 1 assuming that (a) $N = 4$, (b) N approaches infinity.

P3.13. Consider an $N \times N$ output-queued time-slotted packet switch of slot duration T. Arrivals are independent in time and uncorrelated among inputs, and the input activity factor is p. Each output buffer can store an arbitrarily large number of packets. There are K strong communities of interest: each of the first group of N_1 inputs communicates only with the first N_1 outputs, each of the second group of N_2 inputs communicates only with the second group of N_2 outputs, . . . , each of the last group of N_k inputs communciates only with the last group of N_k outputs, and $N_1 + N_2 + \dots + N_k = N$. Within each community, arrivals are equally destined for any output. (a) Find the average delay experienced by packets within group J, $1 \leq J \leq K$, including service time. (b) Show that, for a given activity factor, the average delay within any group is always lower than that experienced for the same activity factor in the absence of communities of interest (that is, when each input is equally likely to be destined for any of the N outputs).

P3.14. For an output-queued $N \times N$ time-slotted packet switch, each input is either busy or idle. If busy, the activity factor is p; if idle, the activity factor is zero. If busy, the packets produced by the source connected to that input are independent in time, and arrivals are uncorrelated among inputs. Each new arrival is equally likely destined for any output. Each output buffer can store an arbitrarily large number of packets. The slot duration is T. (a) For $p = 0.5$, find the maximum number of inputs that are permitted to become busy such that the average delay, including service time, is less than $10T$. (b) Repeat part a for $p = 0.75$. (c) Repeat part a for $p = 1.0$.

P3.15. Consider a 4 × 4 time-slotted packet switch with modified input queuing. At a given moment, the pattern of packets stored at the inputs is given by the matrix

$$\mathbf{C} = \begin{bmatrix} 3 & 0 & 2 & 3 \\ 4 & 4 & 0 & 0 \\ 0 & 2 & 4 & 2 \\ 1 & 2 & 2 & 3 \end{bmatrix}$$

where $C_{i,j}$ = number of packets stored at input i destined for output j; $i = 1, 2, 3, 4$; $j = 1, 2, 3, 4$. Suppose that there are no new arrivals until after all of the packets in **C** have been scheduled and delivered. (a) What is the minimum number of time slots needed to deliver the packets of **C**? (b) Produce a nonconflicting schedule which delivers the packets of **C** within the smallest possible number of time slots.

P3.16. Consider an $N \times N$ time-slotted packet switch, where N is an integer power of 4. The switch is composed of 4 × 4 switching elements, each of which can produce any of 4! input-to-output permutations. (a) What is the minimum number of such switching elements needed such that the $N \times N$ switch is rearrangeably nonblocking? (b) How many β elements are needed to produce one 4 × 4 switching element? (c) Repeat part a under the assumption that each 4 × 4 switching element can produce only two input-to-output permutations.

P3.17. The scheduling algorithm for a modified input-queued space division packet switch always produces a diagonal containing one packet for output 1 whenever there are one or more packets present anywhere in the system bound for output 1. Arrivals are equally likely bound for any output, are independent in time, and are uncorrelated among inputs. The input activity factor is p, and the slot duration is T. Find the average delay, including service time, of a packet bound for output 1.

P3.18. The following situations apply to an 8 × 8 Banyan switch comprised of 2 × 2 switching elements. For each, determine if the input–output combination can be supported and, if so, show a permissible setting for each β element.

	(a) Output	(b) Output	(c) Output	(d) Output	(e) Output
Input 1	5	7	8	2	2
Input 2	2	6	7	1	Idle
Input 3	1	idle	6	4	idle
Input 4	4	4	5	5	4

(continued)

	(a) Output	(b) Output	(c) Output	(d) Output	(e) Output
Input 5	8	3	4	Idle	5
Input 6	7	idle	3	3	Idle
Input 7	3	Idle	2	6	6
Input 8	6	1	1	8	7

P3.19. For the combinations of Problem 3.18, which can be supported if each internal link is operated at twice the rate of each external link?

P3.20. The operating environment of an $N \times N$ Knockout switch contains soft communities of interest, where N is even. For this switch, one particular output is three times more likely to be accessed by each of the first $N/2$ inputs as it is to be accessed by each of the second $N/2$ inputs. The average rate of packet arrivals for that output is one per time slot and, in a given time slot, packet arrivals are uncorrelated among inputs. Find the packet loss rate arising solely from the Knockout mechanism as a function of L (the maximum number of simultaneous arrivals per output which can be stored), and plot for the large switch asymptotic regime as N becomes unbounded.

P3.21. Let b be a Poisson-distributed random variable with mean B. Find its generating function.

P3.22. Each output buffer of an $N \times N$ Knockout switch can store an arbitrary large number of packets. The Knockout concentrator can save up to L simultaneous arrivals per time slot. Let the number of arrivals to a particular output in any given time slot be a Poisson random variable with mean value, B, and let arrivals in different time slots be independent. Find expressions for the mean and variance of the number of cells stored in the buffer for that output.

P3.23. Repeat Problem 3.22, except that the number of arrivals per time slot is Bernoulli distributed with mean B and maximum value N.

P3.24. For Problem 3.22, find the probability at the end of any time slot that the number of cells in the buffer is equal to (a) 0; (b) 1; (c) 2.

P3.25. Consider a large Knockout switch. The activity factor on each input is 90%. Arrivals are independent in time and uncorrelated among inputs. Each arrival is equally likely to be destined for any output. Design the output module (number of buffers per output, depth of each buffer) such that the overall lost cell rate is 10^{-8}, with the loss from the Knockout concentrator contributing less than 1% to the total. You may use curves from the text to help in the design.

P3.26. The sorting network of an 8×8 Batcher–Banyan switch is programmed such that contending packets destined for output No. 3 have priority access to the recirculating buffers. Arrivals are independent in time and uncorrelated among inputs. The activity factor is 100% on inputs 0–3 and 60% on inputs 4–7. Each arrival is equally likely to be destined for any output. There are 8 recirculation ports on the sorting network. (a) Find the loss rate for packets arriving for output No. 3. (b) Among arrivals for output No. 3, priority for the recirculating buffers is given to the oldest group of packets. Find the average time spent in the system for cells that are not lost to buffer overflow.

P3.27. A particular Tandem Banyan switch has a sufficient number of stages, L, such that, for the input activity factor maintained by the admission controller, essentially every arriving packet is delivered to its intended output module. The number of packets arriving for a given output No. J in consecutive time slots can be represented as a string of independent identically distributed random variables with the following statistics:

$$P(\text{no arrivals}) = 0.25$$
$$P(1\ \text{arrival}) = 0.60$$
$$P(2\ \text{arrivals}) = 0.10$$
$$P(3\ \text{arrivals}) = 0.05$$

Assuming that each output buffer can store an arbitrarily large number of packets, find (a) the average time spent in the switch by packets arriving for output No. J, (b) the probability that there are no packets in output No. J's buffer at the beginning of a time slot, (c) the probability that there is 1 packet in output J's buffer at the beginning of a time slot, and (d) the probability that there are 2 packets in output J's buffer at the beginning of a time slot.

P3.28. Each input and output link of a 32 × 32 shared memory switch operates at a speed of 155 Mbit/sec. The cell size is 53 bytes. (a) What is the required data rate of the high-speed bus, and how large is each mini time slot? (b) The high-speed bus is implemented as eight parallel lines. What is the clock rate of each such line? (c) For each of its cycles, the shared memory must read and write one 8-bit word. Assuming equal time for the read and write operations, what is the maximum allowable read time?

P3.29. The buffer of an 8 × 8 shared memory switch can store 32 cells. The input activity factor is 50%, each arrival is equally likely to be destined for any output, and arrivals are independent in time and uncorrelated among inputs. The buffer space is equally and rigidly partitioned among outputs. (a) Compute the cell loss probability. (b) Repeat part a except that, now, the input activity factor is 90%.

P3.30. A shared memory switch has a buffer capable of storing an arbitrarily large number of cells. Each arrival is equally likely to be destined for any output, and arrivals are independent in time and uncorrelated among inputs. The input activity factor is p. There are three classes of arrivals. The first class has highest service priority, that is, any class 1 cells in the buffer intended for a given output will be sequentially served to that output before any class 2 or class 3 cells are served. Class 2 has second highest service priority, and class 3 has lowest service priority. All classes have equal arrival rates, with different classes arriving independently of each other. The number of switch ports is sufficiently large that the number of arrivals per time slot destined for any given output may be approximated by a Poisson random variable. (a) Find and plot the average time spent in the system for class 1 arrivals as a function of p; (b) repeat for class 2 arrivals; (c) repeat for class 3 arrivals.

References

1. H. Ahmadi and W. E. Dinzel, A survey of modern high performance switching techniques, *IEEE J. Selected Areas Commun.* **SAC-7**(7), Sept. 1989.
2. F. Tobagi, Fast packet switch architectures for broadband ISDN, *Proc. IEEE* **78**(1), Jan. 1990.
3. J. Y. Hui, *Switching and Traffic Theory for Integrated Broadband Networks*, Kluwer Academic, 1990.
4. *IEEE J. Selected Areas Commun.*, issue on Broadband Packet Communications, **SAC-6**(9), Dec. 1988.
5. *IEEE J. Selected Areas Commun.*, issue on Switching Systems for Broadband Networks, **SAC-5**(8), Oct. 1987.
6. D. M. Dias and M. Kumar, Packet switching in N log N multistage networks, GLOBECOM '84 Conf. Rec., Nov. 1984.
7. Y. S. Yeh, M. G. Hluchyj, and A. S. Acampora, The knockout switch: A simple modular architecture for high performance packet switching, *IEEE J. Selected Areas Commun.* **SAC-5**(8), Oct. 1987.
8. J. S. Turner, Design of an integrated service packet network, *IEEE J. Selected Areas Commun.* **SAC-4**, Nov. 1986.
9. A. Huang and S. Knauer, Starlite: A wideband digital switch, GLOBECOM '84 Conf. Rec. Nov. 1984.
10. M. G. Hluchyj and M. J. Karol, Queueing in high performance packet switching, *IEEE J. Selected Areas Commun.* **SAC-6**(9), Dec. 1988.
11. M. J. Karol, M. G. Hluchyj, and S. Morgan, Input vs. output queueing on a space division packet switch, *IEEE Trans. Commun.* **COM-35**(12), Dec. 1987.
12. K. Y. Eng, M. G. Hluchyj, and Y. S. Yeh, A knockout switch for variable length packets, *IEEE J. Selected Areas Commun.* **SAC-5**(9), Dec. 1987.
13. V. E. Benes, *Mathematical Theory of Connecting Networks and Telephone Traffic*, Academic Press, New York, 1965.
14. C. Clos, A study of non-blocking switching networks, *Bell Syst. Tech. J.* **32**(3), March 1953.
15. D. Cantor, On non-blocking switching networks, *Networks* **1**, Dec. 1971.
16. D. C. Opfermen and N. T. Tsao-Wu, On a class of rearrangeable switching networks, Part 1: Control algorithms, *Bell Syst. Tech. J.* **50**(5), May 1971.
17. C. L. Wu and T. Y. Feng, Universality of the shuffle-exchange network, *IEEE Trans. Comp.* **30**(5), May 1981.
18. K. E. Batcher, Sorting networks and their applications, Proc. 1968 Spring Joint Comput. Conf.
19. J. S. Turner, Design of a broadcast packet switching network, *IEEE Trans. Commun.* **36**(6), June 1988.
20. T. T. Lee, A modular architecture for very large packet switches, *IEEE Trans. Commun.* **30**(7), July 1990.
21. F. Tobagi, T. Kwok, and F. Chiussi, Architectures, performance, and implementation of the tandem Banyan fast packet switch, *IEEE J. Selected Areas Commun.* **SAC-9**(8), Oct. 1991.
22. S. Shaikh, M. Schwartz, and T. Szymanski, A comparison of the Shufflenet and Banyan topologies for broadband packet switches, IEEE INFOCOM '90, Conf. Proc.
23. R. Melen and J. S. Turner, Nonblocking multirate networks, *SIAM J. Comput.* **18**(2), April 1989.
24. A. S. Acampora and B. R. Davis, Efficient utilization of satellite transponders via time-divison multibeam scanning, *Bell Syst. Tech. J.* **57**(8), Oct. 1978.
25. I. S. Gopal, G. Bongiovanni, M. A. Bonucelli, D. Tang, and C. K. Wong, Optimal switching algorithm for a SS/TDMA system with variable bandwidth beams, *IEEE Trans. Commun.* **COM-30**(10), Oct. 1982.

26. I. S. Gopal, D. Coppersmith, and C. K. Wong, Minimizing packet waiting time in a multibeam satellite system, *IEEE Trans. Commun.* **COM-30**(2), Feb. 1982.
27. Giacopelli *et al.*, Sunshine: A high performance self-routing broadband packet switch architecture, *IEEE J. Selected Areas Commun.* **SAC-9**(8), Oct. 1991.
28. A Pattavina, Nonblocking architectures for ATM switching, *IEEE Commun. Mag.* **31**(2), Feb. 1993.
29. *IEEE J. Selected Areas Commun.*, issue on Large Scale ATM Switching Systems for B-ISDN, **9**(8), Oct. 1991.
30. Y. Sakurai, N. Ido, S. Gohara, and N. Endo, Large-scale ATM multistage switching network with shared buffer memory switches, *IEEE Commun. Mag.* **29**(1), Jan. 1991.
31. J. M. Wozencraft and I. M. Jacobs, *Principles of Communication Engineering*, Wiley, New York, 1965.

4

Metropolitan Area Networks

4.1. Distributed Queue Dual Bus

As discussed in the introductory chapter of this book, the earliest need identified for a Metropolitan Area Network was to provide connectivity over an extended geography among devices connected to different Local Area Networks (see Figure 1.6). For this application, it was envisioned that the MAN would consist of a single high-speed multiple access channel, and that a single access station on each of the interconnected LANs would serve as that LAN's gateway to its corresponding MAN access station. The traffic presented by a LAN to a MAN access station would be aggregated traffic, that is, time-multiplexed traffic generated by a multitude of devices connected to that LAN, needing to be transferred across the MAN to receivers connected to remote LANs. In effect, each LAN serves as a statistical multiplexer for locally generated traffic destined for other LANs, and this statistically multiplexed (or aggregated) traffic is the load presented to the MAN through each LAN's corresponding MAN access station.

Because a MAN is to carry the aggregated traffic offered from several (or many) LANs, the bit rate of the MAN shared channel is substantially greater than that of a LAN's. Also, the MAN might be expected to provide service over a much greater region than that served by a LAN, implying larger geographical distances. As explained in Section 2.6, under these conditions, a distributed bus operating the CSMA/CD protocol would not be very efficient.

An alternative approach known as Distributed Queue Dual Bus (DQDB) was created specifically to redress the inefficiency of CSMA/CD when operated on a geographically long but at high data rates. The basic scheme appears in Figure 4.1. A DQDB system contains two unidirectional buses which carry information in opposite directions. As we shall see, unlike CSMA/CD, the stations of a DQDB system *must* be linearly ordered along a bus (tree and broadcast star physical topologies are inapplicable). A "frame generator" at the head of each bus generates markers which delineate the boundaries of consecutive 125 μsec frames. Each frame is subdivided into fixed-length time slots. The media access protocol pro-

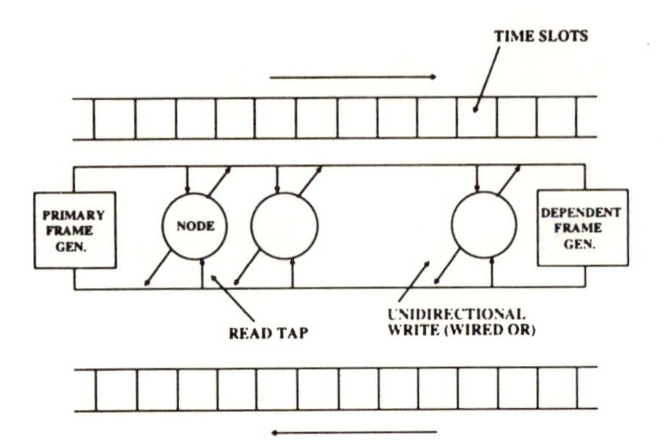

Figure 4.1. Block diagram and operating principles of a DQDB system.

vides an orderly procedure where by active access stations can contend for the right to place fixed-length packets, which we shall call cells, neatly into the DQDB time slots, one cell per slot. Frames and time slots on the upper (lower) bus propagate to the right (left). Relative to the time slot duration, the propagation delay along the buses might be quite large.

Each access station is provided with two passive taps into each bus: a read tap which "eavesdrops" on all bus activity, and a write tap which allows the station to ship cells in the direction of propagation of the corresponding bus. In addition to other functions, the read tap allows the access station to maintain synchronization with the DQDB frame markers; by measuring time elapsed from the local appearance of each frame marker, access stations can establish the time slot boundaries. The read tap also permits a station to recognize its own identity in the destination field of any cell, thereby permitting that cell's payload to be copied into the station's receiver buffer. Other functions of the read tap will be described below.

Any cell written into a time slot of either bus is terminated at the end of that bus; cells on the upper (lower) bus do *not* loop around and reappear on the lower (upper) bus. Frames generated by the frame generators are empty frames, the time slots of which may become filled as the frames propagate along their respective bus.

Access to each bus is mediated by two control bits contained within each time slot: the busy bit and the request bit. For now, we can regard these as occupying the first two bit positions in each time slot. Later, we shall study in greater detail the contents of each time slot.

Before attempting to place an information cell into any time slot, an access station must first determine whether or not the cell's busy bit has been set.

Adequate delay is inserted on the bus between a station's read and write taps (by including an extra length of transmission medium, if necessary) to permit the processing of information observed by the read tap prior to making a write decision. With DQDB, cell collisions cannot occur; a station may not access a time slot known to contain an active cell as determined by the status of that time slot's busy bit. Unlike CSMA/CD, where local determination of an idle bus cannot guarantee against subsequent collision with a packet earlier placed on the bus by an upstream station, the time-slotted nature of DQDB precludes such collisions; although subsequent time slots as seen by a station's read tap may be filled with active cells earlier placed onto the bus by an upstream station, lack of activity in a given time slot as indicated by its busy bit guarantees that that particular time slot may be written to.

The request bit is used, essentially, to make a time slot reservation. It is extremely important to note that the act of reserving a time slot on either bus requires that the request bit of a time slot on the *other* bus be used. Information contained in the cells on either bus is totally independent from information appearing in the cells of the other bus, and the busy bit in any time slot indicates whether or not the payload of that time slot is filled. However, the request bit of any time slot has absolutely nothing to do with the status of that time slot's busy bit and payload; as long as a request bit has not been written into a given time slot by an upstream station, then that request bit can be written by another station, independently of whether or not that cell's payload is filled. Thus, in effect, each time slot contains the equivalent of two independent fields, the busy bit/payload field and the request bit field, and these operate independently.

Each station is equipped with two counters, one to control access to each bus. Let us focus attention on the counter which controls access to the upper bus, whose frames propagate to the right. Each time a station observes a request bit on the lower bus (implying that that request originated somewhere to that station's right), the counter is incremented by unity. Every time that station observes an idle time slot propagating to its right, it decrements the counter by unity, but the value in the counter is never allowed to become negative. We see that if each idle time slot propagating past a given station satisfies a request which was previously generated anywhere to that station's right, then the value contained in a station's counter represents the instantaneous number of *outstanding* requests which were generated to that station's right.

The following simple access rule can now be adopted, and is illustrated in Figure 4.2 which shows a blowup of the access station apparatus and operations required to place a cell onto the upper bus. If a station wishes to transmit a cell anywhere to its right, it must first place a request bit into the empty request field of a time slot propagating to the left. At the time that the request is made, the station observes the value in its counter, and allows an equal number of empty time slots to propagate to its right. The station will then write its waiting cell into the

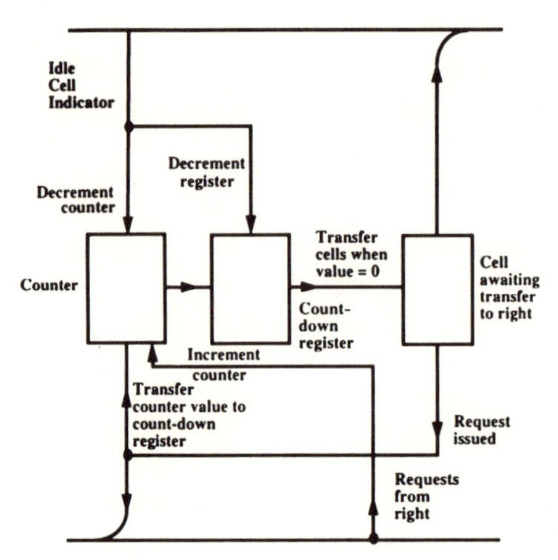

Figure 4.2. Blowup of the access station apparatus and operations required to transfer a cell to the right.

very next idle time slot propagating to its right. During this process, the counter continues to maintain a record of the number of unanswered requests so that the access process can be repeated when the station again wishes to transmit a cell to its right. In effect, the value stored in a station's counter when that station issues a request is transferred to a count-down counter which is decremented by each idle time slot propagating to the right; the station can write a cell when the count-down register reaches zero. At any time, each station can have at most one outstanding request; consequently, only one count-down register is needed per station (actually, two: one for each bus, but for the time being, we are considering access only to the upper bus). Each newly generated cell which arrives during the period of an outstanding request is stored in that station's transmit buffer until a request bit for that cell is written.

If each station obeys these rules and defers transmission of a cell until all requests previously observed to have been generated to that station's right are satisfied, then, during periods of heavy load, each empty time slot which is allowed to pass to the right during the count-down interval will subsequently be filled with a cell (corresponding to one of the outstanding requests) written by a downstream neighbor. The bus utilization efficiency, therefore, can approach 100%; if there is enough traffic, each and every time slot will be filled. At light load it is possible for a slot to unnecessarily remain unfilled, especially if the bus is very long. A station near the right of the bus may make a request which is actually satisfied before that request has propagated past stations to the left of the bus. A time slot will nonetheless be left unfilled by these other stations to satisfy this already satisfied request. Fortunately, if the applied traffic is light, most of the time slots are unfilled because of inadequate demand, and these few additional unfilled slots are of little or no consequence. What the DQDB media access protocol does ensure is that no time slots will be wasted when they are needed, that is, when efficient reduction in the backlog of cells awaiting transmission requires that every time slot be used.

Another consequence of the DQDB media access protocol is that the system behavior will approximate that of a First-In-First-Out Queue (FIFO), since, if we observe the sequence of cells arriving at the bus terminus, this will coincide with the sequence of request bits arriving at the terminus of the opposite bus. This may not, however, coincide *exactly* with the absolute time sequence of request generations because of propagation delay along the bus. Continuing our focus on access to the upper bus, we observe that a station near the right may make a request prior to a station near the left, but that the request may not yet have propagated past the latter station's lower read tap when that latter station makes its own request. The station near the left of the bus will therefore transmit its cell before permitting an idle cell to pass as needed to serve the earlier-generated request. Although the utilization efficiency can approach 100%, the system is not a perfect FIFO, and cells are only approximately served in their absolute order of arrival.

Access to the lower bus is controlled by an identical scheme: requests to send cells to the left on the lower bus are made using the request fields of time slots on the upper bus, and a second counter located within each access station tracks the number of unsatisfied requests which were generated by stations to the left, thereby allowing each station which newly generates a request to pass a number of unfilled time slots on the lower bus equal to the number of requests outstanding at the time the new request was generated.

Under unusual traffic conditions, the DQDB media access scheme may be unfair, allowing stations near the end of a bus to seize a disproportionate share of time slots. Referring to Figure 4.1, suppose a station near the right of the system develops a long, continuous stream of cells that it wishes to transmit to its right. Since this station has the opportunity to write into the request fields on the lower bus before stations appearing to its left, it might be able to write into each and every request field, thereby denying access to the upper bus by any station to its left. In particular, suppose all stations other than one near the right of the bus are idle, and that one active station generates a long sequence of cells for transmittal to its right. Since the time slots seen by that station on the upper bus are idle, that station can write a request field into every time slot on the lower bus, since idle time slots immediately appear on the upper bus to satisfy these requests (remember that a station can have only one outstanding request at any time). Assuming that no other station generates a request during the elapsed time until the first request generated by the only active station has propagated down the entire length of the lower bus, it then becomes impossible for any other station to make a request for the upper bus as long as that one initially active station remains so, because every request field will be written by that active station. Under these conditions, that station has sole access to the bus for as long as it wants. Other pathological situations arising from situations similar to the above, but involving combinations of active stations, may also develop and deny fair access to the system.

A simple modification to the DQDB media access protocol can prevent any such unfairness. Known as bandwidth balancing, this modification prohibits any station from writing into more than some number, K, of consecutive open time slots. Now, continuing the above example, after the one active station has written into K consecutive slots, it must allow one slot to remain unfilled. Since no cell was written to that unfilled slot, the active station cannot immediately place another request onto the opposite bus. Thus, one available request field propagates to its left. A station anywhere to the left of the initially active station can now place a request and, obeying the media access rules, write a cell into the appropriate open time slot propagating to its right (at that point on the bus, the time slots have not yet been filled by the heavy active user located toward the right). Thus, all right-bound time slots propagating past the heavy user are no longer empty; that user cannot write into an already-busy time slot, preventing it from writing into at least one request field, thereby allowing users to its left access to yet another time slot, etc. The heavy user must now refrain from writing request fields by (1) the "no more than K" rule and (2) active time slots written into by stations to its left. After several round-trips, a steady state is approached in which all users enjoy equal access to the DQDB system. The penalty for bandwidth balancing, however, is reduction of bus utilization efficiency. Since no station can access more than K consecutive time slots, some slots will unavoidably be wasted, even if there is only one heavy active user. The parameter K affects the bus utilization efficiency and must be chosen, for a particular bus length, to provide the desired balance between bus utilization efficiency and time to steady state.

Although DQDB stations must be linearly ordered along the dual buses, the actual physical topology can be that of a loop, as shown in Figure 4.3a. The loop topology is needed for reliability. Each DQDB access station is equipped with head-end capability and the means to serve as both the primary and the secondary frame generators. Thus, under normal conditions, both counterrotating frames (one on each bus) are generated by the same station. Each station is also equipped with bypass switches, capable of closing the bus through the station, thereby bypassing the head-end circuitry. These bypass switches are activated in the event of a network failure. Suppose that the cabling is severed, as shown in Figure 4.3b. Upon seeing this failure, the master station of Figure 4.3a informs the stations immediately to the left and right of the failure to assume primary and secondary head-end responsibilities. At the same time, it activates its own bypass switches. The healed configuration of Figure 4.3c results, and operation of the DQDB system is unaffected by the failure. However, unlike the configuration of Figure 4.3a in which both frame generators were physically present in the same access station, the head ends of Figure 4.3c are geographically separated. In the event of a second cabling failure at some other point, bypass switches cannot close the

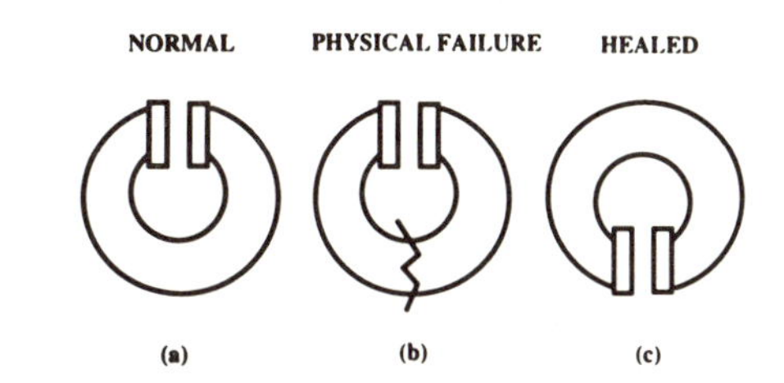

Figure 4.3. DQDB reliability features.

loops through the new head ends (in fact the first failure was a severed cable between the two new head ends). Thus, the DQDB bypass switches protect only against single failure events. Although service is not interrupted by a single failure, that failure must be physically repaired prior to any second failure to prevent an interruption in service caused by that second failure.

DQDB is intended to provide LAN-to-LAN connectivity over extended ranges at bus speeds as high as 155 Mbit/sec (since there are two independent buses, a total maximum capacity of 310 Mbit/sec results). Although this may be adequate for data-only LAN interconnects on a private or virtual-private network basis, the capacity is too low to meet the needs of a multimedia, broadband network. However, when viewed as a MAN subnetwork, DQDB systems may serve as geographically distributed traffic gatherers and local routers and, when several DQDB subsystems are combined as shown in Figure 4.4 with a cell-based packet switch of the type described in Chapter 3, a broadband network with adequate capacity for a multimedia public offering will result. Here, the access stations (not shown) of each DQDB subnetwork terminate transmission facilities which lead to customer premises, other DQDB subsystems, or ports on the packet switch (shown as a multiport bridge, to use the DQDB jargon). As noted in Figure 4.4, DQDB can appear both as public MAN subnetworks and as private backbone networks. By means of such a public network, traffic generated, say, on a LAN may proceed through a first DQDB subnetwork which sends it to the multiport bridge. After being shipped by that switch to a second DQDB subnetwork, it is further sent by that second subnetwork to an access station which serves a host computer in some remote location. We also see that traffic which enters (leaves) the public network always does so by means of an initial (terminal) DQDB subnetwork but, once inside the public network, may be routed by the DQDB subnetworks and the multiport bridge to existing telecommunication facilities (e.g., the public circuit-switched network and/or the narrowband public packet-switched network). Thus, traffic may be transported over extended distances by means of existing facilities, thereby making use of those facilities, but the data rate must then be substantially reduced from that of the DQDB access speed (which may be as high as 155 Mbit/sec).

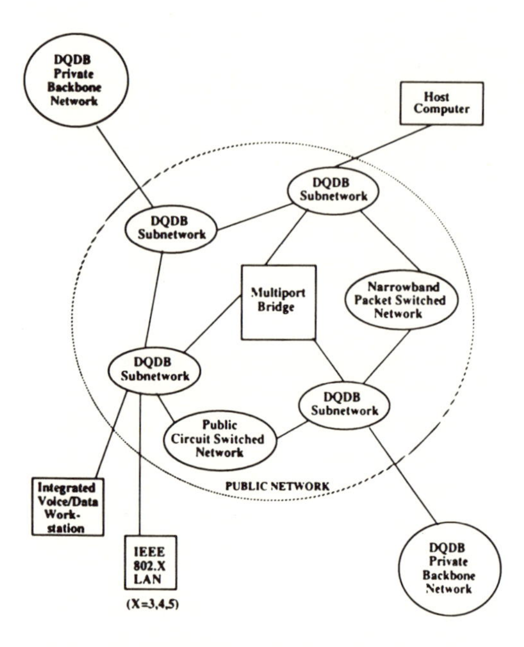

Figure 4.4. The use of DQDB with a multiport bridge (cell switch) to create a broadband network.

4.2. DQDB Segmentation, Reassembly, and Protocol Data Units

The basic unit of information to be transferred across the user interfaces of a DQDB subnetwork is known as a Media Access Control (MAC) Service Data Unit. This may be regarded as a fully formatted block of data, of variable length, generated by a user attached to one access station and delivered to a user attached to another access station. The operations which are performed on the MAC Service Data Unit within a DQDB access station are shown in Figure 4.5. These processes are essentially those of segmentation (creating fixed-length cells which can be written onto the DQDB buses) and reassembly (re-creation of a MAC Service Data Unit at its receiver from a sequence of arriving cells). For convenience, we shall describe the processes which occur at the transmitting access station; the inverse of these processes occur in reverse order at the receiving access station.

The arriving MAC Service Data Unit is first *encapsulated* by header, trailer, and pad fields to create the Initial MAC Protocol Data Unit (IMPDU). The MAC Service Data Unit format remains wholly intact during this process. The IMPDU header and trailer formats are described later.

The IMPDU is then segmented into segmentation units, each of which (with the possible exception of the last) contains precisely 44 octets. Each segmentation

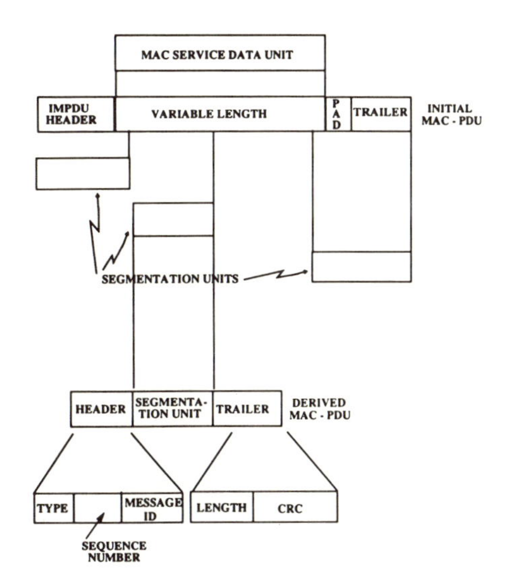

Figure 4.5. DQDB segmentation process.

unit is then encapsulated by a two-octet header and a two-octet trailer to create the Derived MAC-PDU (DMPDU). Each DMPDU therefore contains 48 octets; any segmentation unit containing fewer than 44 octets has sufficient padding attached such that its DMPDU contains 48 octets.

The DMPDU header contains a 2-bit Type field, a 4-bit Sequence Number field, and a 10-bit Message ID field. The Type field denotes whether a particular DMPDU is (1) the Beginning of Message (BOM, the first DMPDU of a new MAC Service Data Unit), (2) the End of Message (EOM, the last DMPDU of a particular MAC Service Data Unit, (3) a Continuation of Message (COM, neither the first nor the last DMPDU of a particular MAC Service Data Unit), or (4) a Single Segment Message (SSM, the only DMPDU of a short MAC Service Data Unit). The Sequence Number field allows the DMPDU of a given MAC Scrvicc Data to be consecutively numbered modulo-16. This is helpful to the receiver in detecting the absence of any DMPDUs which may have been lost or desequenced within the network. If the DMPDUs created from a common MAC Service Data Unit should arrive at the receiver out of sequence, or if any DMPDUs are determined to be absent, then that MAC Service Data Unit will not be reassembled and delivered to the user attached to that receiving access station. The Message Identifier (MID) field is a 10-bit field which is used to associate DMPDUs belonging to a common MAC Service Data Unit. DMPDUs created from a common MAC Service Data Unit shall bear the same 10-bit number within their MID fields. Using its read tap, each access station interrogates the MID field of every cell appearing on the DQDB buses, and on the basis of this information, makes a copy of all those cells intended for local reception. Essentially, each receiver sorts the cells appearing on the buses, ignoring those associated with MAC Service Data Units intended for other stations and sorting those intended for local reception according to the MAC Service Data Unit to which they belong. Since the DMPDUs associated with different MAC Service Data Units which originate at different Access Stations may arrive at a single intended receiver in some random time-multiplexed sequence, the receiver must sort by MID to direct the DMPDUs to the correct MAC Service Data Units being concurrently reassembled.

The DMPDU trailer contains a 6-bit Length Field which is used for End of Message segments and Single Segment Messages; for such segments, the segmentation unit may contain fewer than 44 octets of real data, with the balance of the 44 octets consisting of bit padding. The Length Field permits the receiver to identify the boundary between legitimate data and padding. The remaining 10 bits of the DMPDU trailer contain a Cyclical Redundancy Check (CRC) to permit detection of bit errors which may have occurred within the DMPDU. If any errors are detected, the DMPDU is discarded and the MAC Service Data Unit of which that DMPDU is a part will not be reassembled and delivered.

We see that the reassembling of a MAC Service Data Unit may be terminated by any of three events which may occur at the DMPDU level: detection of a missing DMPDU, detection of an out-of-sequence DMPDU, or detection of bit errors within a DMPDU. Other events which may occur at the IMPDU level will similarly prevent a MAC Service Data Unit from being delivered to the user. In general, the DQDB system will guarantee some maximum rate of "lost MAC Service Data Units," that is, the fraction of MAC Service Data Units which are not delivered will be less than some specified value (say, 10^{-6} or 10^{-8}, for example). If the user's application can tolerate this lost MAC Service Data Unit rate, then no additional protection is needed; if not, then the user is responsible for layering additional protection via higher-level protocols that are above DQDB. For example, the user might implement some retransmission strategy whereby the transmitter again sends any MAC Service Data Unit for which the receiver has not actively acknowledged a receipt. These higher-layer protocols are *not* a part of DQDB, which merely serves as a carrier for MAC Service Data Units supplied by a user through a DQDB Access Port. The user must decide if higher layers of protection are needed, and embody such protection in the MAC Service Data Units.

The format of the IMPDU appears in Figure 4.6. The information block supplied by the user (MAC-SDU) is encapsulated by (1) a header containing three fields (the Common PDU header, MAC Convergence Protocol Header, and Header Extension fields, of lengths 4, 20, and 0–20 octets, respectively); (2) a 4-octet common PDU trailer; and (3) a 0–3 octet pad field, the sole purpose of which is to ensure that the block between the end of the Header Extension field and the beginning of the Common PDU Trailer field is a multiple of four octets.

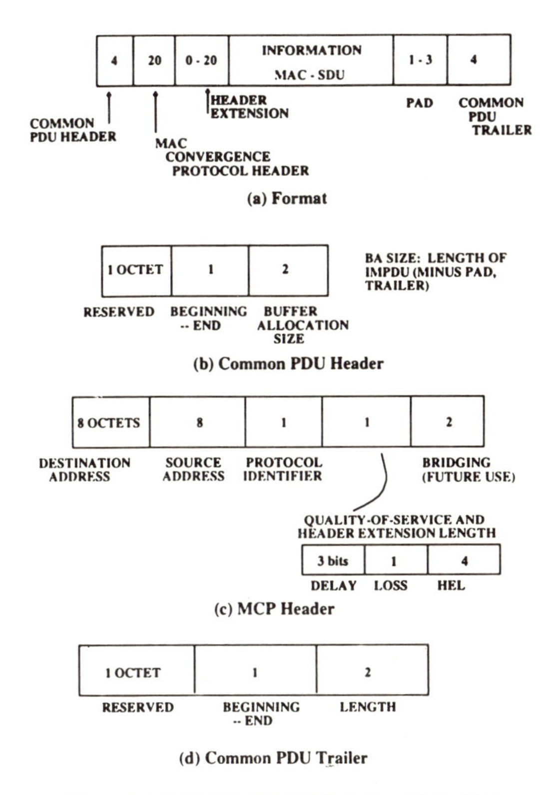

Figure 4.6. DQDB Initial MAC Protocol Data Unit.

The Common PDU Header and the Common PDU Trailer fields are identical. The first octet of each is reserved for unspecified future use. The second octet is the Beginning–End tag. This number, which is of the same value in both the Common PDU Header and Trailer, is of use to a receiver to help ascertain that, during reassembly, the ending of a given IMPDU did not inadvertently become attached to the beginning of a different IMPDU. The Buffer Allocation Size field within the Common PDU Header contains a word which informs the receiver of the overall length of the IMPDU. Occurring within the first segmentation unit of that IMPDU, the Buffer Allocation Size field allows the receiver to reserve sufficient buffer space (when available) to complete the reassembly process. If adequate buffer space to complete the reassembly is unavailable because of prior allocations for IMPDUs currently in the reassembly phase, then the reassembly will not be attempted and the corresponding MAC Service Data Unit will not be delivered, further contributing to the lost MAC-SDU rate. Inclusion of the Buffer

Allocation Size field within the Common PDU Header prevents the receiver from attempting the concurrent reassembly of too large a number of IMPDUs, thereby avoiding the prospect of running out of reassembly buffer space before the reassembly of any was completed. The word contained in the Length field appearing in the Common PDU Trailer must agree with that in the Buffer Allocation Size field and further safeguards against the attachment of the ending of some IMPDU to the beginning of another (an identical function to that performed by the Beginning–End tags in the Common PDU Header and Trailer).

The first and second eight-octet subfields of the MAC Convergence Protocol Header contain the Destination and Source addresses, respectively, of the IMPDU. When preparing the IMPDU, the transmitting station determines these addresses by interrogating the appropriate locations within the MAC Service Data Unit. The MAC-SDU consists of a fully formatted frame of information to be transported by a DQDB system, and therefore must contain both source and destination information. Since the high-layer protocols of the MAC-SDU can be arbitrary, the IMPDU layer of a given DQDB access port must know specifically where to look for source and destination information and must therefore be specific to the higher-layer protocol being run by the user connected to DQDB via that port. Different DQDB access ports may have different IMPDU layers, each specific to the protocol being used by the user attached to a given port. Although the processing done within the IMPDU layer of DQDB may be different for different ports (e.g., as explained above, IMPDU layers specific to different higher-layer user protocols may need to look in different locations to extract source and destination information), the IMPDUs prepared by all DQDB ports are identical in format: destination and source information from their respective MAC-SDUs are mapped into eight-octet DQDB addresses, and the MAC-SDU is totally encapsulated to form the IMPDU, which is then segmented and reformatted as DMPDUs as described above. In a sense, the IMPDU layer of DQDB serves as a terminal adapter, deriving information as appropriate from the MAC-SDU which carries the higher-layer protocols of the user attached to a given DQDB port and formatting a common IMPDU which is independent of the different protocols which may be used by the various DQDB users. In the process, the user protocol-specific MAC-SDU is carried intact by the encapsulation process. Since the IMPDU layer of DQDB must know the protocol being run by the user attached to a given DQDB access port, DQDB provides a location containing the identity of that protocol; this is the third subfield of the MCP Header (Protocol Identifier subfield) and is one octet in length.

The fourth subfield of the MCP Header is also of one octet in length and is further subdivided into three words: (1) a 3-bit word which allows eight levels of priority to be declared with regard to delay; (2) a 1-bit word which allows two levels of priority to be established with regard to loss of an IMPDU; and (3) a 4-bit word which allows up to 16 lengths for the Header Extension field of the IMPDU to be declared. The priority bits are used to help assure quality of service for different types of traffic at the IMPDU level and are used whenever an IMPDU generated within one DQDB system must be delivered to a different DQDB system which is connected to the first via non-DQDB facilities. In such an event, the DMPDUs created from that IMPDU are delivered by the first DQDB system to the gateway DQDB port providing access to the non-DQDB facilities. The IMPDU is reassembled within the DQDB gateway port prior to transmittal over the non-DQDB facilities. Since the non-DQDB facilities may operate at a lower speed than the DQDB systems, congestion may develop within the DQDB gateway port as IMPDUs arrive at a rate greater than that which can be accepted by the non-DQDB system. The priority bits allow the DQDB gateway port to serve the queued IMPDUs in a sequence which best ensures that their respective quality-of-service needs will be met with regard to delay and loss (since the gateway buffer is of finite size, some IMPDU loss will invariably result).

The fifth and final subfield of the MCP Header contains two octets which are reserved for unspecified use in future bridging arrangements among DQDB systems.

The Header Extension field of the IMPDU is a variable length field whose length may assume any of 16 distinct values between 0 and 20 octets. The actual length of the Header Extension field is contained within the 4-bit Header Extension Length word appearing as part of the MCP Header. The Header Extension field is provided by DQDB to enable support of a variety of different services. For example, when DQDB is used to support of a variety of different services. For example, when DQDB is used to support Switched Multimegabit Data Service (SMDS, a connectionless high-speed packet service offered over public facilities to enable LAN interconnection and other applications), the Header Extension field can be used to specify up to four Inter-Exchange Carriers authorized by a customer to carry that customer's traffic; such information must be provided for all SMDS packets originating within the service region of one Local Exchange Carrier and terminating within the service region of another.

Each 48-octet DMPDU created by the DQDB segmentation process is further encapsulated by a 4-octet header, as shown in Figure 4.7, to create a 52-octet DQDB segment. To this is added a 4-octet access control field, which is used to mediate access to the DQDB buses, thereby forming a complete 53-octet DQDB cell. The 4-octet Segment header contains (1) a 20-bit word which can be used to carry the Virtual Circuit Indicator (VCI) for connection-oriented service; (2) a 2-bit word used to declare the type of payload (user data or up to three types of network-generated control data); (3) 2 bits which can be used to help ensure quality of service for different traffic types when DQDB systems are bridged together using DQDB-compatible facilities not requiring IMPDU reassembly within the gateway ports; and (4) an 8-bit CRC to help detect and correct errors occurring anywhere within the segment header.

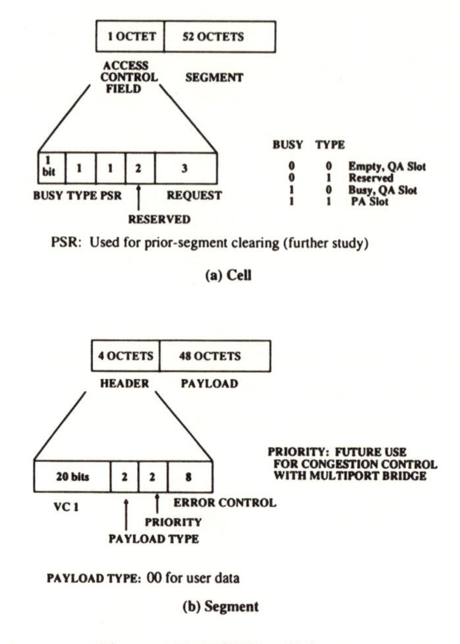

Figure 4.7. DQDB cell format.

The first bit of the 8-bit access control field is the busy bit. A "1" in this bit position is used to declare the presence of a time slot which either is filled with a DQDB cell or is a so-called "prearbitrated" slot which cannot be accessed via the normal DQDB media access protocol; more will be said of this shortly. The busy bit in conjunction with the "type" bit of the access control field can define three types of time slots. If both bits are "0," the time slot is empty and can be accessed via the ordinary DQDB queue-arbitrated media access protocol. A "0" busy bit and a "1" type bit is a prohibited combination. A "1" busy bit and a "0" type bit represents a time slot which was already filled by a DQDB cell via the queue-arbitrated media access protocol. If both bits are "1," the time slot is "prearbitrated."

As originally envisioned, the DQDB access stations would passively attach to the dual buses as previously described. However, referring back to Figure 4.1, we observe that a DQDB system could be also be envisioned in which the access stations actively regenerate cells propagating along the buses. In such an arrangement, the write tap of an upstream station would be connected to the read tap of the next downstream station via a point-to-point transmission system. The read tap would terminate that transmission system, and the access station would regenerate all received signals prior to retransmission from the write tap to the next downstream read tap. This arrangement would allow a DQDB cell to be removed from the bus when it reaches its destination access station; the corresponding time slot would thereby emerge empty from that station's write tap, and could be accessed by another station located farther downstream. The PSR bit in the access control field has been set aside for possible use in conjunction with such prior-segment-release.

The fourth and fifth bits in the access control field are reserved for unspecified future use. The final three bits are used to permit establishment of three request priorities. Recall that, to reserve a time slot on one DQDB bus, a request must first be made in a time slot on the opposite bus. In principle, only one bit in the access control field must be set aside for this purpose. However, each bit of a 3-bit request field can be used to denote a different one of three priorities (the first bit may denote highest priority, and so forth). Now, suppose an access station wishes to transmit one highest-priority DQDB cell. To do so, it must first write a request bit into the high-priority request position. Then, to establish its place in the queue, that station's counter will *ignore* all requests previously generated by upstream stations of priority levels 2 and 3, and will defer transmission only until a number of empty cells sufficient to accommodate those outstanding requests of priority level 1 have passed. Similarly, a priority level 2 request will defer only to those outstanding requests detected at priority levels 1 and 2, as denoted by request bits in positions 1 and 2 of the 3-bit request field, but ignore requests of priority level 3. Finally, a priority level 3 request (lowest priority) must defer to all outstanding requests. We see that use of a 3-bit DQDB request field provides a particularly simple means of allowing urgent messages (perhaps of a network control nature) to be promptly communicated, without having to wait for less urgent messages which were generated earlier to be communicated first.

DQDB prearbitrated slots provide a means for reservation-based circuit switching to be integrated with statistically multiplexed queue arbitrated cells. As we shall see, prearbitrated slots permit bandwidth to be guaranteed to stations offering continuous bit ratc traffic in units of 64 kbit/sec. An example of such continuous bit rate traffic is digital telephony.

As already mentioned, the DQDB frame generator located at the head end of each bus generates frame headers which delineate the boundaries of DQDB frames, each 125 μsec in duration. As shown in Figure 4.8, the frame headers are followed by an empty time interval containing an integral number of DQDB time slots. A pad field is included, if needed, such that the frame is exactly 125 μsec long. A frame generator can declare any of the time slots to be prearbitrated simply by writing "1"s in the busy and type bits of that time slot's access control field. Unlike queue-arbitrated slots which are intended to be filled by DQDB cells comprising a 1-byte access control field, a 4-byte cell header, and a 48-byte information field containing a DMPDU, the prearbitrated slots contain the access

FRAME HEADER	SLOT 0	SLOT 1	.....	SLOT N	PAD

N determined by data rate, slot length, frame header, PAD length
Slots are either isochronous or nonisochronous

(a) Frame of length equal to 125 microseconds

ACF	HEADER	CH 1	....	CH 48

ACF = 1 octet (access control field)
Each channel = 1 octet
1 channel / frame = 64 kb/sec.

(b) Isochronous time slot

(c) Nonisochronous (Queue Arbitrated) time slot

Figure 4.8. DQDB frames, isochronous time slots, and nonisochronous time slots.

control field and cell header but the 48-octet payload consists of up to 48 distinct 1-octet "channel" subfields. Since prearbitrated slots must occur in the same position of each and every DQDB frame, we see that each of the 48 channels of a prearbitrated slot occurs synchronously at a repetition rate of once per 125 μsec. Since each channel contains 1 byte, we note that each channel represents a guaranteed data rate of 1 byte/125 μsec = 64 kbit/sec. Hence, an access station requiring 64 kbit/sec continuous bit rate service would request an allocation of one channel from a prearbitrated slot (allocations are managed by a controller, not shown, which may be part of the head-end station). Once such an allocation has been made, the synchronously occurring 1-byte subfield corresponding to that channel can be assessed only by the station have requested the allocation. Multiple channel allocations can be made, and different channels within the same prearbitrated slot may be allocated to different access stations. When the circuit-switched connection is no longer needed, the now unused channel can be reallocated to some subsequent request. The number of prearbitrated slots created by the frame generator station can be varied to accommodate the instantaneous demand for circuit-switched channels, subject the constraint that the number of queue-arbitrated slots is adequate to handle the demand for non-circuit-switched service. If a free channel is unavailable at the time that a circuit-switched connection is requested, then that request will be blocked.

4.3. Delay Performance of DQDB for Bursty Data Traffic*

Since DQDB is intended to provide LAN–LAN interconnectivity over metropolitan area service regions, we shall study its traffic-handling behavior in the context of a MAN. Arrivals at each DQDB-MAN access station represents the aggregated traffic originating within the corresponding LAN intended for distribution to the other geographically remote LANs, and the DQDB-MAN is assumed to perform like a perfect First-In-First-Out (FIFO) queuing system. We are interested in studying the random delay incurred by a representative packet arriving at any of the LANs as it awaits access to the DQDB-MAN. All LANs are assumed to independently generate inter-LAN traffic having identical traffic statistics.

Two types of traffic will be considered. For the first, we shall assume that each LAN is, itself, a DQDB system, and that each variable length frame of data incident upon each LAN access station is segmented by that LAN into DQDB cells. Since the MAN is also a DQDB system, reassembly of the MAN-SDU is not performed at the LAN/MAN interface, and each DQDB cell arriving at a MAN access station is simply transferred intact across the MAN to the destination LAN/MAN interface, onto the destination LAN, and finally to the receiving access station of the destination LAN where reassembly is performed. For each DQDB cell, the MAN access delay is the time spent in its respective MAN access station awaiting access to the MAN bus.

If the number of interconnected LANs is very large, then, in each DQDB-MAN time slot, we can approximate the total number of new arrivals to all MAN access stations as a Poisson-distributed discrete random variable. The service time is deterministic since, at most, one cell can access each DQDB-MAN time slot. Assuming infinite-length buffers, this model is therefore identical to that previously studied for the short-bus LAN, and the same results apply as were previously derived for the M/D/1 queue. The mean access delay as a function of offered load is therefore as shown earlier in Figure 2.18.

Recall that the short bus analysis of Section 2.11 is valid for arbitrary service time distribution, provided only that the arrival statistics correspond to a Poisson point process. We note here that under the condition of deterministic service time, we can apply the same analysis to compute the exact probability distribution for the total number of cells in the system (this analysis was actually applied in Section 3.5 to compute the buffer overflow probability for the Knockout switch; see Figure 3.19). Also, the analysis can be applied to find the probability distribution for the number of cells in the system and the mean delay for non-Poisson arrivals as long as the number of arrivals in different time slots are statistically independent. In this case, we would need to know the discrete probability distribution for the number of arrivals within a time slot. As an example, if the number of interconnected LANs is equal to N and the probability that a given LAN delivers a cell to its access station within a given time slot is p, and if inter-LAN cell arrivals are independent among the various LANs, then the total number of cell arrivals to the DQDB-MAN within a given time slot is binomially distributed.

In the following [through (4.3.18)], we redevelop some of the results which were previously derived. In so doing, we shall further clarify some of the steps taken in the analysis, present material in the context of a MAN, and provide some additional insight.

As indicated by (2.11.8), the generated function for any nonnegative integer random variable n is defined to be:

$$\psi_n (x) = \sum_{n=0}^{\infty} x^n \, p_n \tag{4.3.1}$$

From (4.3.1), we note that

$$\psi_n' (x) = \sum_{n=1}^{\infty} nx^{n-1} \, p_n \tag{4.3.2}$$

and that

$$\psi_n^{(k)} (x) = \sum_{n=k}^{\infty} n(n-1)(\cdots)(n-k+1)\, x^{n-k} p_n \tag{4.3.3}$$

$$\psi_n(0) = p_0 \tag{4.3.4}$$

$$\psi_n(1) = p_1 \tag{4.3.5}$$

$$\psi_n^k(1) = k!\, p_k, \qquad k = 0, 1, 2, \ldots \tag{4.3.6}$$

Thus, if $\psi_n(x)$ is known, then the exact probability distribution for the nonnegative discrete random variable n can be found.

For the queuing system with deterministic service time, we can directly apply (2.11.15) to find the generating function for the number of cells in the queue, including the one being served:

$$\psi_n(x) = p_0 \, \frac{x\psi_a(x) - \psi_a(x)}{x - \psi_a(x)} \tag{4.3.7}$$

where "a" is the (random) number of arrivals within a given time slot, $\psi_a(x)$ is its corresponding generating function, and p_0 is the probability that $n = 0$. To find the probability that the number of cells in the system is equal to k, we must differentiate the right-hand side of (4.3.7) k times, and evaluate the resulting expression at $x = 0$. The resulting expression will be some rational function of various derivatives of $\psi_a(x)$ evaluated at $x = 0$; since the ith such derivative evaluated at

$x = 0$ is related to the probability that $a = i$ [see (4.3.6)], we conclude that the discrete probability distribution for the number of cells in the system can easily be found from p_0 and knowledge of the discrete probability distribution for the number of arrivals within any time slot. For example, from (4.3.7):

$$p(n = 0) = p_0[-\psi_a(0)/(-\psi_a(0)] = p_0 = p(n = 0) \tag{4.3.8}$$

$$p(n = 1) = \psi_n'(0) = p_o \left[\frac{\{x - \psi_a(x)\}\ \{x\psi'(x) + \psi_a(x) - \psi_a'(x)\}}{\{x - \psi_a(x)\}^2} - \frac{\{x\psi_a(x) - \psi_a(x)\}\ \{1 - \psi_a'(x)\}}{\{x - \psi_a(x)\}^2} \right]_{x=0}$$

$$= p_0 \left[\frac{1 - \psi_a(0)}{\psi_a(0)} \right] = p(n = 0) \left[\frac{1 - p(a = 0)}{p(a = 0)} \right] \tag{4.3.9}$$

Continuing in this fashion, $p(n = k)$ can be found in terms of $p(n = 0)$ and the various $p(a = i)$. Finally, since

$$\sum_{k=0}^{\infty} p(n = k) = 1 \tag{4.3.10}$$

we can find $p(n = 0)$ and express $p(n = k)$ in terms of the various $p(a = i)$, $i = 0, 1, 2, \ldots$. Alternatively, since $p(n = 0) = E\{a\}$ [see (2.11.28), (2.11.29)], we can express $p(n = k)$ in terms of the various $p(a = i)$, and $E\{a\}$.

The average delay for the DQDB-MAN with a finite number of access stations can be found by applying (2.11.17):

$$E\{n\} = p_0 \frac{\psi_a'(1)}{1 - \psi_a'(1)} + p_0 \frac{\psi_a''(1)}{2[1 - \psi_a'(1)]^2} \tag{4.3.11}$$

Since $\psi_a'(1) = E\{a\}$ and

$$\psi_a''(1) = \sum_{a=0}^{\infty} a(a - 1)\, x^{a-2}\, p_a \Big|_{x=0} = E\{a^2\} - E\{a\} \tag{4.3.12}$$

$$E\{n\} = [1 - E\{a\}] \left[\frac{E\{a\}}{1 - E\{a\}} \right] + [1 - E\{a\}] \frac{E\{a^2\} - E\{a\}}{2[1 - E\{a\}]^2}$$

$$= E\{a\} + \frac{E\{a^2\} - E\{a\}}{2[1 - E\{a\}]} \tag{4.3.13}$$

Each time slot, a DQDB-MAN access station will receive a cell with probability p from its respective LAN for transmittal across the MAN. The number of cells arriving in a given time slot among the N access stations is therefore binomially distributed:

$$p(a = k) = \binom{N}{k} p^k (1 - p)^{N-k} \tag{4.3.14}$$

$$E\{a\} = Np \tag{4.3.15}$$

$$E\{a^2\} = N^2p^2 + Np(1 - p) \tag{4.3.16}$$

Here,

$$E\{n\} = Np + \frac{N^2p^2 + Np(1 - p) - Np}{2(1 - Np)} \tag{4.3.17}$$

$$= \frac{Np(2 - p - Np)}{2(1 - Np)}$$

Finally, applying Little's theorem as was done in Section 2.11,

$$\bar{D} = T\left[\frac{2 - p - Np}{2(1 - Np)}\right] \tag{4.3.18}$$

where T is the time slot duration. In (4.3.18), Np must be less than unity; that is, the average rate of cell arrivals must be less than the maximum rate of cell departures (1 per time slot).

Equation (4.3.18) is plotted in Figure 4.9 for various values of N. We see that

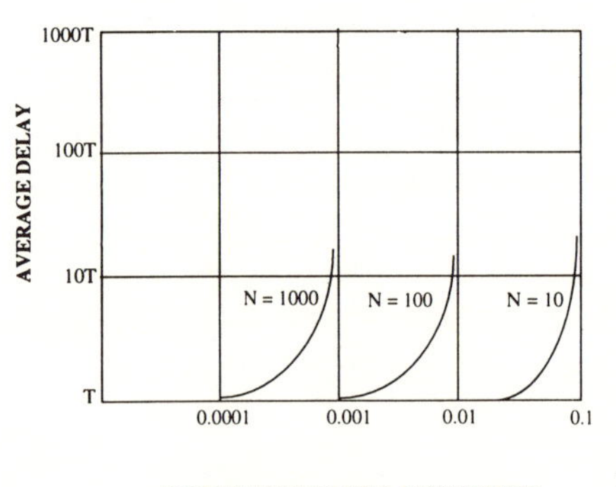

Figure 4.9. DQDB cell access delay versus offered load per user, p.

as N increases, the maximum permissible value of p diminishes almost inversely with N. This unsurprising result simply reflects the fact that as the number of DQDB-MAN access ports increases, the average traffic load that can be offered through each port must diminish since all ports are sharing one fixed-rate shared medium. We shall see in Chapter 7 how the unique opportunities of all-optical networks allows us to modify DQDB and create additional channels on the shared medium as more access stations are added. With optical networks, the load per port does not diminish inversely with N.

Next, suppose that the interconnected LANs are not DQDB systems, but, rather, deliver variable-length MAC-SDUs to their respective DQDB-MAN access ports. Here, we shall again assume that the number of interconnected LANs is large and that in the aggregate, the MAC-SDU arrivals to the DQDB-MAN comprise a Poisson point process with average arrival rate λ. We further assume that if there is only a single MAC-SDU present, then the time needed to transmit all segments of that MAC-SDU across the MAN is an exponentially distributed random variable with mean value T, i.e., we assume that the length of each MAC-SDU is an exponentially distributed random variable, and we ignore the fact that a MAC-SDU is actually transferred as a sequence of discrete DMPDUs. For greater accuracy, we might have modeled the MAC-SDU as containing a random number of DMPDUs drawn from a geometric probability distribution. For long MAC-SDUs, the delay predicted by the more exact discrete time model, wherein the number of DMPDUs contained in a MAC-SDU is a geometrically distributed random variable, is well approximated by the analytically simpler continuous time model, wherein the length of the MAC-SDU is assumed to be an exponentially distributed random variable. Although the DMPDUs of SDUs appearing at different access stations will, in general, appear time mulitplexed in the DQDB-MAN, Little's theorem permits us to find the average delay incurred in transferring a complete MAC-SDU across the DQDB-MAN by modeling the system as a simple single-server queue with Poisson arrivals and exponential service time (M/M/1 queue). Although the performance of a simple M/M/1 queue can readily be studied via the analysis of Section 2.11, it is instructive to illustrate an alternative (and much simpler) methodology which is applicable only to the special conditions which apply: Poisson arrivals, exponentially distributed service time.

The M/M/1 queue is illustrated in Figure 4.10a, where a finite-length buffer capable of storing B-1 messages is assumed. The state of the queue is defined to be the total number of messages in the system, including the one being served. The state increases by unit every time a new message arrives, and diminishes by unity whenever a message has been fully transmitted. Note that, unlike the actual operation of the DQDB-MAN, wherein messages arrive to a distributed queue and are served in a time-shared fashion as DMPDUs from different access stations are time multiplexed onto the bus, our model assumes a single FIFO queue and, once service (transmission) begins on a given message, the server is devoted exclu-

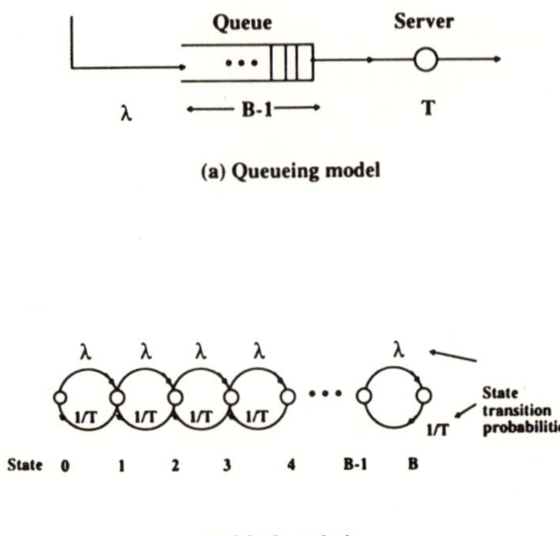

Figure 4.10. The M/M/1 queue.

sively to that message until service is competed. Again, as we will subsequently show, Little's theorem permits the average delay to be found from the simple M/M/1 model.

At any instant in time, the state of the M/M/1 queue can change by, at most, unity, since multiple arrival/departure events cannot occur simultaneously. The system can therefore be modeled by a finite Markov chain as shown in Figure 4.10b. We seek the steady-state probability P_i of finding the system in state i at any randomly chosen point in time.

Consider a point in time $t + \Delta t$, where Δt is a small quantity. The probability of finding the system in state "0" at time $t + \Delta t$ is given by the probability that (1) the system was in state "0" at time t and there were no arrivals in time interval Δt, or (2) the system was in state "1" at time t and there was a departure in time interval Δt. Mathematically,

$$P_0(t + \Delta t) = P_0(t)\,[1 - \lambda\Delta t] + P_1(t)\Delta t/T \tag{4.3.19}$$

where $\lambda\Delta t$ is the probability of an arrival in short time interval Δt and Δt/T is the probability of a departure in interval Δt. Similarly, the system will be in state "i" at time $t + \Delta t$ if (1) the system was in state "i" at time t and there was neither an arrival nor a departure in Δ, or (2) the system was in state "i - 1" at time t and there was an arrival in Δ, or (3) the system was in state "i + 1" at time t and there was a departure in Δt:

$$P_i(t + \Delta t) = P_i(t)[1 - \lambda\Delta t - (\Delta t/T)] + P_{i-1}(t)\,\lambda\Delta t + P_{i+1}(t)(\Delta t/T), \tag{4.3.20}$$

$$1 \le i \le B - 1$$

Finally, the system will be in state "B" at time $t + \Delta t$ (buffer completely filled) if (1) the system was in state "$B - 1$" at time t and there was an arrival in Δt or (2) the system was in state "B" at time t and there was no departure in Δt:

$$P_B(t + \Delta t) = P_{B-1}(t)\,\lambda\Delta t + P_B(t)[1 - \Delta t/T] \tag{4.3.21}$$

After the system has been running for a long time, the state probabilities become independent of time; finding the system in state "i" at time t_1 is no more or less likely than finding it in state i at time $t_2 > t_1$. Thus, in the steady state,

$$P_i(t + \Delta t) = P_i(t) \equiv P_i, \qquad i = 0, 1, \ldots, B \tag{4.3.22}$$

Finally, substituting (4.3.22) into (4.3.19)–(4.3.21), we obtain

$$\lambda P_0 = \frac{1}{T} P_1 \Rightarrow P_1 = \lambda T P_0 \tag{4.3.23}$$

$$\left(\lambda + \frac{1}{T}\right) P_i = \lambda P_{i-1} + \frac{1}{T} P_{i+1} \Rightarrow P_{i+1} = (1 + \lambda T) P_i - \lambda T P_{i-1}, \tag{4.3.24}$$

$$1 \le i \le B - 1$$

$$\lambda\, P_{B-1} = \frac{1}{T} P_B \Rightarrow P_B = \lambda T P_{B-1} \tag{4.3.25}$$

The left-hand column of the above relations reflects the fact that in the steady state, the probability of transitioning out of any given state is equal to the probability of transitioning into that state (i.e., to transition out of state 0, we must first be in state 0 and there must be an arrival; to transition into state 0, we must be in state 1 and there must be a departure).

These equations can be solved recursively in terms of P_0:

$$P_i = (\lambda T)^i\, P_0 \tag{4.3.26}$$

and since $\lambda T < 1$ (traffic intensity < link capacity),

$$\sum_{i=0}^{B} P_i = 1 = P_0 \sum_{i=0}^{B} (\lambda T)^i = P_0 \frac{1 - (\lambda T)^{B+1}}{1 - \lambda T} \tag{4.3.27}$$

$$\Rightarrow P_0 = \frac{1 - \lambda T}{1 - (\lambda T)^{B+1}}$$

and

$$P_i = \frac{(\lambda T)^i\,(1 - \lambda T)}{1 - (\lambda T)^{B+1}}, \qquad 0 \le i \le B \tag{4.3.28}$$

From (4.3.28), for the M/M/1 queue with finite buffer, we can readily compute the probability of buffer overflow (lost packet probability). This is simply the probability that a newly arriving message finds the buffer filled, i.e.,

$$P(\text{loss}) = P_B = \frac{(\lambda T)^B(1 - \lambda T)}{1 - (\lambda T)^{B+1}} \tag{4.3.29}$$

However, this expression, while valid for the M/M/1 queue, cannot be applied to find the lost packet rate of the DQDB system, which would require a much more complicated analysis; the M/M/1 model is useful to us in studying DQDB only with regard to average packet delay, since here we can apply Little's theorem. Furthermore, when studying packet delay, we shall assume that buffers cannot overflow; that is, the DQDB buffers are assumed to be infinite (again, it is only under this condition that Little's theorem can be applied to help us find the average delay). Since $\lambda T < 1$, we find that as B goes to infinity, the state probabilities become:

$$P_i = (\lambda T)^i (1 - \lambda T), \qquad i = 0, 1, 2, \ldots \tag{4.3.30}$$

The mean delay can readily be found from the state probabilities. If there are no messages in the system when a new message arrives, then the average waiting time is T (the average time needed to transmit that message). Similarly, if there are i messages in the system when a new message arrives, the average waiting time is $(i + 1)T$ (the time taken to transmit the i messages already in the system plus the time needed to transmit the newly arriving message). Thus, the average delay or time-in-system may be expressed as:

$$\bar{D} = T \sum_{i=0}^{\infty} (1 + i)\, P_i \tag{4.3.31}$$

and substituting (4.3.30), we obtain

$$\bar{D} = T + T \sum_{i=0}^{\infty} \mathrm{i}(\lambda T)^i/(1 - \lambda T) \tag{4.3.32}$$

$$= T\left[1 + \frac{\lambda T}{1 - \lambda T}\right]$$

General results for the M/M/1 queue are plotted in Figure 4.11 and contrasted with results previously obtained for the M/D/1 queue (Poisson arrivals, deterministic service time).

Now, Little's theorem teaches that, for any arrival statistics and any service discipline, the average number of messages in the system is equal to the product

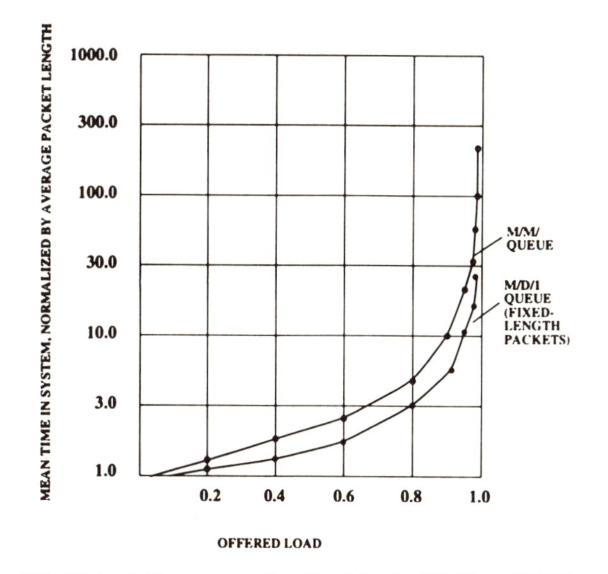

Figure 4.11. Plots of delay versus offered load for the M/M/1 and M/D/1 queues.

of the average arrival rate and the average time spent in the system [see (2.11.33)]. If we define the state of a queuing system to be the total number of messages in the system, then the state equations (4.3.23)–(4.3.25) apply both for a system containing distributed queues (one per access station) and for a system containing one centralized queue. Thus, if no segmentation took place within the access stations but, rather, the access station queues were served in some arbitrary fashion but with whole messages (of exponentially distributed length) being served once a queue had been chosen, then the probability that there are a total of i messages in the system at any time is the same for a system with distributed queues and a system with one centralized queue. The average number of messages in the two systems are therefore identical, and since the average arrival rate for both systems are the same, then the average delay seen by a representative message in either system is the same for both systems. Finally, the fact that the access stations segment MAC-SDUs into DMPDUs, and that DQDB time-multiplexes segments from SDUs arriving at different access stations, does not alter this conclusion. Let us assume that the DQDB cell duration is small compared with the time duration of an SDU, so that we can safely ignore the time granularity caused by bus time slots (i.e., we are approximating a discrete time system by a continuous time system; messages with random lengths which are geometrically distributed in discrete time systems are equivalent to messages with lengths which are exponentially distributed in continuous time systems, as previously discussed). Using this excellent approximation, and observing that the messages in the distributed queue have exponentially distributed lengths, we conclude that the single server of a distributed queue can serve a part of any message, then shift to another message and serve part, and so forth, without affecting the state equations (4.3.23)–(4.3.25); since message lengths are exponentially distributed with mean value T, then the probability that the server is currently serving a message which completes service in the next time increment Δt is simply $\lambda\Delta t$. The state equations (4.3.23)–(4.3.25) can then be solved to yield the probabilities that there are i messages in the system, the service of some of which may already have begun. However, again because messages are exponentially distributed in length with mean T, the expected time to complete service of a message whose service may have already started, conditioned upon service already in progress at any given point in time, is still T as measured from that point in time. Thus, the expected delay experienced by a newly arriving packet finding any combination of i complete or partial messages already in the system is equal to $(1 + i)T$. We conclude that the average message delay for a DQDB system with distributed queues, and with segmented messages served in arbitrary time-multiplexed fashion, is identical to that for the M/M/1 queue, and (4.3.32) applies.

4.4. Blocking Performance of DQDB for Circuit-Switched Traffic*

As previously explained, DQDB offers a circuit-switching feature via prearbitrated cells whereby circuits in multiples of 64 kbit/sec can be exclusively reserved. For circuit-switched service, the quality of service is the call blocking probability, that is, the likelihood that a request for a connection cannot be accommodated. Although the so-called Erlang blocking analysis to be presented is not at all unique to DQDB but, rather, is applicable to a variety of blocking systems, we shall present it in this section as a matter of convenience, and refer back to it in later chapters as appropriate.

The model which we use to study the circuit-switching blocking performance of a DQDB system appears in Figure 4.12a. We assume that there are A prearbitrated time slots, for a total pool of $M = 48A$ circuits, each supporting 64 kbit/sec continuous bit rate service. Furthermore, for simplicity, we shall assume that this number is fixed; prearbitrated slots cannot be converted for queue-arbitrated usage, or vice versa, even if the fixed but otherwise arbitrary split of time slots among pre- and queue-arbitrated slots is mismatched with the relative traffic intensities for the type of service offered by each. Other, so-called movable boundary schemes allow the division of time slots to be adapted in response to the instantaneous demand for each type of service, subject to constraints which ensure that some minimum number of time slots per frame is reserved for each. With movable boundary schemes, such constraints are necessary to prevent one type of

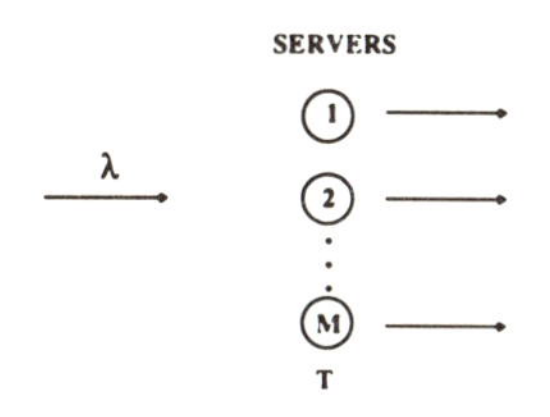

(a) Serving system

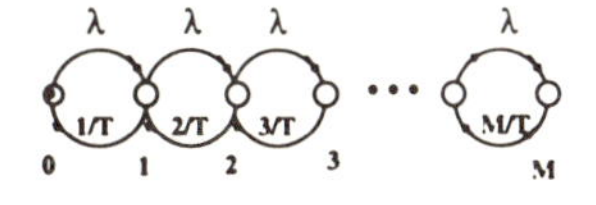

(b) Markov chain

Figure 4.12. The model used to study the blocking performance of a circuit-switched system.

traffic from "starving out" another. For example, suppose that all bursty data sources are temporarily idle and that, at the same time, many requests for 64 kbit/sec digital voice connections are made. In the absence of constraints, all of the time slots may be declared to be prearbitrated to accommodate the influx of voice requests. Unfortunately, the average holding time of each such voice connection may be several minutes and, until some of the voice connections are released, thereby allowing some prearbitrated slots to be converted back into queue-arbitrated slots, bursty data communications will be entirely preempted.

We will confine our attention to a fixed division among pre- and queue-arbitrated time slots, and refer the interested reader to the ample literature which exists on movable boundary techniques. The fixed boundary approach allows us to illustrate and quantify the salient features of blocking systems and, since only circuit-switched connections can compete for circuit-switched minislots, the blocking performance is not influenced by the presence of bursty data traffic.

Requests for circuit-switched connections are assumed to occur at Poisson points in time, with mean arrival rate λ. The duration of each admitted connection is assumed to be much greater than the 125 μsec frame, so that the granularity in holding time (i.e., each connection really lasts an integral number of DQDB frames) can be ignored and the holding time can be regarded as continuous and exponentially distributed with mean value T. Each of the M available circuit-switch minislots per frame can be regarded as a server of one request. Any arrivals which occur when all M servers are busy will be blocked. Blocked calls are assumed to be abandoned.

We define the state of this system to be the current number of busy servers, and construct the $(M+1)$ state Markov chain shown in Figure 4.12b. In a manner analogous to that which was developed when we studied the M/M/1 queue, we note that in any very small time interval Δt, the state of the system can change only by unity (only one request for a new connection or one disconnection can occur in Δt). In time interval Δt, the probability that a system in state i progresses to state $i + 1$ is given by $\lambda \Delta t$ (the probability of an arrival in Δt), $i = 0, 1, \ldots, M - 1$. Also, in time interval Δt, a system in state i can progress to state $i - 1$ if *any* other of the i calls in progress should terminate in Δt, $i = 1, 2, \ldots, M$. Since the probability that any given call terminates in time Δt is simply $\Delta t/T$, and any one of the i calls in progress may be terminated within interval Δt, the probability of a transition from state i to state $i - 1$ in time interval Δt is simply $i\Delta t/T$, $i = 1, 2, \ldots, M$. As was the case for the M/M/1 queue, in the steady state, the probability of transitioning into a state must equal the probability of transitioning out of a state. Thus, if we define P_i to be the steady-state probability of finding the system in state i at time t, we can write:

$$\lambda P_0 = \frac{1}{T} P_1 \Rightarrow P_1 = \lambda T P_0 \tag{4.4.1}$$

$$\left(\lambda + \frac{1}{T}\right) P_1 = \lambda P_0 + \frac{1}{T} P_2 \Rightarrow P_2 = \tfrac{1}{2}[(\lambda T + 1)P_1 - \lambda T P_0] \tag{4.4.2}$$

$$\left(\lambda + \frac{i}{T}\right) P_i = \lambda P_{i-1} + \frac{i+1}{T} P_{i+1} \Rightarrow P_{i+1} = \frac{1}{i+1} [(\lambda T + i)P_i - \lambda T P_{i-1}] \tag{4.4.3}$$

$$\lambda P_{M-1} = \frac{M}{T} P_M \Rightarrow P_M = (\lambda T/M) P_{M-1} \tag{4.4.4}$$

Equations (4.4.1)–(4.4.4) can be solved recursively to yield:

$$P_2 = \tfrac{1}{2}[(1 + \lambda T)(\lambda T)P_0 - (\lambda T)P_0] = \frac{(\lambda T)^2}{2} P_0 \tag{4.4.5}$$

$$P_3 = \tfrac{1}{3}\left[(2 + \lambda T)\frac{(\lambda T)^2}{2} P_0 - (\lambda T)^2 P_0\right] = \frac{(\lambda T)^3}{3!} P_0 \tag{4.4.6}$$

$$P_{M-1} = \frac{1}{M-1}\left[\{(M-2) + \lambda T\}\frac{(\lambda T)^{M-2}}{(M-2)!} P_0 - \frac{(\lambda T)^{M-2}}{(M-2)!} P_0\right] = \frac{(\lambda T)^{M-1}}{(M-1)!} P_0 \tag{4.47}$$

$$P_M = \frac{1}{M}\left[\frac{(\lambda T)^{M-1}(\lambda T)}{M-1}\right] P_0 = \frac{(\lambda T)^M}{M!} P_0 \tag{4.4.8}$$

Finally, since

$$\sum_{i=0}^{M} P_i = 1 = P_0 \sum_{i=0}^{M} (\lambda T)^i / i! \tag{4.4.9}$$

$$P_0 = \frac{1}{\sum_{i=0}^{M} (\lambda T)^i / i!} \tag{4.4.10}$$

The blocking probability, P_{BLOCK} is simply the probability that a newly arriving request finds all circuits busy. Then,

$$P_{\text{BLOCK}} = P_M = \frac{(\lambda T)^M / M!}{\sum_{i=0}^{M} (\lambda T)^i / i!} \tag{4.4.11}$$

Let us assume that we have selected M such that the arrival rate per circuit, λ', is a constant, that is, the total arrival rate $\lambda = \lambda' M$ is directly proportional to M. This would seem to be reasonable; if the aggregate offered intensity rises, we would need to allocate more circuits to handle the additional load. Thus,

$$P_{\text{BLOCK}} = \frac{(\lambda' TM)^M / M!}{\sum_{i=0}^{M} (\lambda' TM)^i / i!} \tag{4.4.12}$$

In (4.4.12), the parameter $\Lambda = \lambda' T = \lambda T/M$ is known as the Erlang load per circuit. The blocking probability is plotted in Figure 4.13 as a function of the Erlang load per circuit (often called the normalized load) for various values of M. Note that, in Figure 4.13, we allow M, in principle, to assume the value of any positive integer; for a DQDB system, of course, M would always be an integral multiple

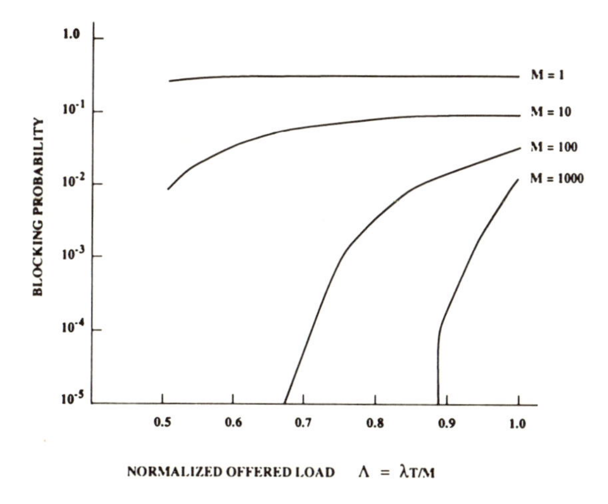

Figure 4.13. Circuit-switching blocking probability versus normalized offered load for select number of servers.

of 48 since each prearbitrated slot can accommodate 48 circuit-switched connections. We notice in Figure 4.13 that for large values of M the blocking probability diminishes rapidly as Λ is reduced below unity. In fact, from (4.4.12) we observe that as M becomes very large, the blocking probability approaches zero for $\Lambda < 1$, and approaches unity for $\Lambda > 1$. This lends the following interesting interpretation. For any number of circuits M, we can find the average number of circuits in use at any given time, E {busy circuits}:

$$E\{\text{busy circuits}\} = \sum_{j=0}^{M} jP_j = \frac{\sum_{j=0}^{M} j(\lambda T)^j/j!}{\sum_{i=0}^{M} (\lambda T)^i/i!} \tag{4.4.13}$$

$$= \frac{\sum_{j=1}^{M} (\lambda T)^j/(j-1)!}{\sum_{i=0}^{M} (\lambda T)^i/i!} = \frac{\lambda T \sum_{j=1}^{M} (\lambda T)^{j-1}/(j-1)!}{\sum_{i=0}^{M} (\lambda T)^i/i!}$$

$$= \lambda T \left[1 - \frac{(\lambda T)^M/M!}{\sum_{i=0}^{M} (\lambda T)^i/i!} \right] = \lambda T[1 - P_B]$$

For large M and $\lambda T < 1$, E {busy circuits} is approximately equal to λT since P_B tends toward 0. Thus, when the blocking probability is very small, the Erlang load corresponds to the average number of circuits which are busy and the Erlang load per circuit represents the fraction of time that any given circuit is busy.

Returning to Figure 4.13, we further observe the principle of trunking efficiency. For a given Erlang load per circuit less than one, the blocking probability reduces rapidly as the number of circuits increases. This is not the result of a greater number of circuits being made available to serve a constant load. Rather, for a large number of circuits, the number of circuits actually in use remains closer to its average value, and the fraction of time that all circuits are busy is very small. For a small number of circuits, the variability in the number of circuits actually in use is much greater, and, for $\lambda' T < 1$, there is greatly increased likelihood of a newly generated request finding all circuits busy.

4.5. Fiber-Distributed-Data-Interface

In contrast to the bus architecture of DQDB, an alternative architecture that has developed for metropolitan area applications is that of a ring. Known as Fiber-Distributed-Data-Interface (FDDI) by virtue of its use of fiber links to interconnect the FDDI access stations, its original intent was to provide high-speed connectivity among mainframes, mass storage devices, and other large data peripherals. FDDI can, however, also serve as the backbone network for interconnection of lower-speed networks. The maximum data rate at which a station can access FDDI is 100 Mbit/sec, although the symbol rate on the ring is 125 Mbaud to accommodate a four-out-of-five code used on the optical medium (each 4-bit unit of information is mapped into a 5-bit word to facilitate physical-level synchronization, bit timing recovery, etc.).

As shown in Figure 4.14a, an FDDI system consists of two counterrotating rings, each supporting an access data rate of 100 Mbit/sec. Each access station intercepts both rings, terminating and electronically regenerating the inbound signals and retransmitting these onto the outbound fibers. Unlike the counterdirective buses of DQDB in which each bus is active, only one FDDI ring is active under normal conditions. The second ring is a "hot standby" used to maintain service in the event of a station or cable failure. Each access station is equipped with a switch which can electrically connect the two rings as shown in Figure 4.14b. Under normal conditions, FDDI data frames are relayed around one ring, undergoing electronic regeneration and processing within each access station. In the event of a cable or station failure, the station to the left and right of the failure will activate their switches, thereby maintaining a closed loop among the nonfailed stations, and service will be uninterrupted.

FDDI is responsible only for the data link and physical layers of the OSI protocol stack and operates by encapsulating arbitrary Layer 3 frames into stan-

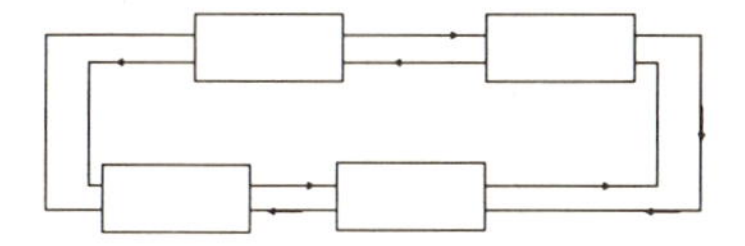

(a) Two counter-rotating rings

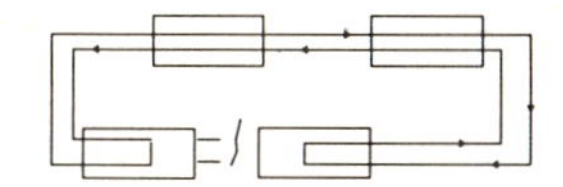

(b) Reconfiguration after cable cut

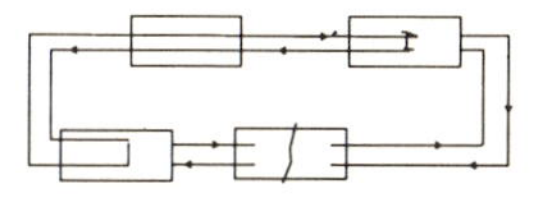

(c) Reconfiguration after station failure

Figure 4.14. FDDI ring architecture and reliability of features.

dard FDDI Layer 2 frames. Unlike DQDB, the FDDI media access protocol does not operate on packet segments. Rather, stations access FDDI with fully formatted Layer 2 frames. Two types of FDDI Layer 2 frames have been defined: data frames and tokens. The token is a unique combination of bits which is relayed among access stations around the ring. In order to transfer a data frame onto the ring, a station must first await receipt of the token. Any station currently holding the token can transfer a number of data frames no greater than that permitted by the Timed Token Rotation Protocol (TTRP) to be described later. FDDI is a single token system; only one token circulates around the ring. After a token-holding station has transmitted its permitted number of data frames, it transfer the token along the ring to its neighbor.

FDDI data frame and token formats are as shown in Figure 4.15. The data frame starts with 16 so-called "idle" symbols (an FDDI symbol is a combination of bits in a 5-bit word). The "idle" symbol preamble is used to permit the next station on the ring to recover bit timing. Since stations can transfer data frames only when holding the token, an appreciable time interval may elapse during which a transmission link interconnecting two adjacent access stations is idle. During this idle time, the receiving end may lose synchronization with the exact frequency and phase of the transmitter's bit clock, and the preamble permits those to be acquired. After the preamble comes the Start Delimiter. This unique com-

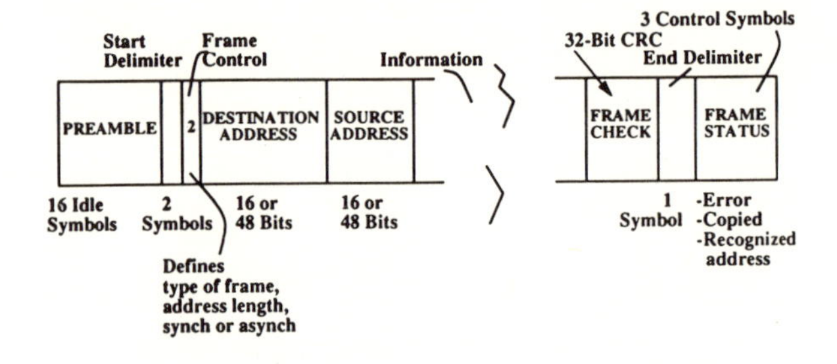

(a) Data frame

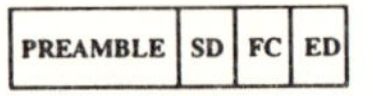

(b) Token

Figure 4.15. FDDI data frame and token formats.

bination of two symbols enables the receiver to unambiguously recognize the official "start" of the data frame. This is followed by a two-symbol Frame Control field which is used to define the type of frame (data or token), the length of the subsequent address fields, and various other types of frame-specific information. Next comes the Destination Address and Source Address fields, each either 16 or 48 bits in extent (presumably to support smaller or larger configurations). As implied, the Destination Address contains the identity of the intended receiving access station, and the Source Address identifies the sending access station. The Source Address is followed by a variable-length Information field; each 4-bit combination of user data is mapped into one of 16 five-bit FDDI symbols, each of which is different from the idle, frame control, frame station, start delimiter, and end delimiter symbols. The Information field is followed by a Frame Check field containing a 32-bit CRC used to detect and correct certain combinations of bit errors which may have occurred within the Information field. Next is the End Delimiter, a unique one-symbol field used to define the "end" of the frame. Finally, the three-symbol Frame Status field is written by the receiver to inform the transmitter of the status of the received frame. In FDDI, a data frame can be removed or stripped from the ring only by the station having placed the frame onto the ring. As a frame propagates around the ring, each station checks the destination field and, if the frame is determined to be destined for that station, a local copy of the frame is prepared. The receiving station checks the CRC and writes into the frame status field to advise the sending station as to whether or not the frame was correctly copied. The frame is then further relayed around the ring until it arrives back at its source. As frames are relayed around the ring, each access station checks the Source Address as well as the Destination Address fields, and when a station finds a frame with its own Source Address, it removes the frame from the ring by not relaying it any farther. It then checks the Frame Status field to see if the frame was correctly copied by the destination station. In this fashion, the Frame Status field provides an active acknowledgment (ACK) or negative acknowledgment (NCK) back to the transmitting station. The appropriate response to an ACK or NCK is the responsibility of higher-layer protocols. The maximum size of an FDDI Layer 2 frame is 4500 octets.

The FDDI token is a simplified version of the data frame. It, too, starts with a 16-symbol preamble to facilitate receiver synchronization, followed by the two-symbol Start Delimiter. The Frame Control field for the token contains a unique combination of two symbols which define this frame to be the token. Finally, the one-symbol end delimiter defines the end of the token.

Access to FDDI to mediated by the TTRP. This is a strategy for limiting the Token Holding Time (THT) of any station in possession of the token such that, for each station, the maximum time interval between consecutive receipts of the token is bounded by twice some universally agreed-upon Target Token Rotation Time (TTRT). Under steady-state conditions, each station receives the token approximately every TTRT (or more frequently if the ring is lightly loaded with traffic). The TTRP allows synchronous traffic with guaranteed bandwidth needs to be integrated with asynchronous traffic not needing a guaranteed bounded latency. Unlike DQDB in which prearbitrated time slots ensure reserved bandwidth for circuit-switched traffic, the TTRP only ensures that each station can transmit some allocated number of FDDI frames at least once in every time interval equal to 2 × TTRT (again, this is "worst case"; the TTRP "tries" to circulate the token to each station at least once every TTRT).

To accommodate its synchronous traffic needs, an access station may request a transmission time allocation. If its request is granted, then each time that station receives the token, it can transmit data frames until its time allocation is reached; under certain conditions, it can continue to transmit for a longer interval if it has additional queued frames awaiting transmission. Allocation of transmission time is administered by a network manager, which may be either a processor attached to one of the access stations or some number of distributed processors located within the access stations which cooperate to reach some collective decision.

Let S_i be the transmission time allocated to station i, and let D_p be the round-trip delay associated with transferring the token once around a totally idle ring (that is, if the ring presents no traffic whatsoever, then D_p is the sum of the round-trip propagation delay plus the sum of the processing delay incurred in each access station). Since each station can hold the token for a period of time at least equal to S_i, and the TTRT is some mutually agreed-upon number, it follows that, for an N-station system,

$$\sum_{i=1}^{N} S_i + D_p \leq \text{TTRT} \tag{4.5.1}$$

Clearly, if the sum of all time allocations plus the fixed delay D_p exceeds TTRT, then it is impossible for each station to receive the token within one TTRT. Thus, when deciding whether or not a new request for an allocation can be satisfied, the network manager checks to see if the granting of that request will cause inequality (4.5.1) to be violated; if so, then the request must be denied [actually, the American National Standards Institute FDDI standard adds the transmission time of the largest permitted FDDI frame, 4500 octets, to the left side of (4.5.1); inclusion of this additional term would obscure an understanding of the TTRT and will be ignored in our treatment]. As long as the allocation side (4.5.1) is not violated, then, as we shall see, station i can, on average, hold the token for a fraction of time at least as great as S_i/TTRT, and therefore enjoy a "guaranteed" bandwidth of at least (S_i/TTRT) × 100 Mbit/sec.

The access rule is as follows. Each station is equipped with a timer which measures the elapsed time since that station last received the token (the so-called Token Rotation Time, TRT). Then, that station may hold the token for a period of time (the Token Holding Time, THT) equal to

$$\text{THT} = \begin{cases} S_i, & \text{TTRT} - \text{TRT} < S_i \quad (4.5.2a) \\ \text{TTRT} - \text{TRT}, & \text{TTRT} - \text{TRT} > S_i \quad (4.5.2b) \end{cases}$$

The station must release the token after a time interval equal to THT or after it has transmitted all of its queued data frames, whichever occurs first. Equation (4.5.2a) states that if the TRT is within that station's allocation of the TTRT, or if the TRT exceeds the TTRT, then that station can only transmit its allocation (it is always permitted to transmit its allocation whenever it receives the token, even if this will cause the elapsed time since the token was received at subsequent access stations to exceed the TTRT). Equation (4.5.2b) states that if the difference between the TTRT and the TRT is greater than that station's allocation, then the station may transmit its allocation and then transmit additional queued data frames (if it has any) until the TRT plus the total time which that station has held the token equals the TTRT; at that time, the station must release the token.

Now, since (1) each station can always hold the token for a time interval equal to S_i to transmit its allocation, (2) no station can transmit any traffic for which no allocation was made if such transmission will cause the elapsed time since it last saw that token until it releases the token to exceed the TTRT, and (3) the sum of the allocations made by the network manager plus the fixed round-trip token rotation time D_p must be less than the TTRT, it follows that the average fraction of time during which station i can hold the token must always be at least S_i/TTRT, and the average data rate or "bandwidth" guaranteed to station i, B_i, will always be at least:

$$B_i \geq (S_i/\text{TTRT}) \times 100\text{Mbit/sec}, \qquad i = 1, 2, \ldots, N \tag{4.5.3}$$

If the allocation rule (4.5.1) is satisfied with equality (only allocated transmissions can be made for all stations), then (4.5.3) also becomes an equality for every station. However, if there is some "unallocated" capacity [(4.5.1) is an inequality], then stations can make "unallocated" transmissions and their access "bandwidth" can exceed their allocated "bandwidth."

It also follows that under transient conditions, the maximum access delay experienced by any station is equal to 2 × TTRT. (It is important to recognize that, in the FDDI context, maximum access delay is defined to be the maximum elapsed time between consecutive receipts of the token, or the maximum elapsed time between the beginnings of two consecutive transmissions. The maximum access delay is most definitely *not* the maximum delay experienced by a newly appearing data frame as it awaits transmission within its access station's transmit buffer. This latter delay is a random variable whose time is determined by instantaneous traffic-loading conditions and although its average value is bounded if the total external load presented to the network is less than the usable capacity of the network, its extreme value can, in general, be unbounded.) Assume that the fixed round-trip delay D_p is zero (small ring, no processing delay) and suppose that, for some reason, all stations are temporarily idle (they are not even transmitting their allocations). Under these temporary conditions, the TRT seen by each station is zero. Now, suppose that each station suddenly "springs to life" and generates a number of data frames equal to its allocation plus a good number more. Finally, suppose that the allocation rule (4.5.1) is satisfied with equality. Then, the station holding the token when all stations suddenly "spring to life," in accordance with (4.5.2b), transmits for a period of time TTRT (remember that its TRT is zero and, when it springs to life, a great backlog of frames instantaneously appear; after it has transmitted for the full TTRT, it must release the token, even though its backlog is not depleted). The TRT seen by the next and all subsequent stations is now at least TTRT, the amount of time for which the first station has held the token. Since each station can always transmit its allocation, and (4.5.1) is satisfied with equality, an *additional* period of TTRT elapses before the initiating station again receives the token. Since it held the token for TTRT and a period of time equal to TTRT has elapsed before it gets the token back, the *total* elapsed time between consecutive receipts of the token at that station is 2 × TTRT. Thus, in addition to a guaranteed bandwidth given by (4.5.3), the TTRP also guarantees a maximum access delay to each station of 2 × TTRT.

An example illustrating the operation of the TTRT Protocol appears in Figure 4.16. For this example, there are four access stations, each enjoying an allocation of 2 (in arbitrary units of time). The token propagation time D_p is assumed to be zero, and after a prolonged period of ring inactivity, all stations "spring to life" at a moment when station No. 1 is in receipt of the token. The TTRT is 12 units.

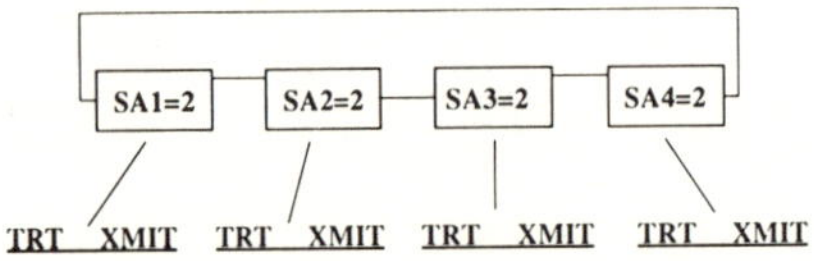

SA1=2		SA2=2		SA3=2		SA4=2	
TRT	XMIT	TRT	XMIT	TRT	XMIT	TRT	XMIT
0	2.10	12	2	14	2	16	2
18	2	8	2.2	10	2	10	2
10	2	10	2	8	2.2	10	2
10	2	10	2	10	2	8	2.2
10	2	10	2	10	2	10	2
8	2.2	10	2	10	2	10	2
10	2	8	2.2	10	2	10	2

ASSUMES TOKEN ROTATES AS FAST AS POSSIBLE

Figure 4.16. Example illustrating the operation of FDDI's Time Token Rotation Protocol.

When the system springs to life, the amount of time which has elapsed since station No. 1 last saw the token is zero (TTRT = 0). That station therefore transmits its allocation of 2 units, plus 10 more unallocated units. Since it has now held the token for the TTRT (12 units), it then releases the token to station No. 2. The TRT seen by station No. 2 is 12 units (the amount of time during which station 1 has held the token). Since its TRT equals the TTRT, it is only permitted to transmit its allocation of 2 units. It then releases the token to station 3 which can only transmit its allocation of 2 units since its TRT equal 14 units (the total time for which stations 1 and 2 have held the token). When station No. 4 receives the token its TRT is 16 units, and it can only transmit its allocation of 2 units before transferring the token back to station No. 1. Station No. 1's TRT is 18 units, and it transmits its allocation of 2 units and then transfers the token to station No. 2, whose TRT is only 8 units (since station No. 2 last received the token, each station including itself held the token for two time units). Station No. 2 can therefore transmit its allocation of 2 units, plus 2 additional unallocated units, before the sum of its TRT and THT equals the TTRT; it must then release the token to station No. 3. Continuing in this fashion, we see that for this example, each station receives the token approximately once every 10 time units; when it gets the token, each station transmits its allocation, and occasionally is allowed to transmit a little bit more. Then, the actual bandwidth enjoyed by each station is slightly greater than $(2/10) \times 100$ Mbit/sec = 20 Mbit/sec, which exceeds its allocated (guaranteed) bandwidth of $(2/12) \times 100$ Mbit/sec = 16.6 Mbit/sec. Furthermore, the greatest access delay is 18 time units, less than the guaranteed maximum of 24 time units (note that the sum of the allocations is less than the TTRT, resulting in the greatest access delay being smaller than $2 \times$ TTRT and the actual bandwidth being greater than the "guaranteed" bandwidth).

Notice from (4.5.3) that, for a larger value of TTRT, each station will enjoy a constant guaranteed bandwidth if its station allocation increases proportionately, as permitted by the allocation rule (4.5.1): if, for example, TTRT should double, then each station's allocation can double without violating (4.5.1), and a constant guaranteed bandwidth results. Large TTRTs are desirable since, as TTRT grows larger, the allocation inefficiency represented by the constant round-trip delay D_p diminishes (if D_p is comparable to, or large compared with, TTRT, then the fraction of TTRT which can be allocated diminishes and the total of the allocated bandwidth guaranteed to each station will correspondingly diminish). Larger TTRTs are also consistent with greater ring utilization efficiency, as shown in Section 2.9; in general, the efficiency improves as the ratio of data frame size to round-trip delay D_p increases, and for large TTRT, all time intervals except D_p (which is a constant) scale up accordingly. However, the maximum access delay also scales directly with TTRT, and bounded latency requirements for traffic receiving a TTRP allocation may mandate that a small TTRT be used. It has been suggested that, for most applications and over a wide range of FDDI system size, a TTRT in the neighborhood of 8 msec should be used. With this value for TTRT, the bandwidth allocation efficiency for all configurations smaller than the largest permitted FDDI installation (500 access stations, 100 km of cabling) is greater than 75%, and the maximum access delay of an allocated transmission is 16 msec, adequate for real-time traffic such as voice.

4.6. Delay and Blocking Performances of FDDI*

For small round-trip fixed delay (D_p approximately equal to 0), we can characterize the delay performance of an FDDI system by modeling the system as either an M/M/1 or an M/D/1 queue. As previously discussed when studying the delay performance of DQDB, Little's theorem allows us to model the distributed queues which exist in the various access stations as if all data frames awaiting transmission were stored in one centralized queue, at least insofar as the average delay is concerned. Furthermore, the average delay is independent of the sequence in which queued packets are served. Thus, three different FDDI systems—one which permits each station to transmit one data frame every time the token is received (each station has an allocation of unity, and TTRT is N units); a second which permits each station to transmit all of its queued packets whenever it receives the token (no allocations; TTRT equal to infinity); and a third system with some specified allocations and specified TTRT (corresponding to some traffic-dependent queuing discipline in which the sequence of a packet service is traffic-pattern dependent)—all produce the same average delay. If we model data frame arrivals as a Poisson point process, and if $D_p = 0$, then the average delay for arriving frames with exponentially distributed lengths, or with constant deterministic length, can be found by applying the results for the M/M/1 queue and M/D/1 queue, respectively.

Similarly, for some given TTRT, if we fix the total amount of allocable transmit time at some value S_T less than TTRT [to conform with (4.5.1)], and if transmit time is assumed to be allocable in units of b seconds, where S_T is an integral multiple of b, we can model FDDI allocations as an Erlang blocking system with $M = S_T/b$ servers: if requests for an allocation of b units arrive in accordance with a Poisson point process, and if the holding time for each request is random with exponential probability distribution, then the Erlang blocking formula for an M-server system can be applied to find the probability that a request for allocation cannot be served. (By setting S_T at a fixed value less than TTRT, we are implying the use of a fixed-boundary scheme; moving-boundary schemes are beyond our present scope.)

4.7. Problems

4.1. A DQDB system contains five stations. The stations are sequentially numbered, with station No. 1 positioned at the left of the linear dual bus system. The propagation delay between any two adjacent stations is exactly one time slot. Suppose that at time $t = 0$, all stations are empty, and that time slot boundaries are time aligned at the various access stations. Consider now access to the upper bus only, on which the direction of propagation is to the right. During the first time slot, segmentation units are generated by each of the four access stations. During the second time slot, segmentation units are generated by stations 1 and 3 only. During the third time slot, segmentation units are generated by stations 2 and 3 only, and during the fourth time slot, segmentation units are generated by stations 1 and 4 only. No segmentation units are generated in subsequent time slots. Draw a timing diagram showing the sequence of segmentation units appearing on the upper bus, with each labeled by originating access station. Justify your answer by drawing an appropriate set of supporting timing diagrams.

4.2. A DQDB Service Data Unit contains 981 bytes. The IMPDU header extension field contains 12 bytes. Find (a) the number of padding bytes needed in the IMPDU; (b) the number of DMPDUs created; (c) the total number of bytes needed to transfer the SDU over a DQDB system, and (d) the DQDB overhead for this SDU, expressed as a percentage of total number of bytes needed.

4.3. Each access station of a DQDB system uses a single shared buffer to store arriving segmentation units prior to reassembly into IMPDUs. An IMPDU is reassembled only when all of its segmentation units have arrived at the buffer. The shared buffer of a given access station can store 320 segmentation units and, at a certain point in time, is empty. At that point in time, segmentation units from four new IMPDUs start to arrive. Suppose that each IMPDU contains 100 segmentation units, and that arriving segmentation units are perfectly interleaved, i.e., the sequence of segmentation unit arrivals is from IMPDU 1, then 2, then 3, then 4, then 1, and so forth, until all segmentation units from all IMPDUs have arrived. (a) Suppose that the Buffer Allocation field is ignored by the reassembly machine, with all incoming segmentation units simply stored until an IMPDU has been reassembled or the buffer overflows. How many IMPDUs will be reassembled? (b) How many IMPDUs will be reassembled if the Buffer Allocation field is properly used?

4.4. Each access station of a large DQDB system contains a buffer capable of storing an arbitrarily large number of cells. Consider only the upper bus. At the access stations, the pattern of cell arrivals for a given output is such that, for that output, the number of cell arrivals in consecutive time slots can be modeled as a set of independent identically distributed Poisson random variables, each with mean value p (the input activity factor). The DQDB system operates as a perfect first-come–first-served distributed queue, with appropriate deference to each of the three DQDB access priority classes. Each arriving cell is of priority class 1 with probability 25%, class 2 with probability 50%, and class 3 with probability 25%, and the priority of each cell is totally independent of the priority of any other cell. Find and plot the mean delay for each priority class as a function of p.

4.5. At time $t = 0$, an M/M/1 queuing system is empty. The system is capable of storing up to two requests, including the one being served. The arrival rate is λ, and the mean service time is T. Let $P_j(t)$ be the probability that there are j requests in the system, including the one being served. (a) Find and plot $P_0(t)$, $P_1(t)$ for $\lambda T = 0.25$. (b) Repeat for $\lambda T = 0.5$. (c) Repeat for $\lambda T = 0.9$.

4.6. Each access station of a large DQDB system generates SDUs which are sufficiently long that the number of segmentation units contained in each may be modeled as an independent exponentially distributed random variable (the discrete time DQDB system may be modeled as a continuous time system). Let T be the amount of time needed to transmit an SDU containing the average number of segmentation units over one of the DQDB buses. The arrival process at each station is Poisson with arrival rate $2\lambda/N$, where N is the number of access stations, and each arrival is equally likely to be destined for either bus. The holding buffer at each station can store an arbitrarily large number of segmentation units. All segmentation units of any given SDU are of either access class 1 or access class 2, where class 1 SDUs have service priority over class 2 SDUs. The class of any SDU is independent of the class of any other SDU, and the two classes are equally likely to occur. Ignore all DQDB overhead. (a) Find and plot as a function of λT the average time spent in the system for each new class 1 SDU arrival. (b) Repeat for each new class 2 arrival.

4.7. An M/M/1 queuing system has arrival rate λ, mean service time T, and can store an arbitrarily large number of requests. Find and plot as a function of λT the variance of the time spent in the system.

4.8. N sources are statistically multiplexed in a single-server queuing system. Each source produces Poisson arrivals, and the service time of each new arrival is an independent exponentially distributed random variable of mean value T. The arrival rate of each source is $0.1/T$. It is required that 99% of all new arrivals have departed with a period of time equal to $10T$. Use the result of Problem 4.7 and the Chebychev inequality to find a conservative admission control policy, that is, to find a lower bound on N such that the 99% delay objective is met.

4.9. Consider an integrated video/data DQDB system. Suppose that each 125 μsec DQDB frame consists 10 time slots, and that each video connection requires allocation of one time slot. New video connections are given priority access to unallocated time slots, and only those slots remaining unallocated to video are available to serve data. Requests for new video connections are Poisson distributed and the holding time for call connection is an exponentially distributed random variable. In the aggregate, the rate of new requests for video to either of the buses is 6/hr, and the average holding time for a video connection is 1 hr. In the aggregate, the rate of generation of data packets is 8000/sec, and, on average, each packet is of length such that three time slots are needed to complete its transmission. (a) What fraction of time will there be insufficient unallocated time slots to serve the needs of the data traffic? (b) What is the video connection blocking probability?

4.10. Again consider the system of Problem 4.9, but for a different time slot allocation rule. Video again has priority, except that some maximum number of time slots per frame is absolutely reserved for data. (a) What is this minimum number if the effective bandwidth reserved for data must always be greater than or equal to the average arrival rate of data information? (b) What is the new video connection blocking probability?

4.11. A connection-oriented DQDB system admits new calls until the average delivery time for data packets exceeds some threshold. Each bus operates at a speed of 100 MB/sec. Packet arrivals from each active connection are Poisson distributed with rate 10 packets/sec, and the length of each packet is an exponentially distributed random variable with mean value equal to 100,000 bits. Ignoring the cell granularity (or, equivalently, treating the system as running in continuous time), find the maximum number of connections which can be admitted to each bus such that the mean packet delay (including service time) is less than 50 msec.

4.12. For the system of Problem 4.11, call attempts are Poisson distributed with an average rate of 1/min. The holding time of each call is exponentially distributed with mean value equal to 80 min. For the same QOS objective, find the call blocking probability.

4.13. The Target Token Rotation Time of an FDDI system is 20 msec, the data rate is 100 Mbit/sec, each packet contains 10,000 bits, and both the round-trip propagation delay and the round-trip token processing delay are negligible. There are N access stations, and the allocation to station i is S_i packets, $i = 1, \ldots, N$. (a) Find the maximum possible value for the allocations summed over all access stations. (b) Find the maximum station access delay, that is, the maximum time interval between consecutive token arrivals at any station.

4.14. Packet arrivals to each access station of a 100-station FDDI system are Poisson with mean value λ. The number of bits in each arrival is an exponentially distributed random variable with mean value $M = 100{,}000$ bits, which translates into a mean service time of 1 msec for the data rate of this particular system. Each station has an allocation of 2 msec. Find the mean packet delay (including service time) and plot as a function of λ.

4.15. For the system of Problem 4.14, let there be two priority access classes. Each packet is equally likely to belong to either class, independently of all other packets. Find and plot as a function of λ the mean delay for (a) high-priority packets and (b) low-priority packets.

References

1. *IEEE Commun. Mag.,* special issue on Metropolitan Area Networks, **28**(4), April 1988.
2. IEEE Standard 802.6: Distributed Queue Dual Bus (DQDB) Subnetwork of a Metropolitan Area Network, Final Draft 15, IEEE Computer Society, Washington, D.C.
3. M. W. Garrett and S.-Q. Li, A study of slot reuse in dual bus multiaccess networks, *IEEE J. Selected Areas Commun.* **SAC-9**(2), Feb. 1991.
4. E. L. Hahne, A. K. Choudhury, and N. F. Maxemchuck, DQDB networks with and without bandwidth balancing, *IEEE Trans. Commun.* **COM-40**(7), July 1992.
5. F. E. Ross, An overview of FDDI, *IEEE J. Selected Areas Commun.* **SAC-7**(7), Sept. 1989.
6. F. E. Ross, FDDI—A tutorial, *IEEE Commun. Mag.* **24**(5), May 1986.

7. W. E. Burr, The FDDI optical data link, *IEEE Commun. Mag.* **24**(5), May 1986.
8. F. E. Ross and J. R. Hamstra, Forging FDDI, *IEEE J. Selected Areas Commun.* **SAC-11**(2), Feb. 1993.
9. American National Standard, Fiber Distributed Data Interface (FDDI), Token Ring Media Access Control (MAC), ANSI X3.139–1987.
10. F. E. Ross, Get ready for FDDI-2, *Net. Manag.* July 1991.
11. R. Jain, Error characteristics of Fiber Distributed Data Interface (FDDI), *IEEE Trans. Commun.* **COM-38**(8), Aug. 1990.
12. *Digital Tech. J.,* issue on FDDI, **3**(2), Spring 1991.
13. R. Jain, Performance analysis of FDDI, *Digital Tech. J.* **3**(3), Fall 1991.
14. *IEEE Lightwave Trans. Syst.,* special issue on FDDI, **2**(2), May 1991.
15. R. B. Cooper, *Introduction to Queueing Theory,* Elsevier/North-Holland, Amsterdam, 1981.
16. D. Bertsekas and R. Gallager, *Data Networks,* Prentice–Hall, Englewood Cliffs, N.J., 1987.
17. F. P. Kelly, *Reversibility and Stochastic Networks,* Wiley, New York, 1979.

5

Broadband ISDN and Asynchronous Transfer Mode

5.1. Broadband ISDN and ATM: Preliminaries

Much of the material we have covered thus far is related to the notion of Broadband Integrated Services Data Network (B-ISDN), a vision for the evolution of local and wide-area telecommunication networks based on the concepts of universal interfaces, bandwidth-on-demand, hardware-based packet interconnects, and multimedia traffic integration. In the B-ISDN vision, all types of traffic presented by end-users of the telecommunication network would be formatted (either by the user or at the user interface to the network) into short, fixed-length packets containing an information field and a routing field (header). This would necessitate segmentation and reassembly of larger blocks of data and continuous bit rate traffic. The network interfaces would operate at access speeds of 155 and 622 Mbit/sec, and information would be carried in the Asynchronous Transfer Mode (ATM).

ATM is a means of transferring information which is quite analogous to that of DQDB. Like DQDB, the basic building block is a 53-octet cell consisting of a 5-byte header and a 48-byte payload. On all ATM transmission facilities, ATM cells from different sources are asynchronously multiplexed into fixed-length time slots with rigidly defined boundaries occurring synchronously in time, each time slot carrying one ATM cell. With ATM, as shown in Figure 5.1a, it is the appearance of cells associated with the same information source–destination pair (designated C1, C2, etc.) which occurs asynchronously; the transmission links are operated synchronously at both the bit and time-slot levels, and a single ATM cell cannot overlap two consecutive time slots. This is to be contrasted with the Synchronous Transfer Mode (STM) shown in Figure 5.1b in which time on a transmission link is divided up into fixed-length frames (delineated by frame marker boundaries), which are further divided up into fixed-length time slots. With

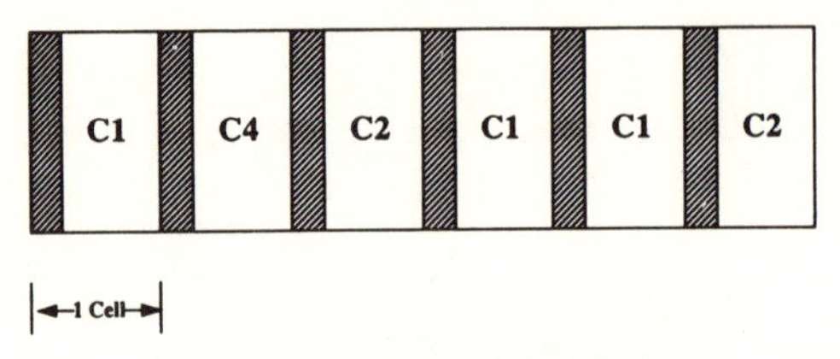

(a) Asynchronous Time Multiplexing

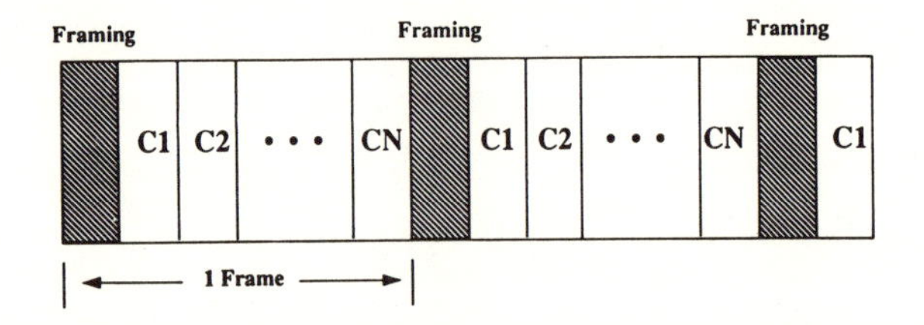

(b) Synchronous Time Multiplexing

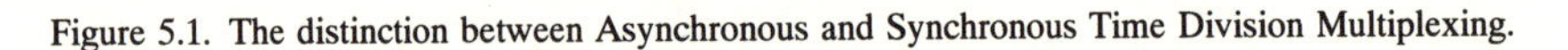

Figure 5.1. The distinction between Asynchronous and Synchronous Time Division Multiplexing.

STM, cells associated with a given source–destination pair are always transmitted during the same time slot of each frame and therefore occur synchronously with time. STM multiplexing is done via reservation (circuit-switching); ATM multiplexing is statistical in nature.

A reference architecture for B-ISDN is shown in Figure 5.2. In this model, ATM cells are generated on customer premises, that is, the ATM Adaptation Layer (AAL) which converts arbitrarily formatted information supplied by the user into an ATM cell stream, and vice versa, is located on customer premises in a Customer Premises (CP) node. More will be said of the AAL later. For now, it is sufficient to note that differently formatted user information signals will require different AALs; the AAL therefore serves the function of a terminal adapter, converting the information which is supplied by the user in some specified form into the universal ATM format expected by the network. In this regard, the AAL should be viewed as performing a service identical to that of the LAN Terminal Adapter shown in Figure 2.1, and is but one example of a familiar concept initially arising within the context of an earlier type of network reappearing once again in the context of a broadband network. As with the earlier LAN concept, each user of a broadband network must declare its signal format, and an appropriate AAL must be provided at the user's CP node. If the user's equipment natively produces ATM cells (i.e.,

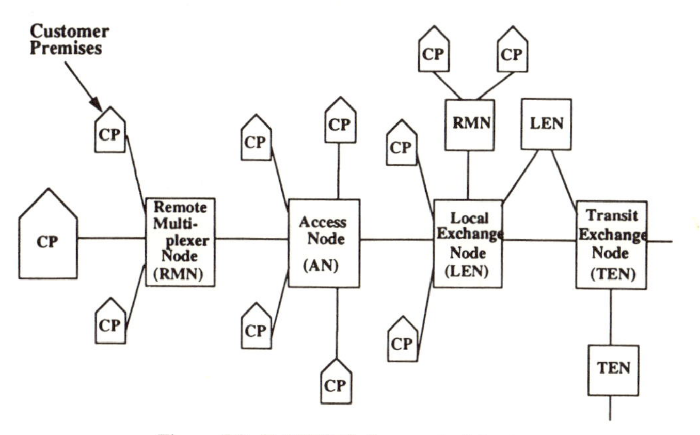

Figure 5.2. B-ISDN Reference Architecture.

if the AAL is performed internally to the equipment), then no additional AAL would be needed in the CP node to connect that user to the B-ISDN network.

In the reference model (Figure 5.2), ATM cells generated by some multitude of relatively low-bandwidth CP nodes are carried by fixed-point transmission facilities, and flow into a Remote Multiplexer Node (RMN) where they are statistically concentrated into an ATM stream with higher time slot occupancy. Several such concentrated links next converge on an Access Node and are further statistically multiplexed with themselves and with ATM cell streams produced by higher-bandwidth CP nodes. This highly multiplexed ATM stream is carried to the Local Exchange Node (LEN), an ATM switch. Other possible interfaces to the LEN may be RMNs and very large CP nodes. Finally, the ATM cells not routed back to the community of users served by a given LEN are sent either to other LENs or to larger, Tandem Exchange Nodes (TEN, a large ATM switch). As shown, TEN interfaces exist only to other TENs and LENs, and not to Access, Remote Multiplexer, and Customer Premises nodes.

In the B-ISDN reference model of Figure 5.2, the various transmission links shown serve as the carriers for ATM cells. Each transmission link consists of a transmitter and receiver as shown in Figure 5.3a. The transmitter generates fixed-length transmission frames which are delineated by framing sequences (shaded areas) used for transmission system synchronization and various transmission system housekeeping functions (overhead). ATM cells which arrive at the transmitter are placed onto the transmission link during the time intervals between framing sequences. These time intervals may not contain an integral number of ATM time slots, in which case some ATM cells might actually be split and carried in two consecutive frames. At the receiver, the framing sequences are removed, and any ATM cells which were split at the transmitter are reattached. The important point to note is that ATM cells arrive at the transmitter and ATM cells depart from the receiver; the actual format used by the transmission system may be peculiar to that transmission system, as long as any manipulation done to the ATM cells at the transmitter is completely undone at the receiver.

Similarly, the ATM switches and multiplexers appearing in Figure 5.2 may manipulate the format of the arriving ATM cells as long as any such manipulation is undone at the output. For example, as shown in Figure 5.3b, the ATM switch fabric is surrounded by line interface cards serving any of several functions. The line interface cards may, for example, be used to delay the arriving ATM cells such that cells appear time-aligned at the input to the switch. The line interface cards may also contain switch routing tables, each of which is responsive to the virtual circuit identifiers of the arriving ATM cells. Here, at call setup time, a routing table is provided with the switch output port number to which all cells bearing the newly assigned virtual circuit identifier are to be sent. Next, the line interface appends this port number to the ATM cell in a new field appended to the cell header. The switch fabric then operates on this new "output port" field to relay the cells to their correct outputs. The output line interface cards then remove the output port field, which is no longer needed, thereby re-creating standard ATM cells. Again, the point to be noted is that, just like the transmitter of a transmission system, the input

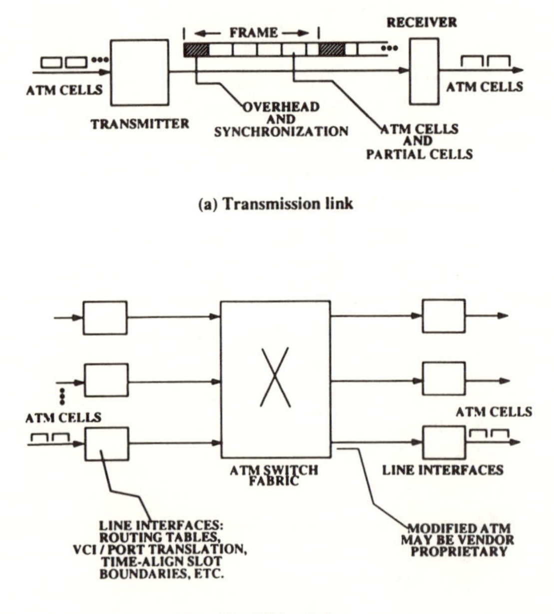

Figure 5.3. Transmission and switching system interfaces for ATM.

line interface to an ATM switch may actually modify the format of the ATM cells if such modification simplifies the relaying of the cells by the switch fabric, as long as the modification is totally reversed by the output line interface to re-create standard ATM cells. The external interfaces to the switching and transmission systems appearing in Figure 5.2 are therefore standard ATM, allowing transmission and switching systems with different proprietary internal interfaces to be used at different locations within the B-ISDN reference model.

B-ISDN is a connection-oriented network, and each CP node may support many concurrent virtual circuits. Although the actual ATM cells are carried over virtual circuits, the service seen by any particular user may be either connectionless or connection-oriented, as noted earlier in Chapter 1 and will again be illustrated in Section 5.6 in the context of an ATM LAN.

5.2. B-ISDN Protocol Reference Model

Protocols relevant to the operation of a B-ISDN CP node appear in Figure 5.4. As shown, the ATM Adaptation Layer for such a node may need to support several types of services. User services involve Connection-Oriented and Connectionless Variable Bit Rate services, possibly some other types of Variable Bit Rate services, and Constant Bit Rate services. The Variable Bit Rate services would support nonpersistent types of traffic of various peak data rates, such as bursty data traffic, image files, large data base file transfer, packet video, and packet voice. The Constant Bit Rate services would support persistent types of traffic characterized by constant data rate over a prolonged period, such as digital video and/or 64 kbit/sec digital voice. In addition to user information, the CP node AAL must support control information as yet another Variable Bit Rate service.

Presented across the user or control interfaces to the AAL are information signals, the formats of which may be different for different CP nodes. As mentioned earlier, it is the responsibility of the AAL to convert these signals into a standard format suitable for ATM prior to sending these signals into the ATM network, and to reconstruct the signals from ATM cells arriving from the network prior to presentation to the user or control interfaces. With regard to the OSI Protocol Reference Model, the AAL, coupled with the ATM and Physical Layer shown in Figure 5.4, perform services of layer 1 and some of the services of layer 2. The boundary between the user/control interfaces and the AAL is therefore somewhere within layer 2 of the OSI model.

The AAL is divided into two sublayers. The Convergence Sublayer performs an encapsulation/deencapsulation function for user and control data. The encapsulated data frame created by the Convergence Sublayer is known as a Convergence Sublayer Protocol Data Unit (CS-PDU) and is quite similar in format to the IMPDU of DQDB. In the transmit direction, the Segmentation and Reassembly

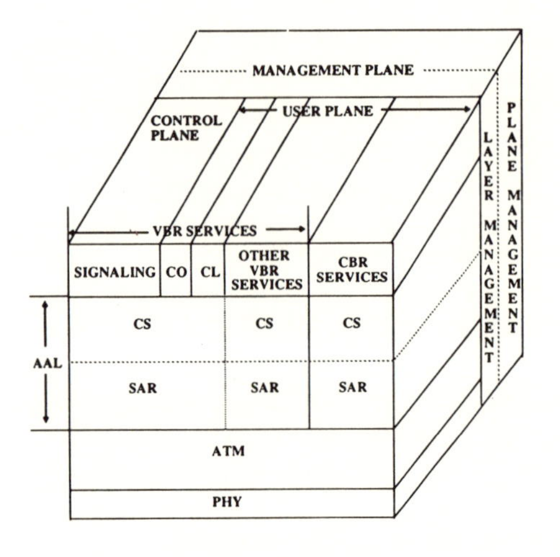

CS: CONVERGENCE SUBLAYER
SAR: SEGMENTATION AND REASSEMBLY SUBLAYER
AAL: ATM ADAPTATION LAYER
CO: CONNECTION-ORIENTED
CL: CONNECTIONLESS
VBR: VARIABLE BIT RATE
CBR: CONTINUOUS BIT RATE

Figure 5.4. B-ISDN Protocol Reference Model.

sublayer creates 48-byte segments from each CS-PDU to form Segmentation and Reassembly Protocol Data Units (SAR-PDUs); in the receive direction, these are reconnected to form CS-PDUs. More will be said later of the CS-PDUs and SAR-PDUs. The ATM layer is responsible for attaching/stripping the 5-byte header from each SAR-PDU, thereby forming the 53-octet ATM cell. The physical layer is responsible for placing these cells onto the transmission link medium. Note that the ATM network, consisting of Remote Multiplexer, Access, Local Exchange, and Tandem Exchange Nodes, operates only on the ATM cell headers; the 48-byte cell payloads are neither processed nor even read by the ATM network entities. The principal (but not the only) function of the ATM cell header is cell relaying; more will be said later of the ATM cell header.

The B-ISDN Protocol Reference Model also includes a plane responsible for management of all user and control layers within the CP node and is involved, for example, in call setup procedures. The layer management entity of the management plane interfaces each layer of the control and user planes and is responsible for providing instructions to those layers (either for local management purposes or for transmittal to the management plane of distant CP nodes), and for accepting replies from those layers (either locally generated or generated within the management plane of some distant CP node).

5.3. Call Setup Procedures

ATM cells will be carried in B-ISDN on the basis of connection-oriented virtual circuits. Unlike real circuit-switched connections, a virtual circuit does not receive an exclusive allocation of network resources at the time that the connection is made. Rather, many virtual circuits statistically share some aggregated pool of network resources, thereby avoiding waste of reserved (but possibly unused) resources. For example, to do its job, a user's data device may need a multiplicity of connections to be concurrently established, each to a device attached to a distinct port on the network. In operation, that user might generate consecutive bursts of high-speed information, each intended for a different receiver. Rather than reserving an exclusive high-speed physical connection to each receiver, which would be used only sporadically, the user of a virtual circuit-oriented network would be allocated a number of concurrent virtual circuits, each to a different one of its cooperating receivers. Then, to access a particular virtual circuit, the user would place that virtual circuit's identifier in the appropriate field of the information packet. At call setup time, the routing tables of the network's self-routing switches would be provided with relaying instructions such that all packets associated with a particular virtual circuit flow over the same route (or path) from source to destination. A user could send each of a sequence of packets to a different destination by writing the appropriate virtual circuit identifier into each packet's routing field. The network would then demultiplex and sort that transmitter's packets by virtual circuit, sending each over a different route to its intended destination. On each link of a packet's journey, other statistically multiplexed packets will appear, belonging to different virtual circuits.

Both permanent and switched virtual circuits will appear in B-ISDN. Each CP node may have one or more permanent virtual circuits to each of several other CP nodes; these might be used, for example, to allow that node to offer connectionless service to its user, as was described in Chapter 1. Rather than requiring a call setup procedure to transfer a datagram over the network, the CP node would inspect the destination field of a user-generated datagram and route the ATM cells produced from that datagram to their destination by means of a preexisting permanent virtual circuit to that destination. Permanent virtual circuits might also be used to carry signaling and control information among CP nodes, network switches, management entities, and call processing units.

Users can also request establishment of new virtual circuits on a demand-assigned basis. The decision as to whether or not such a request can be granted is made by the call processor which, among other things, must ascertain that the traffic expected to be offered over that newly requested virtual circuit will not cause the quality of service enjoyed by existing virtual circuits to degrade below some guaranteed level. Remember that many virtual circuits statistically share the network links and switches; the traffic of a newly admitted virtual circuit will cause additional delay and packet loss to other existing virtual circuits as the overall traffic load increases. It is the job of the call processor to make admission decisions in such a fashion as to honor as many requests as possible, consistent with maintaining the quality of service guaranteed to each. More will be said about admission control in Section 5.7. Since the admission decision is based on the collective traffic generated by all virtual circuits which share a common facility, it is important that virtual circuits which are no longer needed be disconnected.

B-ISDN call setup procedures are illustrated in Figure 5.5, where numbered arrows correspond to the sequence of events taken among the involved protocol entities. The process starts with a request by the user for establishment of a new virtual circuit. This request arrives at the user plane of that user's CP node, and a signal is sent to the management plane (event 1). The management entity instructs the control plane to prepare a message for the call processor (event 2), and that message is segmented and shipped to the call processor as a sequence of ATM cells over a permanent virtual circuit (event 3). After processing that request, the call processor makes an admission decision and, if affirmative, assigns a virtual

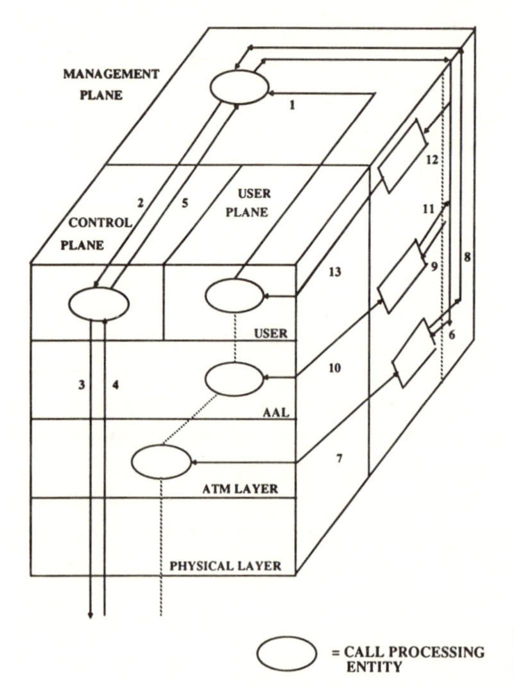

Figure 5.5. Illustration of call setup procedures and signaling in B-ISDN.

circuit number. (If the decision is to admit, the call processor must also select a path and update the routing tables of the ATM multiplexers and switches along that path, and must further inform the receiving CP node of the newly assigned virtual circuit number.) The admission decision and any virtual circuit number is then delivered as a sequence of ATM cells over the network back to the originating CP node (using another permanent virtual circuit), where the cells are reassembled into a message delivered to the control plane (event 4). The control plane then relays this reply to the management plane (event 5). The remainder of Figure 5.5 is based on an affirmative reply; if the request is blocked, the management plane would simply advise the user using a path (not shown) running parallel to, but in the opposite direction to, that of event 1. If the decision is to admit, the management plane instructs the ATM layer management entity (event 6), which supplies the ATM layer with the virtual circuit number (event 7); as we shall see, most of the bits of the ATM cell header appended to SAR-PDUs by the ATM layer are allocated to the virtual circuit number. An acknowledgment is then sent back to the management plane (event 8); the AAL is then advised of the new connection via its layer management entity (events 9 and 10); an acknowledgment is again sent to the management plane (event 11), and the user is advised via its layer management entity that the request has been granted and that communication over its newly established virtual connection can begin (events 12 and 13).

5.4. Virtual Channels and Virtual Paths

Each Virtual Circuit is itself composed of two subunits: the Virtual Path and the Virtual Channel. As shown in Figure 5.6a, each physical link in the network can contain several virtual paths, and each virtual path can contain several virtual channels. The virtual path is that component of a virtual circuit which exists between two CP nodes, and defines the route that a particular virtual circuit will take between the transmitter and receiver. In essence, user ports are interconnected by virtual paths, so called because different portions of a virtual path may be carried over different physical links, and several virtual paths between distinct source–destination pairs may share the same physical links. The virtual paths sharing the same physical link with a particular virtual path in one portion of the network may be different from those sharing the same physical link with the same particular virtual path in another portion of the network. At call setup time, a virtual path is established between the corresponding transmitter and receiver CP nodes if no prior virtual path existed between that same transmitter/receiver pair. As shown in Figure 5.6b, each user port can have a virtual path to each of several other user ports if it has virtual circuits established with those ports. It is important to note that, in general, user ports are not fully connected by virtual paths. In fact, the number of virtual paths which may be established by any user is limited by the

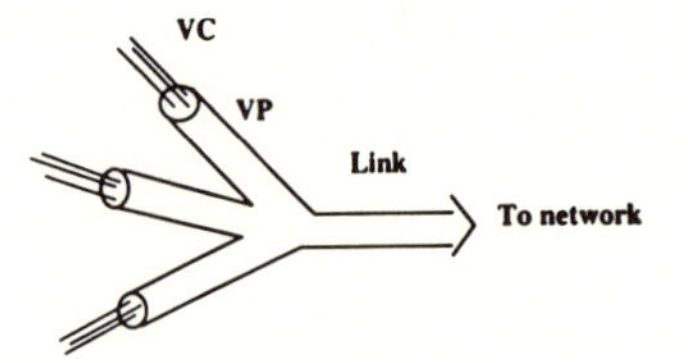

(a) Multiplexing hierarchy in ATM

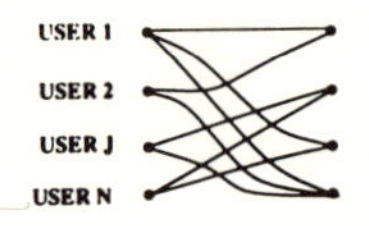

(b) A possible set of user-to-user virtual path connections

Figure 5.6. Virtual channels, virtual paths, and physical links.

number of bits allocated to the virtual path identifier in the ATM cell header. A virtual path connection diagram such as shown in Figure 5.6b is useful when making admission decisions by means of two simplified algorithms to be described at the end of this section (the second and third techniques to be mentioned).

Virtual Channels are used to distinguish among several virtual circuits which may be concurrently established between two users, all of which share the same Virtual Path. Thus, some multitude of Virtual Channels between two users can be multiplexed onto the same Virtual Path, as shown in Figure 5.6a. Each Virtual Path, then, carries some number of multiplexed Virtual Channels between the same user ports. A field within the ATM cell header bears the Virtual Channel number, and the virtual circuit consists of the combination of Virtual Path and Virtual Channel number.

As noted above, the number of Virtual Paths which may be established by any user (in both the transmitted and receive directions) is limited by the size of the Virtual Path field in the ATM cell header. The number of Virtual Channels which may be multiplexed onto any Virtual Path is similarly limited by the size of the Virtual Channel Identifier field in the ATM cell header. To permit the network to support many more virtual circuits than that which may be supported by a given user port, it is necessary that Virtual Channel and Virtual Path identifiers be reused in geographically different portions of the network. When doing so, it is essential that different virtual circuits never bear the same virtual circuit identifier on the same physical link; if this were to occur, those virtual circuits would become indistinguishable. However, two virtual circuits can bear the same virtual circuit identifier on distinct links, maintaining their distinguishability by virtue of the different physical links that they exist upon. Moreover, it is possible for the same virtual circuit to be assigned different virtual circuit identifiers on different physical links, implying the need for virtual circuit identifier translators which receive translation instructions at call setup time. For example, an ATM switch could relay the cells of two different virtual circuits bearing the same identifier, but arriving on different input ports, to different switch output ports; here, the routing tables have been instructed to relay cells on the basis of their virtual circuit identifier and port of arrival. If those same two virtual circuits were destined for the same physical output, the virtual circuit identifier of one or the other would need to be translated prior to relaying to enable the two to be distinguished on the output. This situation is depicted in Figure 5.7, which also illustrates the multiplexing/demultiplexing of Virtual Channels onto Virtual Paths at the user ports.

Thus, in general, a given virtual circuit is defined by a sequence of identifier numbers, one applicable on each physical link along the path from transmitter to receiver. As long as the correct identifier translation is performed from link to link,

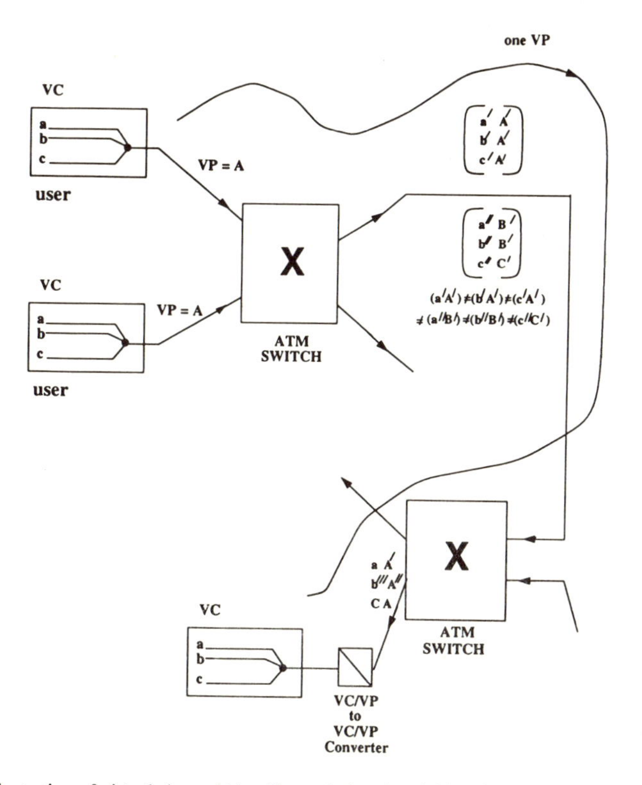

Figure 5.7. Illustration of virtual channel identifier and virtual path identifier translation and renumbering in different parts of the network.

and no two virtual circuits bear the same identifier on the same physical link, there is no risk that virtual circuits can be misidentified. For example, suppose that, at call setup time, a new virtual circuit is established from port A to port B along a path involving four physical links and three ATM switches. The sequence of identifiers assigned by the call processor to this new virtual circuit is 10, 31, 10, 8 (for simplicity, we have not separated the identifier into Virtual Path and Virtual Channel identifiers). Along the first link (from port A to the first switch), the new virtual circuit is identified by the number 10. At the first ATM switch, cells arriving from port A bearing identifier number 10 are relayed to the second link, and the identifier fields in the cell headers are converted from number 10 to number 31 (the virtual circuit identifier translators were programmed to do so at call setup time). At the second ATM switch, cells arriving from the first switch bearing identifier number 31 are relayed to the third link, and their identifier number is converted from 31 to 10. Finally, at the third ATM switch, cells arriving from the second switch bearing identifier number 10 are relayed to port B along the fourth link, and their identifier number is converted from 10 to 8.

Dividing the virtual circuit identifier into Virtual Path and Virtual Channel identifiers can also be used to simplify call setup procedures. In B-ISDN, three types of multiplexing techniques might be envisioned. For the first type, the sum of the peak data rates needed by all virtual circuits multiplexed onto a common link may exceed the peak rate of that link. By this we mean that if all virtual circuits are active, the collective rate at which information arrives from the individual virtual circuits, to be statistically multiplexed onto the common physical link, exceeds the capacity or bandwidth of that link; the queue in the buffer feeding that link would therefore begin to build. Of course, the sum of the average rates of the virtual circuits must always be less than the link rate, or queues will build forever. Of the three techniques, this first approach makes maximum utilization of statistical multiplexing, and the links can be operated most efficiently, but in response to each and every request for a virtual circuit, the call processor must check the resulting quality of service on each physical link for each other virtual circuit multiplexed onto that link. Note that the notion of Virtual Path does not exist with this first multiplexing technique.

For the second technique, Virtual Channels are first multiplexed onto Virtual Paths. Each Virtual Path is allocated a fixed bandwidth, meaning that, for each physical link along the length of the Virtual Path, ATM cells associated with that Virtual Path can flow at some maximum guaranteed data rate, a situation which is essentially equivalent to having a single dedicated transmission link of the specified bandwidth connecting the transmitting to the receiving end of the Virtual Path. In practice, the buffers feeding each physical link serve the Virtual Paths statistically multiplexed onto that physical link in such a sequence as to effectively guarantee the bandwidth of each Virtual Path and, on each link, the sum of the bandwidths assigned to the Virtual Paths sharing that link cannot exceed the bandwidth of that link. No Virtual Path can transmit to a physical link at a rate greater than its bandwidth allocation. Thus, congestion cannot occur within the network since, under worst-case conditions when all Virtual Paths on a given physical link are active at their allocated data rates, that link has the capacity to meet that peak demand. Virtual Channels are then multiplexed onto Virtual Paths, and on any Virtual Path, the sum of the peak rates of its Virtual Channels may exceed its allocated bandwidth. The number of Virtual Channels which may be statistically multiplexed onto a given Virtual Path is dependent on the bandwidth of the Virtual Path and the quality-of-service needs of the Virtual Channels (delay, lost packet rate, etc.). Under no circumstance can the sum of the average data rate among the Virtual Channels sharing a given Virtual Path exceed the bandwidth of that Virtual Path. Thus, statistical multiplexing only occurs among those Virtual Channels sharing a common Virtual Path, and queues can only develop at the input ports of the network where Virtual Channels are statistically multiplexed onto Virtual Paths. To decide on the admissibility of a newly requested Virtual Channel, the call processor merely needs to confirm the resulting quality of service for those Virtual Channels sharing the same Virtual Path as that which would carry the new request. The existence of an appropriate Virtual Path is ascertained by checking a diagram such as shown in Figure 5.6b, modified to show the bandwidth of each established Virtual Path. This obviously simplifies call setup procedures relative to the first technique, but wastes physical link bandwidth since no statistical multiplexing of the Virtual Paths is possible.

The third technique represents a compromise. As was the case with the second technique, each Virtual Path is allocated a bandwidth, and is never allowed to transmit onto a physical link at a rate greater than its allocation. Again, the sum of the bandwidths of the Virtual Paths sharing the same physical link *cannot* exceed the bandwidth of that link; statistical multiplexing of Virtual Paths onto physical links, again, is not permitted. Again, Virtual Channels are then statistically multiplexed onto Virtual Paths; the sum of the peak rates of all Virtual Channels multiplexed onto a common Virtual Path can exceed the peak bandwidth allocation of that Virtual Path. To determine admissibility of a new request for a Virtual Channel, the call processor needs to confirm the quality of service for those Virtual Channels sharing the same Virtual Path to be used by the new request and, if the guaranteed quality of service can be maintained, the new request is admitted (again, a diagram similar to that of Figure 5.6b is used to ascertain the existence of a Virtual Path and to determine its bandwidth). However, with the third technique, if a request for a new Virtual Channel cannot be accepted onto the existing Virtual Path, the call processor will attempt to allocate additional bandwidth to that Virtual Path, but must ensure that the quality of service guaranteed to all Virtual Paths sharing any common physical links can be maintained. Only if such a guarantee cannot be maintained will the new request be blocked. Since most requests will be admitted to their respective Virtual Paths without requiring that additional bandwidth first be allocated, it is only rarely that additional bandwidth must be sought for an existing Virtual Path, an event which requires the complex task of confirming the quality of service of all other Virtual Paths sharing any common physical link with the Virtual Path for which additional bandwidth is being sought. Thus, the call setup simplicity of the second technique has almost been achieved, but with a substantially improved link utilization efficiency which approaches that of the far more call-setup-intensive first technique. For large networks, the third technique is therefore preferred, and its call setup algorithm is depicted in Figure 5.8.

5.5. Function of the ATM Adaptation Layer

A primary function of the AAL is segmentation and reassembly of higher-level blocks of information. The procedure and protocols used are similar to those used for DQDB, and are illustrated in Figure 5.9a for bursty data traffic; similar operations are performed for continuous bit rate traffic. (Note that the procedures depicted in Figure 5.10 formally correspond to so-called AAL-4, the adaptation layer intended for connectionless data. Similar procedures, not shown, apply for AAL-3 and AAL-5, both intended for connection-oriented data. The interested reader is referred to the abundance of literature depicting the precise procedures associated with each of these other two data-oriented adaptation layers, and/or those associated with AAL-1 and AAL-2, the adaptation layers intended for continuous bit rate traffic.) Although operations and formats shown in Figure 5.10a are similar to DQDB, the names given to the various entries encountered are different. We shall describe the operations necessary to transfer a block of data from a transmitting CP node; the reverse operations would be performed at the receiving end.

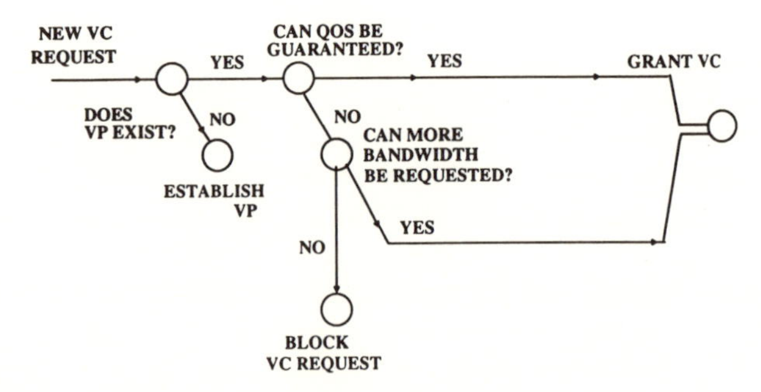

Figure 5.8. Call setup algorithm for a multiplexing technique which prevents statistical multiplexing of virtual paths but permits additional bandwidth to be allocated to a virtual path.

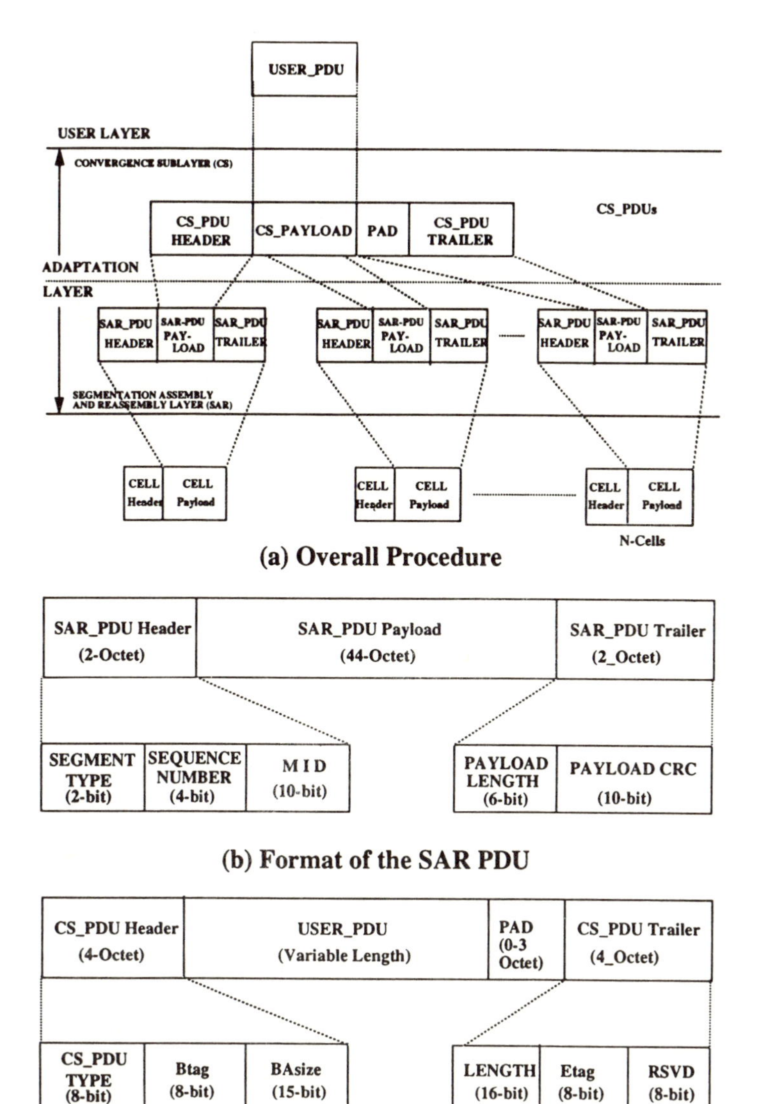

Figure 5.9. Processing associated with the ATM Adaptation Layer, and the resulting data formats (shown for AAL-4 only).

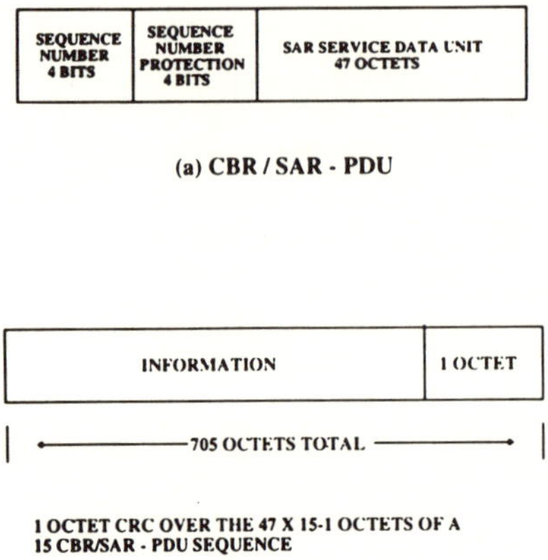

SEQUENCE NUMBER 4 BITS	SEQUENCE NUMBER PROTECTION 4 BITS	SAR SERVICE DATA UNIT 47 OCTETS

(a) CBR / SAR - PDU

INFORMATION	1 OCTET

|←——— 705 OCTETS TOTAL ———→|

1 OCTET CRC OVER THE 47 X 15-1 OCTETS OF A 15 CBR/SAR - PDU SEQUENCE

(b) Exemplary CBR/CS - PDU

Figure 5.10. AAL data formats for constant bit rate service.

A user having a block of data to be transferred across the ATM network presents this to the AAL in the form of a User Protocol Data unit, a block of information fully prepared by that user conforming with some arbitrary user-specified protocol. As with the IMPDU sublayer of DQDB, the Convergence Sublayer of ATM must know the protocol of the user PDU and might even need to extract information from some of the protocol fields, but always leaves the User PDU intact, merely encapsulating it with a Convergence Sublayer header, trailer, and pad, to form the CS-PDU. The Segmentation and Reassembly Sublayer of the AAL divides the CS-PDU into 44-octet segments, each of which comprises the payload of one SAR-PDU (the last segment may need padding to form a full 44-octet payload). The SAR-PDU consists of a 2-byte header, 44-octet payload, and 2-byte trailer.

The format of the SAR-PDU for AAL-4 is identical to that of a DQDB-Derived MAC-PDU and is shown in Figure 5.9b. The segment type (BOM, EOM, COM, SSM) field, sequence number field, and Message ID field in the SAR-PDU header are identical in form and function to their respective fields in the DMPDU header and will not be described further; the reader may wish to refer back to Section 4.2. The same is true for the Payload Length and Payload CRC fields appearing in the SAR-PDU header.

The format of the CS-PDU is shown in Figure 5.9c and is similar, but not identical, to the DQDB Initial MAC-PDU. The CS-PDU header begins with an 8-bit type field, which is used to declare the type of service being provided to that CS-PDU (e.g., the "all zeros" word denotes connectionless service). This is followed by an 8-bit Beginning tag (B tag) field, containing a word which must agree with that contained in the End tag (E tag) field in the CS-PDU trailer; the functions of the B tag/E tag fields in the ATM are identical to those of their respective fields in the IMPDU for DQDB. Also, the 16-bit Buffer Allocation field in the CS-PDU header and the 16-bit Length field in the CS-PDU trailer serve the same functions as their respective fields in the IMPDU of DQDB. The last 8-bit field in the CS-PDU trailer is reserved for unspecified future use. As in DQDB, the function of the CS-PDU Pad field is to ensure that the overall length between the end of the header and beginning of the trailer is an integral multiple of 4 bytes.

The CS-PDU and SAR-PDU for Constant Bit Rate service are different in format from those used for Variable Bit Rate service, and are shown in Figure 5.10. For CBR service, the CBR bit stream is divided in the Convergence Sublayer into segments which are 704 octets in length, and the CS-PDU consists of this 704-octet segment plus a 1-octet Cyclical Redundancy Check to detect the presence of errors in the 704-octet information field. Each 705-octet CS-PDU is divided into fifteen 47-octet segments, each one of which is the payload of one CBR SAR-PDU. The CBR SAR-PDU consists of the 47-octet payload plus a 1-octet header. The header consists of a 4-bit sequence number field which allows the segments to be sequentially numbered from 0 to 15, plus a 4-bit error-correction field to protect the sequence number.

For both VBR and CBR service, the 48-octet SAR-PDUs are supplied to the ATM layer which adds the 5-byte cell header. The format of the ATM cell header added at a User–Network Interface (UNI) access port appears in Figure 5.11. This starts with a 4-bit Generic Flow Control (GFC) field which might be used, for example, to mediate access to a shared channel serving several CP nodes. We note that the B-ISDN Reference Architecture of Figure 5.2 shows each CP node served by a dedicated transmission facility, in which case no media access control field is necessary. However, B-ISDN may admit CP node access on a shared-channel contention basis, and, in such a case, the GFC field might be used to carry busy and request bits if a DQDB-like media access protocol were used. The GFC field may also serve other functions associated with controlling traffic flow within the network. Some of these (policing a source to be sure that its usage does not violate that represented at call setup time; restricting new call arrivals during periods of congestion; etc.) are described in Chapter 6. The GFC field is followed by an 8-bit Virtual Path Identifier field. Thus, each CP node can establish a Virtual Path for

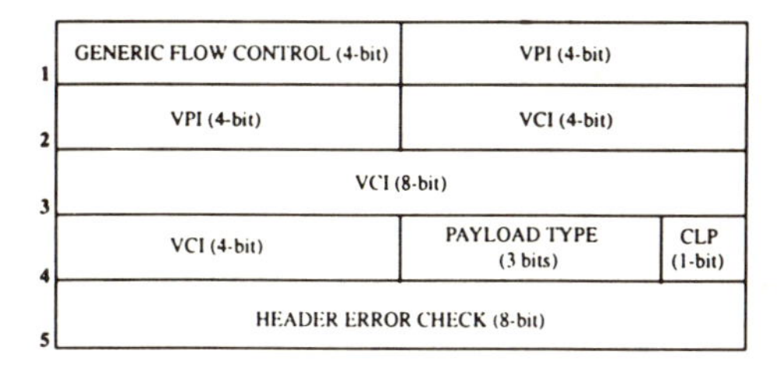

Figure 5.11. Format of the ATM cell header appearing at the user–network interface.

each of 256 destination ports. Next is the 16-bit Virtual Channel Identifier field; each Virtual Path can therefore support 65,536 Virtual Channels between its transmit port and receive port. A 3-bit Payload Type field is next, which is used to declare whether the cell contains customer traffic or any of seven possible types of network-generated control traffic. Finally, a 1-bit Cell Loss Priority field is provided such that one of two priority levels can be established with regard to cell loss. Presumably, this field might be interrogated during periods of congestion to be sure that higher priority cells are not discarded by a buffer on the verge of overflow. The final 8-bit field in the cell header is used to detect the presence of errors occurring anywhere within the cell header.

In addition to segmentation and reassembly, the AAL must merge ATM-formatted control signals into the user-generated ATM stream, detect the appearance of higher-level information blocks arriving at the user interface, detect the boundaries of reassembled information blocks and deliver these across the user interface, and detect bit errors and missing cells of arriving signals. Also, for CBR service, the AAL must compensate for the variable delay of the segments delivered by the ATM network (caused by statistical multiplexing), and for recovering the bit timing clock of the CBR source. CBR data must be delivered continuously across the user interface at a CBR receiver, and if the next-needed cell arrives too late, an unacceptable gap in delivery would occur. Also, the receiver's bit clock must be synchronized with the transmitter's, such that the reconstructed CBR signal has exactly the same bit rate as the original. For example, if the transmit clock for digital voice is exactly 64,000 kbit/sec, then the reconstructed bit stream at the receiver must be delivered across the user interface at exactly this rate. If the receiver clock runs at 63.9999 kbit/sec, information will be delivered more slowly than it is generated, and the delivery delay would increase forever with time. Similarly, if the receiver clock runs at 64.0001 kbit/sec, information will be delivered at a rate faster than that at which it is generated, and gaps in delivery would unavoidably result.

As an example, compensation for the variable delay of the ATM network and synchronization of the receiver's clock might both be accomplished through the use of apparatus such as shown in Figure 5.12. Here, we show arriving cells temporarily stored in a cache memory, the primary function of which is to accumulate a sufficient supply of cells prior to delivering these to the reassembly machine at a constant rate such that the likelihood that the cached supply is depleted before a new cell arrives is acceptably small. The cache memory will thereby have effectively smoothed out the random jitter between the arrival times of adjacent ATM cells comprising the same CBR signal. Also shown on the buffer are two level markers along with a derived clock which controls the rate at which cells (or bits) are read from the cache memory. The objective is to maintain the number of cached cells between the low- and high-level markers. If the number of cells should ever fall below the low watermark, a control signal will be

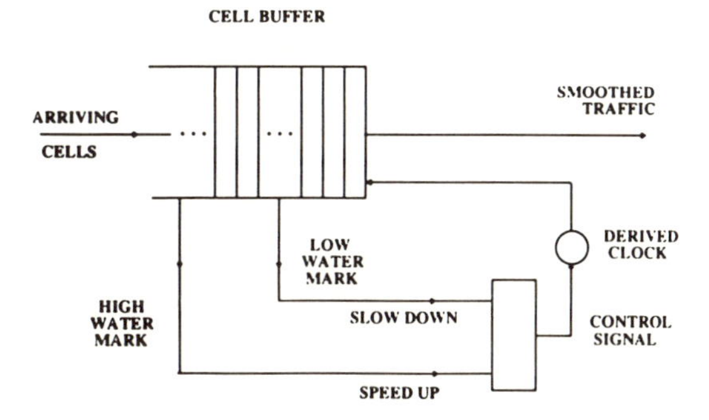

Figure 5.12. A possible means to compensate for the variable delay introduced by an ATM network.

generated which instructs the derived clock to "slow down," thereby enabling a greater supply of cached cells to redevelop. Alternatively, if the level of cells should ever exceed the high watermark, then this would indicate that the derived clock is running too slowly, and another control signal would be generated to instruct the derived clock to speed up. In effect, the average frequency of the derived clock becomes synchronized with that of the transmit clock, which has the additional benefit of further smoothing the intercell arrival jitter.

5.6. ATM LAN

As mentioned in Section 5.1, the function provided by the AAL in an ATM network setting is similar to that provided by a Terminal Adapter in a LAN setting. To carry this analogy with LANs one step further, we note that there is no inherent reason why ATM could not be used as the universal format for a LAN. Here, the terminal equipment to be attached to the network would contain the AAL internally such that its physical connector natively carries ATM cells. The so-called ATM LAN, shown in Figure 5.13, would contain ATM switches, ATM statistical multiplexers (one output module of the Knockout switch, for example, might serve as a statistical multiplexer), and fixed-point transmission links to connect each native ATM terminal to either a statistical multiplexer or an ATM switch, and to appropriately interconnect the statistical multiplexers and the ATM switches.

Operation might appear as shown in Figure 5.14. Here, the AAL contained within each terminal connects variable-length Link Level (Layer 2) packets into ATM cells. For example, the AAL might convert IEEE 802.2 Link Level packets (the common Link Level format for all IEEE 802 Medium Access Control protocols such as 802.3—CSMA/CD and 802.5—Tóken Ring) into ATM cells. The ATM network, using permanent virtual circuits to fully interconnect the user ports,

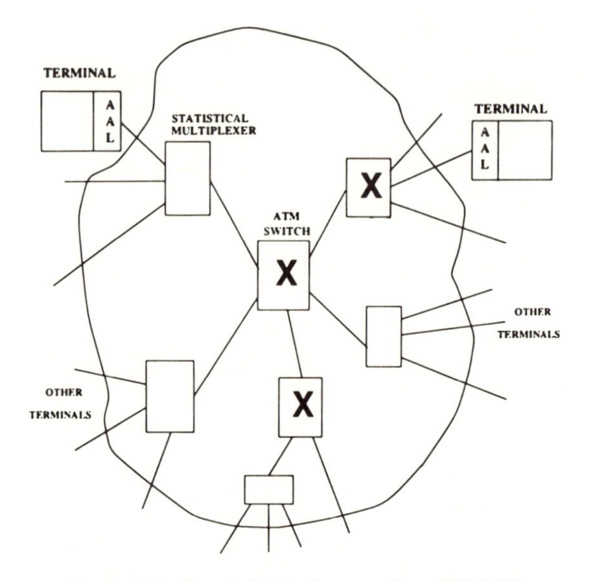

Figure 5.13. Generic block diagram of an ATM LAN.

then emulates the shared LAN medium (e.g., a bus for 802.3; a ring for 802.5) by broadcasting each arriving cell to each output. (In this case, the permanent virtual circuits are broadcast circuits causing each cell, as it arrives at the input to a switch or demultiplexer, to be reproduced and carried to all outputs. This capability is readily provided by ATM switches which support multicast and broadcast services, such as the Knockout and Shared Memory switches, and is accomplished by writing the appropriate information into the routing tables and virtual circuit identifier translators of the switches.) At each output, the Link Level packet is reassembled and its destination field checked for a local match. The reassembled Link Level packet is then ignored by all terminals except those intended to receive that packet. In this fashion, ATM replaces the need for a MAC sublayer, a connectionless service is provided to the users, and the ATM network is invisible to the Link Layer and all higher protocols. Thus, no changes are needed to the software of the higher protocol layers or to the application software if a terminal originally equipped to operate over a "traditional" LAN (e.g., 802.3 or 802.5) should subsequently be attached to an ATM LAN; the AAL provides the requisite signal reformatting and protocol/application software transparency.

In the above example of an ATM LAN, the broadcast capability of an ATM network was exploited to provide connectionless service over a connection-oriented network in such a way as to totally avoid the need for changes to the higher-layer LAN protocols. An alternative way to support connectionless service

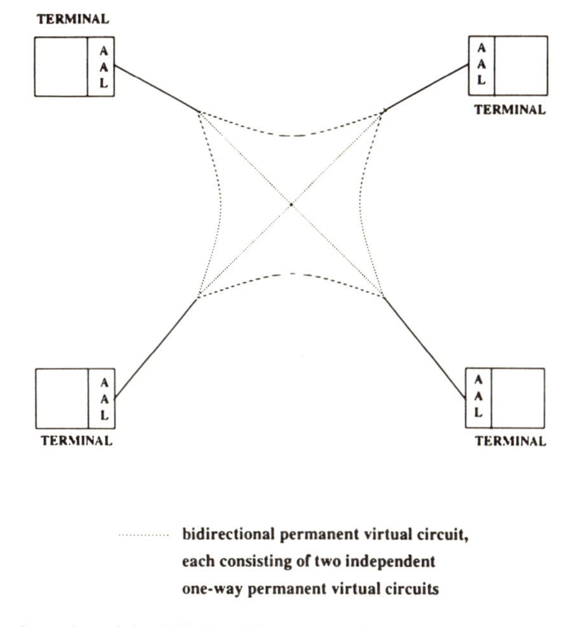

Figure 5.14. Operation of the ATM LAN by means of broadcast permanent virtual circuits.

over a connection-oriented network was described in Chapter 1 and was briefly mentioned again in Section 5.3. Although that earlier technique was also based on the use of permanent virtual circuits, each arriving packet was sent over a specific virtual circuit which delivered that packet only to its intended receiver. While such an approach could, in principle, be used as the basis for an ATM LAN, the lack of a broadcast capability as seen at the Link Level would require modification to the higher levels of some LAN protocols. Alternatively, the use of ATM as the basis for some LANs might encourage modification to higher-layer LAN protocols to better exploit the connection-oriented nature of ATM; this might facilitate support for connection-oriented and/or continuous bit rate services, as well as more readily enabling the integration of multimedia traffic.

Ultimately, the viability of the ATM LAN is dependent on its cost/performance competitiveness with more "traditional" types of LANs. As wide-area ATM networks become more prevalent, the convenience of natively attaching an ATM LAN to a wide-area ATM network, the ability to directly connect a terminal natively equipped with an AAL to either a local or a wide-area network, and the inevitable reduction in cost of ATM equipment might tend to enhance the competitiveness of the ATM LAN.

5.7. Problems

5.1. A certain transmission system operates at a clock speed of 100 MHz. Each frame is exactly 425.06 μsec long, and starts with a frame marker containing 106 bits. In addition to frame markers, the transmission system carries only ATM cells, and no bit positions are ever idle. It so happens that, for a particular frame, say, frame No. J, the time interval immediately after the frame marker contains a complete ATM cell. In which subsequent frame will this again occur?

5.2. A given physical link of an ATM network is shared by three virtual paths. Each VP is allocated for its exclusive use one-third the bandwidth of that physical link. No statistical multiplexing of VPs onto physical links is permitted anywhere in the network. Each time slot on the physical link is of length T seconds. Several virtual channels are statistically multiplexed onto each VP such that, at the input to each of the VPs, the probability that a cell is present in any T-second time slot is p and cell arrivals are independent in time. Find and plot as a function of p the average cell waiting time in the statistical multiplexer at the head end of any VP, ignoring all time granularity smaller than $3T$.

5.3. Consider an ATM network for which the delay performance is dominated by one heavily used physical link. During every time slot, each virtual channel generates a cell with probability $p = 0.1$, independently of every other VC, and cell arrivals are independent in time. In one system, no statistical multiplexing of virtual paths is permitted, but VCs can be statistically multiplexed onto VPs, and three VPs equally share the bandwidth of the critical physical link. In the second system, VPs are not used and VCs are statistically multiplexed onto each physical link. It is required that the average delay of the ATM cells be less than $10T$. (a) For the first system, how many VCs can flow over the critical link without violating the QOS objective? (b) Repeat part a for the second system. (c) Compare the results of parts a and b and qualitatively discuss your findings.

5.4. The delay performance of a particular ATM network is dominated by one critical physical link, which is shared by three fixed-bandwidth virtual paths. Statistical multiplexing of VPs onto physical links is not permitted. The statistical multiplexer at the head end of every VP can store an arbitrarily large number of cells. Each time slot, the probability of a cell arriving on the first, second, and third VP is 0.1, 0.2, and 0.3, respectively, and cell arrivals are independent in time. (a) Find the allocation of link bandwidth among the three VPs such that the average cell delay is the same for all VPs. This problem may be solved graphically, if desired, by first plotting the average delay versus fractional bandwidth for each VP. (b) What is the average delay for the allocation found in part a? (c) Repeat parts a and b if the probability of a cell arriving on the first, second, and third VP is 0.05, 0.05, and 0.5, respectively. (d) Compare and discuss the results of parts a–c.

5.5. A certain virtual path in an ATM network has a fixed bandwidth of 15.5 Mbit/sec. The clock frequency of each physical link is 155 Mbit/sec, producing an ATM slot time $T = (424)/(155 \text{ Mbit/sec})$. Each virtual circuit statistically multiplexed onto that VP produces a cell in any given time slot with probability $p = 1/100$. Cell arrivals are independent in time and independent among VCs. The statistical multiplexer can store an arbitrarily large number of cells. (a) How many VCs can be statistically multiplexed onto that VP such that the average time that a cell waits in the statistical multiplexer is less than $100T$? (You may ignore time granularity smaller than $10T$.) (b) Arrivals of new requests for VCs are Poisson with rate 1/min, and connection duration is an exponentially distributed random variable with mean value of 1 hr. Find the call blocking probability such that the delay QOS objective is achieved.

5.6. An N-port ATM Local Area Network uses one $N \times N$ centralized switch with output queuing. Each of the N input/output ports attaches to one of the N switch I/O ports. Each output queue can store an arbitrarily large number of cells, and all virtual connections are fully broadcast in nature. Variable length packets are segmented into ATM cells for transfer over the ATM LAN. Each packet contains a large number of cells, so that the cell granularity can safely be ignored. Packet arrivals are Poisson with arrival rate λ per input port. The length of each packet is an exponentially distributed random variable and if all cells were to be transferred contiguously, the mean occupancy time on any I/O link would be T seconds. (a) For $N = 10$, find and plot as a function of λT the mean packet delay, defined to be the time interval between generation of the lead cell from that packet at its source and delivery of the last cell of that packet to its destination. (b) Repeat part a for $N = 100$. (c) What do you conclude about the per-port traffic-handling capability of such an ATM LAN when used in this broadcast fashion?

5.7. N virtual channels are statistically multiplexed onto one virtual path. Each input link to the statistical multiplexer carries one VC and operates at 155 Mbit/sec, and, each time slot, a virtual channel generates an ATM cell with probability p. Cell arrivals are independent in time and independent among inputs. The slot time is T and the guaranteed QOS requires that the mean cell delay within the statistical multiplexer be less than $10T$. (a) For $p = 0.1$, find and plot the bandwidth needed on the VP as N varies from 1 to 10. (b) If the physical links carrying the VP can operate at a maximum speed of 155 Mbit/sec, what is the maximum permissible value of N? (c) Repeat part a for $p = 0.1$.

References

1. J.-Y. Le Boudec, The asynchronous transfer mode: A tutorial, *Comput. Net. ISDN Syst.* **24,** 1992.
2. S. E. Minzer, Broadband ISDN and asynchronous transfer mode (ATM), *IEEE Commun. Mag.* **7**(9), Sept. 1989.
3. *IEEE Net. Mag.,* special issue on Broadband Networks, **2**(1), Jan. 1989.
4. M. De Prycker, ATM switching on demand, *IEEE Net. Mag.* **6**(2), March 1992.
5. T. Aoyama, I. Tokizawa, and K. Sato, ATM VP-based broadband networks for multimedia services, *IEEE Commun. Mag.* **31**(4), April 1993.
6. M. De Prycker, Evolution from ISDN to BISDN: A logical step toward ATM, *Comput. Commun.* **12**(3), June 1989.
7. S. Ohta, K. Sato, and I. Tokizawa, A dynamically controllable ATM transport network based on the virtual path concept, IEEE GLOBECOM '88 Conf. Rec., Hollywood, Fla.

8. M. J. Rider, Protocols for ATM access networks, IEEE GLOBECOM '88 Conf. Rec., Hollywood, Fla.
9. CCITT Recommendation I.321, BISDN Protocol Reference Model and its Application.
10. CCITT Recommendation I.121, Broadband Aspects of ISDN.
11. CCITT Recommendation I.150, BISDN ATM Functional Characteristics.
12. CCITT Recommendation I.361, BISDN ATM Layer Specification.
13. CCITT Recommendation I.362, BISDN ATM Adaptation Layer (AAL) Functional Description.
14. CCITT Recommendation I.363, BISDN ATM Adaptation Layer (AAL) Specification.
15. I. Gopal, R. Guerin, J. Jannello, and V. Theoharakis, ATM support in a transparent network, *IEEE Net. Mag.* **6**(6), Nov. 1992.
16. *Int. J. Digital Analogue Cabled Syst.,* special issue on Asynchronous Transfer Mode, **1**(4), 1988.
17. A. Hac and H. B. Mutlu, Synchronous optical network and broadband ISDN protocols, *IEEE Comput. Mag.* Nov. 1989.
18. I. M. Leslie, D. R. McAuley, and D. L. Tennenhaus, ATM everywhere, *IEEE Net. Mag.* **7**(2), March 1993.
19. A. G. Fraser, Early experiments with asynchronous time division networks, *IEEE Net. Mag.* **7**(1), Jan. 1993.
20. Z. Wang and J. Crowcroft, SEAL detects cell misordering, *IEEE Net. Mag.* **6**(4), July 1992.
21. S. C. Farkouh, Managing ATM-based broadband networks, *IEEE Commun. Mag.* **31**(5), May 1993.
22. A. R. Modarressi and R. A. Skoog, Signaling system No. 7: A tutorial, *IEEE Commun. Mag.* **28**(7), July 1990.
23. M. Kawarasaki and B. Jabbari, B-ISDN architecture and protocol, *IEEE J. Selected Areas Commun.* **SAC-9**(9), Dec. 1991.
24. E. Biagioni, E. Cooper, and R. Sansom, Designing a practical ATM LAN, *IEEE Net. Mag.* **7**(2), March 1993.
25. J. Burgin and D. Dorman, B-ISDN resource management: The role of virtual paths, *IEEE Commun. Mag.* **29**(9), Sept. 1991.

6

Issues in Traffic Control and Performance Management

6.1. The Importance of Traffic Control and Performance Management

In this section, we will discuss various issues associated with maintaining the quality of service expected by the different types of telecommunication traffic which may be present in an integrated, multimedia network. The ATM network is expected to support several types of variable and continuous bit rate traffic having a broad range of bandwidth needs. Both low-rate voice and high-rate video services must be supported, along with transfer of short-duration bursty data files and long-duration bursty image files. The average and peak rates of the bursty traffic may range over a wide extent. Different types of traffic will have different delivery needs (allowable delay or latency, permissible frequency with which information might be misdelivered or lost). A network which is optimized for one type of traffic might be providing a quality of service which is inappropriate for other types of traffic.

The topics of traffic control and performance management seek to address and resolve the various (and often conflicting) issues which arise in the high-speed multimedia traffic environment of broadband networks. This is an advanced and sophisticated topic, and a rigorous treatment is well beyond our present scope. In keeping with the introductory nature of this text, the various issues and solutions will be described on a physical, intuitive basis and the presentation will be kept entirely qualitative. Extensions of the analytical methodologies developed elsewhere in this text can be applied to study and compare various approaches to traffic control and performance management. The problem set at the end of this chapter attempts to illustrate some of the analytical difficulties encountered.

There are various schools of thought concerning the need for, and importance of, traffic control/performance management in the broadband environment. Two extreme opposite schools of thought hold that (1) there is enough capacity

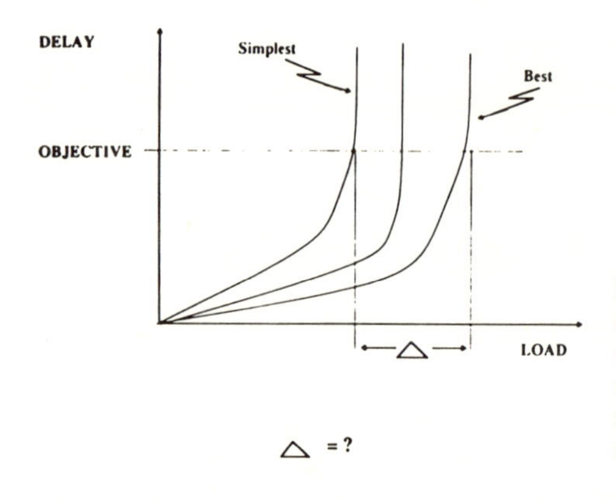

Figure 6.1. Qualitative comparison of two different performance management algorithms. Although both produce the same quality of service objective, the volume of traffic which can be handled by each is different.

in the broadband environment that simple control strategies will suffice and sophisticated strategies are of academic interest only and (2) congestion control is *the major* issue in broadband networks and sophisticated new strategies are absolutely essential to guarantee quality of service with reasonable resource utilization. As diametrically opposed as these two philosophies may appear, both are based on the following observation. In the statistically multiplexed environment of broadband networks, *any* desired set of quality-of-service metrics (mean delay, 95th percentile of delay, average lost cell rate, virtual circuit call blocking probability, etc.) can *always* be achieved by allocating more network resources (more transmission links operating at higher speeds, larger smoothing buffers, etc.) and/or reducing traffic intensity on the network (intentionally reducing the resource utilization efficiency), subject to fundamental constraints (the speed of light, for example, is not infinite, so that the communication latency between any two remote access ports cannot be zero). The difference between these two philosophies is illustrated, for example, by the qualitative curves shown in Figure 6.1. These curves are meant to show the mean delay experienced by a representative message as a function of offered traffic load for a network with fixed resources. Two traffic control policies are reflected. The first is a simple policy which simply admits a volume of traffic sufficiently low as to achieve the desired delay objective by blocking new requests as appropriate. The second is a complex strategy which seeks to maximize the load which can be carried while satisfying the mean delay objective, and might involve alternate routing around congested switches and links, along with selectively discarding low-priority cells from buffers to reduce the waiting time for the remaining cells. Presumably, both strategies permit the same average delay objective to be achieved, but the second does so while handling a greater volume of traffic. The issue concerns the size and importance of the difference between the load-carrying capabilities of the two approaches. The first school of thought holds that the relative differences may not be large and that, in any event, there is so much available capacity that even large differences are entirely acceptable. The second school of thought holds that the relative differences may approach 100% of the capacity available under the best strategy, that is, the load which can be carried by the simple approach may be only a tiny fraction of that possible with the best strategy, and that such a difference is intolerable and cannot be ignored. No doubt, the truth lies somewhere in between.

Yet a third school of thought holds that in the broadband multimedia environment, there are simply too many variables and unknowns to permit the *a priori* development of sound policies, and that the network must be equipped with a system of monitoring points and knowledge bases which permit it to try new policies while adaptively learning from its prior mistakes. As the network gains operational experience, it will learn to recognize current traffic patterns and make decisions (admission, rerouting, cell discard, etc.) which were successful in the past with regard to maintaining the quality of service (QOS) under similar traffic conditions. In this case, the traffic control/performance manager resides in management entities attached to the geographically distributed access ports and shares local information and data bases over the network to reach collective decisions.

There are various time scales associated with traffic control/performance management. The finest granularity involves control operations that may be exercised at the ATM cell level. Next is control that may be exercised over blocks of data generated by bursty sources. Call or admission control is next in the control granularity hierarchy and involves decisions as to whether or not a newly requested virtual circuit will be admitted to the network. On an even longer time scale are operations to increase the pool of, or redeploy, network facilities. These control operations can be exercised within the network, or within dedicated processors which are attached to the network access ports. Some traffic-related QOS metrics which these control procedures seek to guarantee include average cell delay, cell delay variation (especially important for smoothing at the receiver), cell delay extremes (the tails of the cell delay probability distributions), average lost cell rate, average lost packet rate, frequency and duration of high congestion periods, lost packet probability and cell delay statistics during periods of high congestion, and virtual circuit call blocking probability for each of several different types of connections. Other important QOS metrics which are not related to traffic volume (and which will not be further considered) are service availability, mean time between failures, failure interval statistics, and mean time to repair. We will start our treatment with a qualitative description of admission control.

6.2. Admission Control

Admission control involves a set of procedures which are intended to limit the load on network access and internal links such that the QOS guaranteed to each of the existing connections is maintained. The question to be asked is, "given existing admitted connections, can a path be found for a newly requested connection such that the QOS guarantees are maintained for all connections?" If not, then the new request must be blocked.

To answer the above question, the call controller must have access to a set of statistical descriptors for the traffic to be carried over the requested connection. Is it VBR or CBR? If VBR, what is the data rate when active? What are the on-and-off period statistics, and what are their dependencies? What is the average rate? What are the burst and idle length probability distributions? What is the on/off ratio? Are several types of traffic involved? Answers to questions such as these must be known if the call processor is to assess the effect of the newly requested connection's traffic on the QOS for existing connections. Generally, detailed traffic descriptors are unavailable (for example, how does one characterize the traffic generated by an advanced workstation in a multiprocessing environment, and how does this vary with application?). Although admission control is potentially a very powerful performance management tool (especially in the multimedia environment), lack of accurate traffic descriptors may compromise its potential effectiveness. Clearly, if delay is of no consequence and very large buffers are available to smooth over the statistical variations of burst patterns, then knowledge only of the long-term average traffic expected from a newly requested connection would be needed to make the admission decision, since cell loss could not occur as long as, for each buffer, the average traffic intensity is less than the transmission rate of the output link serving that buffer. Unfortunately, delay *is* an issue and internal buffers *are not* infinite.

Another potential inhibitor of effective admission control is the computation required to determine the equivalent capacity needed of a virtual path or physical link to accommodate the new request. Assuming that the traffic descriptors for a newly requested connection are known, it remains to be determined how much reserve capacity must be present in a virtual path or physical link to accommodate the needs of that new connection without excessively penalizing existing connections. As is typically the case, the amount of "bandwidth" needed by the requested connection is not a constant dependent only on the traffic descriptors of that connection, but is further dependent on the nature and volume of traffic presented by other virtual channels sharing the same virtual path and physical links. The equivalent capacity of a virtual path or physical link is the smallest capacity needed to assure QOS for all virtual connections sharing that virtual path or physical link if the new request should be admitted. Since the equivalent capacity is dependent on the traffic of all connections, it is often difficult to compute the required reserve capacity and to determine whether or not sufficient reserve exists.

Some simple examples will be offered to illustrate the difficulties associated with determination of equivalent capacity. For these examples, we will assume that all virtual connections offer bursty traffic characterized by the same traffic descriptors. The source of traffic for each virtual connection will be represented by a two-state model. This model, shown in Figure 6.2a, is commonly used to represent bursty data sources and to study many aspects of traffic control/performance management. The traffic source is represented by two states: on and off. When a source is on, it generates information at a constant rate R bits/sec; when off, it generates no information. If the source is in its "on" state during a particular ATM time slot, then (1) it remains in its "on" state and generates an additional cell in the next time slot with probability p or (2) it turns "off" with probability $(1 - p)$ in the next time slot. If the source is in its "off" state during a particular time slot, then (1) it remains in its "off" state for the next time slot with probability q or (2) it turns "on" and generates an ATM cell in the next time slot with probability $(1 - q)$. Although ATM cells are generated continuously while the source is in its "on" state, the data rate is R bits/sec, which may be less than the data rate S of the ATM access port. Thus, the continuous stream of ATM cells generated by an "on" source appears at the input to the network as a quasiperiodic sequence of ATM cells with average arrival rate of 1 cell every S/R time slots. When in the "off" state, the rate of ATM cells generated by a source is zero cells per time slot. Since, in the steady state, the probability of transitioning into a state must equal the

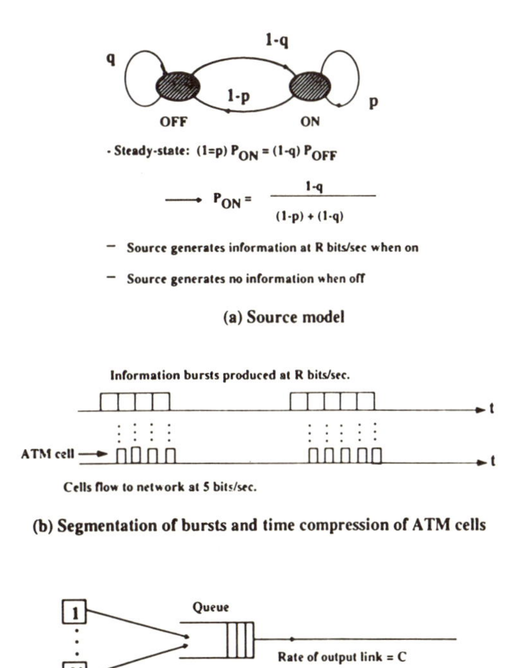

(a) Source model

(b) Segmentation of bursts and time compression of ATM cells

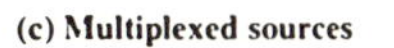

(c) Multiplexed sources

Figure 6.2. On–off source traffic model, production of ATM cells, and model used to compute equivalent capacity needed to serve multiplexed sources.

probability of transitioning out of a state, we conclude that $(1 - p)\,P(\text{on}) = (1 - q)\,P(\text{off})$, where $P(\text{on})$ and $P(\text{off})$ are the probabilities, respectively, of finding the source in the "on" and "off" states. Since the source is either "on" or "off," $P(\text{on}) + P(\text{off}) = 1$, and it follows that $P(\text{on}) = (1 - q)/(2 - p\text{-}q)$.

Consider now the multiplexing of N such sources onto a common virtual path as shown in Figure 6.2b. The input port to that virtual path contains a buffer to statistically smooth the arrivals, storing those cells which cannot immediately be placed onto the finite capacity output link for subsequent delivery. We note that if the size of the buffer is infinite, then the minimum required rate of the output link must be equal to the average rate from the N sources if cells are not to build up forever within the buffer. The average rate per source is $R \times P(\text{on})$; the required capacity of the output link is therefore $N \times R \times P(\text{on})$, and as long as the output link is run at a rate greater than this (equivalent capacity), finite delay with no cell loss is assured. A system with a very large (infinite) input buffer needs the smallest equivalent capacity since the "on" and "off" periods of all sources are statistically smoothed. However, for a finite size buffer, the equivalent capacity needed to maintain some specified level of packet loss because of buffer overflow must be greater. If only one virtual channel exists, the required equivalent capacity could be as high as R if the buffer size is small since, when active, that single source generates information at R bits/sec and, if the output runs at a lower rate, unacceptable buffer overflow might result. If the buffer size is small, then the equivalent capacity needed to permit multiplexing of N sources while producing an acceptably small buffer overflow probability might be as high as $N \times R$ (equivalent capacity equal to N times the peak rate of the source). Under these conditions, if the equivalent capacity were any smaller, then if all of the sources should concurrently become active, cells would temporarily stream into the buffer at a rate greater than that at which they are removed and, since the buffer size is small, unacceptable cell loss might occur. Thus, for a reasonably sized buffer, the equivalent capacity needed to multiplex N sources is between $N \times R \times P(\text{on})$ and $N \times R$.

In general, as additional virtual channels are admitted, the equivalent capacity increases, but at a rate which scales nonlinearly with N; if the sources are independent (i.e., there is no correlation between their "on" and "off" states), then equivalent capacity grows more slowly than linearly with N since, as more sources are added, the likelihood of all being simultaneously active diminishes. Thus, for one source, the equivalent capacity is $\alpha_1 R$ where $\alpha_1 = 1$; for two sources, the equivalent capacity is $2\alpha_2 R$ where $0 < \alpha_2 < 1$; for three sources, the equivalent capacity is $3\alpha_3 R$ where $0 < \alpha_3 < \alpha_2 < 1$; for N sources, the equivalent capacity is $N\alpha_N R$ where $0 < \alpha_N < \alpha_{N-1} < \ldots < \alpha_3 < \alpha_2 < 1$. Furthermore, as the buffer size becomes larger, α_N approaches $P(\text{on})$ as shown from the infinite buffer example described above for which equivalent capacity $= NRP(\text{on})$.

A qualitative plot of equivalent capacity versus number of multiplexed sources might appear as shown in Figure 6.3, drawn for a fixed QOS objective. We conclude that even for identical bursty sources characterized by a very simple two-state traffic model, equivalent capacity is a complicated function of many parameters: *R*, *P*(on), *N*, buffer size. Curves of this type are, nonetheless, needed by the admission controller such that, for a fixed-bandwidth virtual path, a decision can be reached concerning the admissibility of a new request. Here, the equivalent capacity is fixed and is equal to the bandwidth of the virtual path; the equivalent capacity curve then provides the maximum number of sources, or virtual channels, that can be multiplexed onto that virtual path. Also, through use of an equivalent capacity curve, the admission controller can determine, as appropriate, the amount of additional virtual path bandwidth needed to accommodate a new request, and inquire as to whether or not this additional capacity is available on each physical link associated with that virtual path.

Equivalent capacity curves for identical on–off sources multiplexed onto a common virtual path appear in Figure 6.4. Shown on these graphs is the exact equivalent capacity needed to produce an overflow probability of 1×10^{-5} from a 3-Mbit buffer as a function of the source activity *P*(on), produced from extensive computer simulations. Also shown are the results of an analysis based on a simplified fluid flow approximation in which data sources are replaced by "water hoses" which are either off or on. When on, water flows at a constant rate. If the channel outlet rate is less than the instantaneous sum of the input rates, then water will build up in the buffer. If any water is present, it flows out of the buffer at a rate equal to the channel outlet rate. If no water is present and the instantaneous sum of the input rates is less than the channel outlet rate, then water flows out at the arrival rate. (This fluid flow approximation is often applied to study problems of this type.) The final curve corresponds to the equivalent capacity with an infinite buffer, that is, $N \times R \times P(\text{on})$. For Figure 6.4a, there are 50 identical sources, each

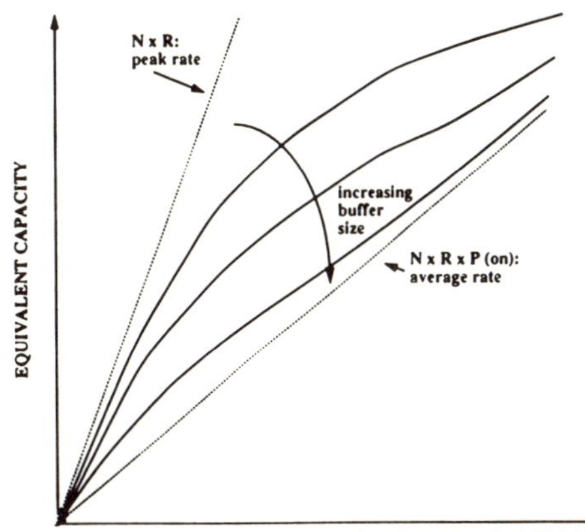

Figure 6.3. Qualitative example of the various factors which influence equivalent capacity.

of 4 MB/sec peak rate, and the average burst length is 100 msec. For Figure 6.4b, there are five sources, each of 40 MB/sec peak rate and 10 msec burst period. From curves such as these, it is apparent that the required equivalent capacity is much higher than the mean bit rate when the sources are relatively inactive, that is, when P(on) is low, but is only slightly higher than the mean bit rate for very active sources. This is intuitively pleasing: as P(on) increases toward unity, the sources behave increasingly like continuous bit rate sources for which the peak and average rates are the same. As P(on) diminishes, the sources become increasingly bursty, and the required equivalent capacity far exceeds the mean bit rate.

When multiplexing different traffic types onto the same virtual path, each possibly characterized by a more complicated source model, the equivalent capacity becomes increasingly difficult to determine. Here, rather than presenting the admission controller with a curve showing equivalent capacity as a function of N, it may be more appropriate to provide the admission controller with a capacity

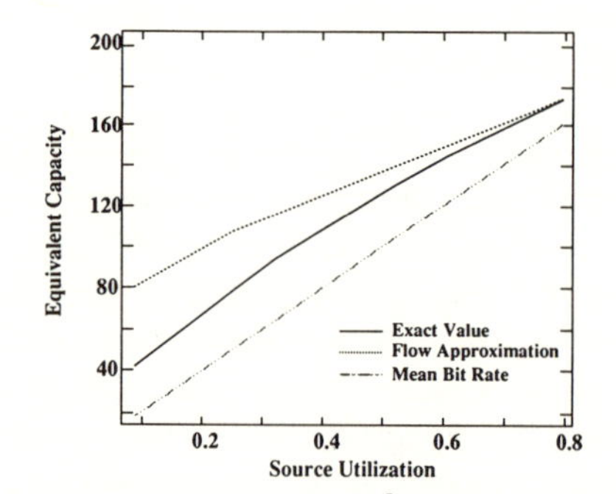

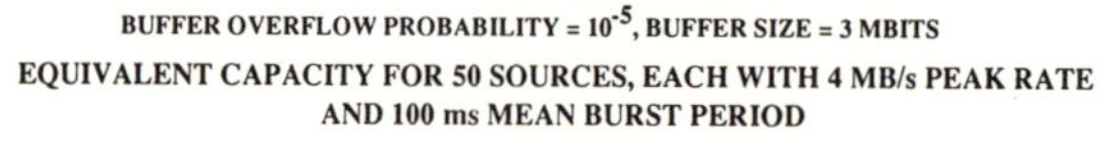

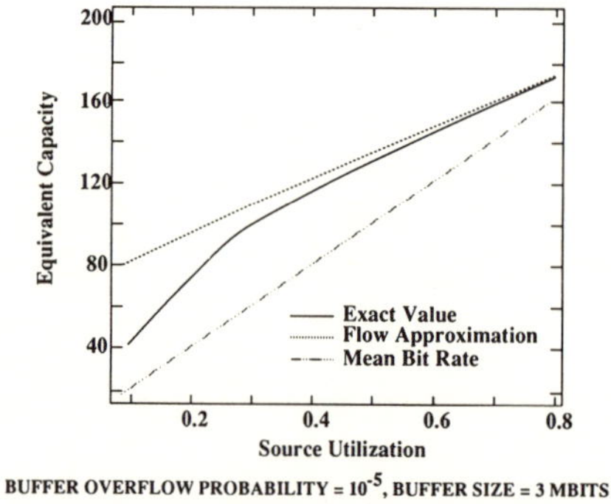

BUFFER OVERFLOW PROBABILITY = 10^{-5}, BUFFER SIZE = 3 MBITS
EQUIVALENT CAPACITY FOR FIVE SOURCES, EACH WITH 40 MB/s PEAK RATE AND 10 ms MEAN BURST PERIOD

Figure 6.4. Some specific examples of equivalent capacity.

region which shows, for a fixed virtual path bandwidth, the number of virtual channels of each type which may be multiplexed onto that virtual path while maintaining adequate QOS. For example, suppose there are two types of traffic sources. Then, the capacity region could be depicted by two-dimensional plots such as is qualitatively shown in Figure 6.5. For a fixed equivalent capacity (virtual path bandwidth), the area beneath the curve shows the permissible number of type 1/type 2 virtual channel combinations which may be multiplexed onto the virtual path. For a fixed equivalent capacity, a new request can be accommodated if the resulting (N_1, N_2) pair is beneath the respective curve. Also, as before, curves such as these can be used by the admission controller when requesting additional bandwidth for the virtual path.

If the number of traffic sources is equal to K, then the capacity region depicts those permissible virtual channel combinations $(N_1, N_2, \ldots, N_K)$ such that the objective QOS can be maintained, where N_i is the number of virtual channels of type i, $i = 1, 2, \ldots, K$. The capacity region is therefore K-dimensional. When deciding upon admissibility for a new connection of type i, the controller must determine if the resulting combination $N_1, N_2, \ldots, N_{i-1}, N_i, N_{i+1}, \ldots, N_K$ is within the capacity region.

Several possible admission control strategies have been proposed to cope with these traffic descriptor and equivalent capacity difficulties. The Class Related Rule seeks to divide virtual channels into a small number of distinct classes with all channels of the same class possessing similar worst-case traffic statistics. Then, when a request for a new connection is made, the call processor needs to know the equivalent bandwidth needed to accommodate one more connection of the class containing the request. Equivalently, the admission controller must have available the capacity region for a number of traffic types equal to the number of classes. Since the number of classes is far smaller than the number of virtual connections, the capacity region is of much smaller dimensionality and the admission decision

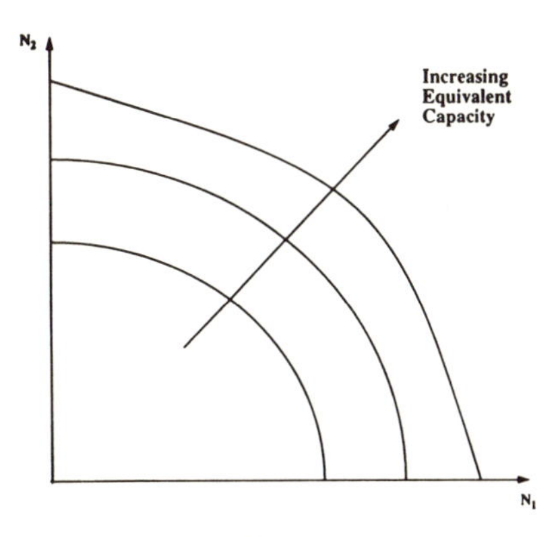

Figure 6.5. Qualitative example of a two-dimensional capacity region.

is simplified, although it still remains a complex function of the number of existing virtual connections of each class.

The QOS Evaluation Method seeks to *estimate* how QOS will be affected if a newly requested virtual connection is admitted, and is based on various bounding and approximation techniques.

The Fixed Boundary Method introduces traffic classes and, within a virtual path, provides for some fixed number of virtual channels for each class such that, as long as the number of virtual channels admitted for each class is less than the fixed number reserved for that class, then QOS for all virtual channels of that class will be met. The admission decision is therefore quite simple: block the request if all virtual channels for that class are currently in use. This method avoids the need to find the capacity region, at the expense of sacrificing network capacity: since fixed boundaries are used for each traffic class, it is not possible to reallocate the unused virtual channels reserved for one class for use by another class which is using all of its reserved virtual channels, and overall utilization efficiency will therefore suffer.

The Measurement Method relies on actual traffic measurements, rather than traffic descriptors. The required equivalent bandwidth is found from actual measurements, and a new request is admitted if adequate reserve bandwidth exists as determined from prior experience.

6.3. Policing

Suppose adequate traffic descriptors are developed and admission decisions are based on availability of virtual path bandwidth adequate to meet the equivalent capacity need. What is to prevent an admitted virtual channel from knowingly, unknowingly, or maliciously violating its declared traffic descriptor, thereby compromising QOS for all users of that virtual path? To prevent such an occurrence, source policing may be needed to ensure that virtual channels do not violate the covenant made at call set up time as manifested in their declared traffic descriptors. Policing is also known as traffic enforcement, usage parameter control, and access control. Various policing strategies based on mean and peak rate enforcement have been proposed. The Leaky Bucket approach is particularly simple to implement and effective for ensuring that no source will inject cells at a rate greater than its declared peak for longer than some limited time period.

To implement Leaky Bucket policing, each source is provided with a counter. Each cell generated by the source causes the value in the counter to increment by unity. Meanwhile, the counter is continually decremented at some constant rate R, except that the integer stored in the counter is never permitted to become negative. Any cell which is generated when the value in the counter exceeds some preset value M is "tagged," and tagged cells are either discarded immediately or given lowest priority and discarded later if congestion should develop internally to the network. With Leaky Bucket, a source cannot supply cells to the network at a rate greater than *R* for any prolonged period since, if cells are generated at a rate greater than *R,* the value of *M* will soon be exceeded and subsequent cells will be tagged (actually, as explained above, the Leaky Bucket can be made to be "lenient" in that it is possible to admit tagged cells to the network if traffic conditions elsewhere are so light that the tagged cells are not offensive to other virtual circuits). Controlling the parameters *M* and *R* can effectively "shape" the nontagged cells such as to conform with the traffic descriptors declared at call setup time. A possible Leaky Bucket "trajectory" is shown in Figure 6.6 which depicts the value in the counter as a function of time for a particular sequence of cell arrivals. As shown, the two cells which arrived when the value in the counter exceeded *M* were tagged and became candidates for discarding. All cells of any cell arrival sequence which maintains a counter value below *M* will be admitted to the network on a nontagged basis.

6.4. Flow Control

Another aspect of traffic control/performance management involves flow control, also known as reactive traffic management. Flow control seeks to recognize congestion within the network as it develops, and then attempts to limit the flow of new traffic into the network accordingly. For example, if the flow control mechanism determines that congestion is developing at a particular ATM switch output, as evidenced by the volume of cells in the buffer feeding that output, then an appropriate signal would be generated to deter all sources with virtual channels or virtual paths through that output from sending any additional cells into the network until the buffer has been cleared out. In effect, flow control seeks to limit the arrival of new information which cannot be delivered with acceptable quality until the congestion which impedes delivery has been cleared out, thereby reducing the length of the congestion period and improving service quality.

Clearly, the utility of such flow control is strongly dependent on link speeds, traffic loads, and delay in propagating the flow control signals back to the traffic sources. If link speeds are low and/or the propagation delay is small, then flow

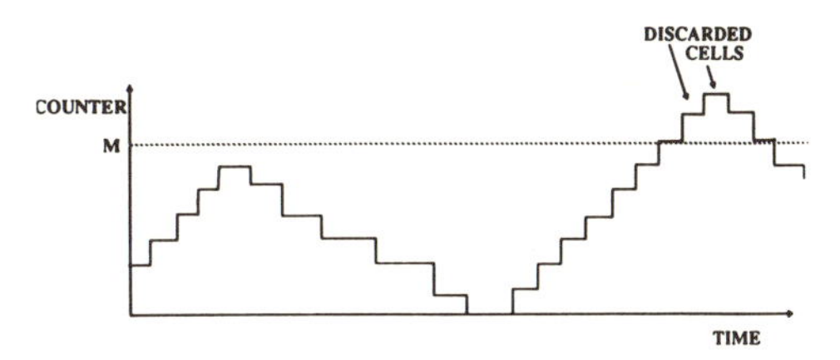

Figure 6.6. An exemplary counter-value trajectory for Leaky Bucket policing.

control can be a very effective technique for maintaining QOS since the volume of new traffic arriving at a congested network node while the control signal was propagating back to the sources is correspondingly small. However, if the link speed is very high (implying that much new traffic might arrive at the point of congestion before the sources have been silenced) or if the propagation time of the control signal is very long (again, much new traffic might arrive), then flow control may become useless. For broadband, wide-area networks, the link speeds are high *and* the propagation delay is long. For example, the link speed for ATM is 155 Mbits/sec, implying a 53-octet cell transmission time of 2.735 μsec. The minimum propagation time across the United States from east to west, as determined by the finite speed of light, is 13.44 msec. Thus, before a source in New York can be silenced by congestion developing at a node in Los Angeles, as many as 13.44 msec/2.735 μsec = 4914 new ATM cells may have been generated by one source alone. Since many sources may be feeding that one congested node, we see that the effectiveness of flow control in maintaining QOS could be severely compromised. Moreover, if done improperly, the additional traffic burden created by the flow control signals could actually exasperate an already high-congestion situation.

Window flow control is a technique not requiring the generation of an explicit congestion-initiated feedback signal. Here, the receiver is responsible for generating a short acknowledgment back to the transmitter whenever a higher-layer protocol data unit is received, and the transmitter only allows some fixed number of unacknowledged PDUs (the window size) to be outstanding on the network. Thus, if congestion should develop within the network, the PDUs will arrive late at their receivers, causing delayed acknowledgments. If, while awaiting the next acknowledgment, a transmitter has already placed a full window of PDUs onto the network, it must cease transmitting any additional PDUs. Thus, the flow of traffic into congested nodes will be curtailed. Although window flow control does not require an explicit congestion-initiated feedback signal, its effectiveness is nonetheless compromised when the link speed is high and/or the propagation delay is long. Under either condition, efficient link utilization and high user throughput mandate the use of a long window. For example, if the window length is too small for the link speed and propagation delay, then after generating a window's worth of PDUs, the source must go idle, even though the network may be uncongested, if the earliest PDU is still in transit to its receiver (which cannot develop an acknowledgment until that PDU has been received). At high speed and long propagation delay, a large window is needed to ensure that the "pipeline remains filled." Unfortunately, the effectiveness of window flow control as a preserver of QOS is compromised as the window length increases because, as before, too much new data may enter the network after the onset of congestion. Use of a congestion-dependent window size may partially alleviate this drawback; here, not only does the transmitter refrain from placing more than one window of unacknowledged data onto the network, but, also, it reduces its window size in response to late-arriving acknowledgments. Thus, as congestion develops, the window size is gradually reduced (thereby preventing a large amount of unwanted data from entering the network) until the backlog clears out.

Rate flow control is a technique which adaptively controls the peak rate at which a source may transmit in response to congestion. Either an explicit congestion-indicating feedback signal, or a late-arriving acknowledgment from a receiver of earlier-transmitted data, can initiate the adaptive rate mechanism. As with window flow control, both the high-speed links and long propagation delay compromise the effectiveness of rate flow control.

Explicit Congestion Notification (ECN) is a possible flow control mechanism for B-ISDN which works to keep the end points informed of congestion conditions. ECN would require use of one bit in each ATM cell header to notify the end points, and can affect both window and rate flow control. As described above, ECN works best when the propagation delay is small (compared with the duration of the congestion). ECN may actually be harmful if the propagation delay is too long since an indication of noncongestion may take so long to propagate back to the source that new congestion has already developed at precisely that time when the source begins transmitting at its maximum rate.

Fast Reservation Protocol (FRP) is another possible B-ISDN flow control mechanism which may be useful if the sources generate very long data bursts containing many ATM cells. Here, a source generating such a long data burst for transmission on a preestablished virtual circuit would first send a special request cell to a fast scheduler, which seeks to reserve bandwidth equal to the peak burst rate along the entire length of the virtual path. (Note that, in the absence of FRP, a source can transmit at will once its virtual channel has been set up, subject to policing and/or other flow control mechanisms, but a specific request for permission to transmit is not needed; with FRP, such a special request is needed.) The fast scheduler would send an acknowledgment only if the reservation is successful. Since, during periods of congestion, no acknowledgment would be sent, the transmitter is inhibited accordingly. Acknowledged bandwidth would be released once the burst transmission is completed. FRP is most appropriate when delay is tolerable (since signaling and reservation are involved) but lost cells are not (no source would be permitted to send a burst to a node whose buffers were about to overflow).

6.5. Priority Control

Priority control is another tool for traffic control/performance management which seeks to improve link utilization by declaring several QOS classes. Within the ATM concentrators and self-routing switches, different classes are handled in different buffers, as shown in Figure 6.7. A priority scheduler then empties these buffers onto the output link in such a fashion as to preserve QOS for each. For example, the priority scheduler would serve the buffer of a traffic class with real-time delivery needs but which could withstand cell loss in such a way as to meet that real-time need, unless prevented from doing so by congestion, in which case cells which would be delivered too late are simply discarded. Similarly, the priority scheduler would last serve the buffer of a traffic class which could tolerate delay but could not tolerate cell loss, unless that buffer was about to overflow, in which case that buffer would receive highest priority. With priority control, the greatest link utilization improvements would be realized when the QOS classes have widely different, complementary requirements.

6.6. Self-Learning Strategies

An example of a self-learning traffic control/performance manager is shown in Figure 6.8. Here, a three-tiered adaptive controller is overlaid on the ATM network. Users are shown as open squares, ATM concentrators and switches as shaded squares, and network control elements as circles. The control elements communicate via observation and control links (which may be part of the ATM network or may involve a totally separate signaling network). The Cell Transfer Control Layer is responsible for categorizing the traffic pattern generated by the users, setting the parameters for window flow control and Leaky Bucket policing accordingly, and declaring a QOS. The Call Control Layer is responsible for admission control and congestion control (e.g., generating ECN signals) in response to the QOS classifications provided by the Cell Transfer Control Layer. The Network Control Layer is responsible for sensing resources that are underutilized over a time period comparable to (or greater than) the holding time of the typical virtual connection, and for dynamically rerouting connections and reassigning link capacity such as to achieve better load balancing (for example, reroute away from links which carry a disproportionate share of virtual connections as a result of virtual connection departures and new assignments). As explained earlier, one school of thought holds that self-learning approaches such as this are the only viable means for effecting traffic control and performance management in the B-ISDN environment.

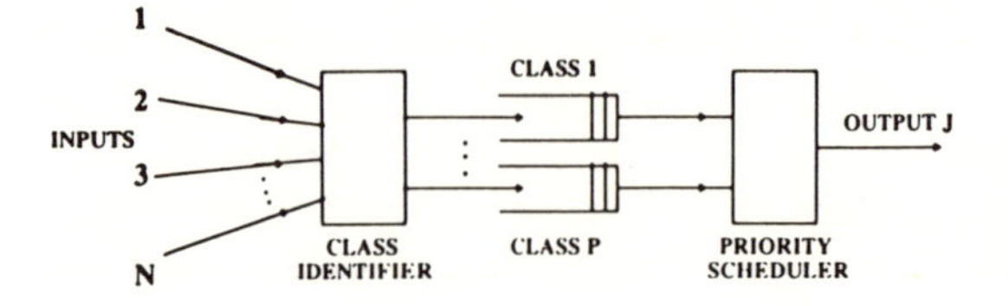

Figure 6.7. Block diagram of a statistical multiplexer using priority scheduling.

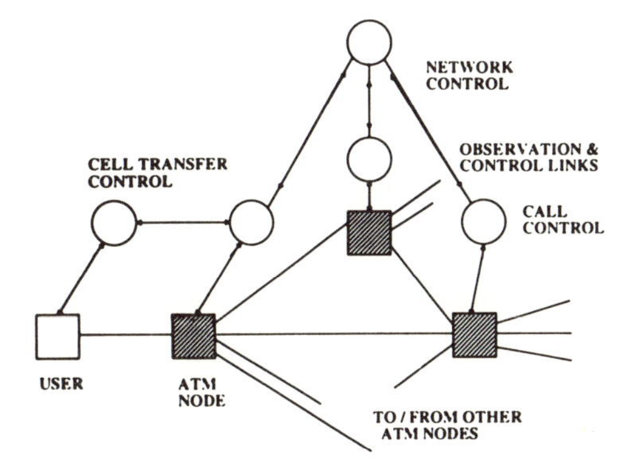

Figure 6.8. Partial block diagram of an ATM network employing a self-learning traffic control/performance management strategy.

6.7. Conclusion

How important are traffic control and performance management in broadband networks? In this chapter we have posed this question, described several conflicting schools of thought, and summarized the various technical aspects to this issue. The simplest approaches based on virtual connection peak bandwidth allocation are unnecessarily wasteful of network resources and limit the applied traffic load to a small fraction of that which the network might potentially handle. Alternatively, the incremental load supportable by very complicated strategies may not warrant the additional implementation complexity.

Clearly, open-loop admission control will be a powerful technique for maintaining adequate QOS in the integrated multimedia environment. Approaches based on the creation of a limited set of traffic classes represent a reasonable compromise between resource utilization efficiency and ease of implementation. Here, each new request for a virtual connection is assumed to exhibit the worst-case traffic pattern for its class. This limits the dimensionality of the capacity region, thereby simplifying the admission decision algorithm, while sacrificing some of the network capacity (within a traffic class, few if any of the virtual connections may exhibit worst-case behavior). Fixed-boundary techniques are certainly simpler to implement but suffer the disadvantage of precluding the reallocation of underutilized resources dedicated to one traffic class to another class which may be overtaxing its fixed resource allocation.

For any admission control strategy, some form of policing is unavoidable since, along a given virtual path, resources are statistically shared among users and violation of its traffic descriptor by any one user adversely affects the QOS enjoyed by all. Leaky Bucket policing is both effective and particularly easy to implement.

In principle, flow control strategies could permit a greater traffic load to be carried by the network. Here, the network is more tolerant of the increased frequency of short-term high congestion resulting from the admission of more connections, maintaining QOS by preventing new traffic from entering the network on existing virtual connections during such high-congestion episodes. The issue which compromises the effectiveness of this technique as applied to broadband networks concerns the propagation delay of the flow-controlling feedback signal and the resulting inability to stop the flow of entering traffic sufficiently soon (i.e., before too great a volume of new traffic has entered the network).

Overall, sacrifice of some potential capacity is the price to be paid for the benefits to be enjoyed from a manageable integrated multimedia network. Fortunately, one of the benefits to be expected from a broadband network is an ample overabundance of capacity such that the sacrifice of some of this capacity to achieve simplification of control over the remainder is indeed a favorable trade-off.

6.8. Problems

6.1. Consider a time-slotted ON–OFF traffic source. If the source is idle in a given time slot, it becomes active in the next time slot with probability $1 - q = 0.01$; if the source is active in a given time slot, it becomes inactive in the next time slot with probability $1 - p = 0.5$. The data rate, when active, is 25 Mbit/sec, and the line speed is 150 Mbit/sec. (a) Find the fraction of time that the source is active; (b) find the average data rate of the source; (c) find the probability that a newly active source remains active for precisely three time slots; (d) find the probability that a newly active source generates three active time slots followed by ten idle slots followed by two active slots followed by a transition to the idle state. (e) For the sequence of part d, draw to scale the sequence of cells appearing at its output, both before and after cell compression to convert the peak rate of the source to the line speed.

6.2. Two classes of traffic are statistically multiplexed onto one virtual path. Each class of traffic can be modeled by an ON–OFF source. For class 1, the On-to-OFF transition probability is $1 - p = 0.09$ and the OFF-to-ON transition probability is $1 - q = 0.01$. For class 2, the ON-to-OFF and OFF-to-ON transition probabilities are 0.4 and 0.1, respectively. When active, class 1 and class 2 sources generate information at peak rates of 5 and 50 Mbit/sec, respectively. The bandwidth of the virtual path is 100 Mbit/sec, the buffer at the head of the virtual path can store an arbitrarily large number of cells, and delay is not a QOS issue. Let n_1 and n_2 be the number of class 1 and class 2 virtual channels, respectively, which are multiplexed onto the virtual path. (a) Find an expression for the capacity region, that is, the permissible combinations of (n_1, n_2) pairs. (b) Plot the result of part a.

6.3. To achieve some specified QOS objectives, the equivalent capacity needed to multiplex n identical sources (each with its own virtual channel) onto a common virtual path is given by:

$$C_{eq} = (1 \text{ Mbit/sec}) \times n + (2 \text{ Mbit/sec}) (1 - e^{-2n})$$

(a) Plot the equivalent capacity as a function of n. (b) If the VP bandwidth is fixed at 10 Mbit/sec, how many VCs can it support and still meet the QOS objective? (c) If two additional VCs must be added to the result from part b, by how much must the VP bandwidth be increased?

6.4. Referring to Problem 6.3, suppose that requests for new VCs are Poisson with arrival rate $\lambda = 3$ calls/hr, and the duration of each call is exponentially distributed with mean value equal to 1 hr. Find the call blocking probability if the VP bandwidth is fixed at 10 Mbit/sec.

6.5. Referring to Problem 6.2, suppose that, for both class 1 and class 2 sources, requests for new VCs are Poisson with arrival rate $\lambda = 8$ calls/hr and that the duration of each call is exponentially distributed with mean value 1 hr. (a) Find the blocking probabilities for class 1 and class 2 calls. (b) Suppose that a fixed-boundary allocation rule is used with the following parameters:

maximum number of type 1 VCs = 40
maximum number of type 2 VCs = 8

Find the blocking probabilities for class 1 and class 2 calls. (c) Graphically superimpose the fixed-boundary allocation rule on the capacity region of Problem 6.2b and show how capacity is "wasted." (d) Although capacity is wasted, do requests for class 1 connections benefit from use of the fixed-boundary rule?

6.6. A Leaky Bucket mechanism is used to police a particular source. Each time slot, the number of cells generated by the source is a binomially distributed random variable with mean value 0.1 and maximum value 3. The service rate is $R = 1$ cell/2 time slots, the threshold value is set at $M = 5$, and the buffer can hold an arbitrarily large number of cells. Find the probability that a particular cell is tagged for violation.

6.7. A source generates Poisson arrivals, and each message length is an exponentially distributed random variable of mean value T. Arrivals enter a single-server queue. The source is flow controlled: the rate of new arrivals from the source is dependent on the number of queued messages in accordance with the relationship $\lambda_n = \lambda_0/(n - 1)$ where λ_n is the average arrival rate when there are n messages in the queue, $n = 0, 1, 2, \ldots$. The queue can store an arbitrarily large number of messages. (a) Find the probability that there are n messages in the queue, $n = 0, 1, 2, \ldots$. (b) What is the overall rate of new arrivals when averaged over the number of messages in the queue? (c) If λ_0 is the "uninhibited" average rate of the source, by what fraction is this "uninhibited" rate reduced by the flow control mechanism?

6.8. Consider a two-class priority scheduler. Class 1 and class 2 are characterized by the following parameters: Poisson arrivals of rates λ_1 and λ_2, respectively, and exponentially distributed message lengths with mean values T_1 and T_2, respectively. There are two buffers, each storing arrivals associated with only one class of traffic. Each buffer can store a maximum of B messages. The priority scheduler always serves the class 1 buffer if there are any class 1 messages present, unless the number of class 2 messages in class 2's buffer is greater than or equal to M, in which case class 2 receives priority. (a) Find expressions for the overflow probabilities of the class 1 and class 2 buffers. (b) For those class 1 messages which are not lost to buffer overflow, find an expression for the average queuing delay. (c) Numerically find values for parts a and b when $\lambda_1 = \lambda_2 = 10$/sec, $T = 10$ msec, $B = 10$, and $M = 8$.

References

1. R. Guerin, H. Ahmadi, and N. Naghshineh, Equivalent capacity and its applications in high-speed networks, *IEEE J. Selected Areas Commun.* **SAC-9**(7), Sept. 1991.
2. J. M. Hyman, A. A. Lazar, and G. Pacifici, Real-time scheduling with quality of service constraints, *IEEE J. Selected Areas Commun.* **SAC-9**(7), Sept. 1991.
3. E. P. Rathgeb, Modeling and performance comparison of policing mechanisms for ATM networks, *IEEE J. Selected Areas Commun.* **SAC-9**(3), April 1991.
4. G. Gallassi, C. Rigolio, and L. Fratta, ATM: Bandwidth assignment and bandwidth enforcement policies, 1989 IEEE GLOBECOM Conf. Rec., Boston.
5. D. Anick, D. Mitra, and M. M. Sondhi, Stochastic theory of a data handling system with multiple sources, *Bell Syst. Tech. J.* **61**(8), Oct. 1982.
6. M. Butto, E. Cavallero, and A. Tonietti, Effectiveness of the leaky bucket policing mechanism in ATM networks, *IEEE J. Selected Areas Commun.* **SAC-9**(3), April 1991.
7. G. M. Woodruff, R. G. H. Rogers, and P. S. Richards, A. congestion control framework for high-speed integrated packetized transport, 1988 IEEE GLOBECOM Conf. Rec., Hollywood, Fla.
8. E. P. Rathgeb and T. H. Theimu, The policing function in ATM networks, 1990 Int. Swit. Symp. Conf. Proc., Stockholm.
9. L. Dittman, S. B. Jacobsen, and K. Moth, Source-independent call acceptance procedures in ATM networks, *IEEE J. Selected Areas Commun.* **SAC-9**(3), April 1991.
10. A. Y.-M. Lin and J. A. Silvester, Priority queueing strategies and buffer allocation protocols for traffic control in an integrated broadband switching system, *IEEE J. Selected Areas Commun.* **SAC-9**(9), Dec. 1991.
11. M. Gerla and L. Kleinrock, Flow control: A comparative survey, *IEEE Trans. Commun.* **COM-28**(4), April 1980.
12. N. K. Jaiswal, *Priority Queues,* Academic Press, New York, 1968.
13. S. Yazid and H. T. Mouftah, Congestion control methods for B-ISDN, *IEEE Commun. Mag.* **30**(7), July 1992.
14. I. W. Habib and T. N. Saadawi, Multimedia traffic characteristics in broadband networks, *IEEE Commun. Mag.* **30**(7), July 1992.
15. D. Hong and T. Suda, Congestion control and prevention in ATM networks, *IEEE Net. Mag.* **5**(4), July 1991.
16. H. Ohnishi, T. Okada, and K. Noguchi, Flow control schemes and delay-loss tradeoffs in ATM networks, *IEEE J. Selected Areas Commun.* **SAC-6**(9), Dec. 1988.
17. M. Wernik, O. Aboul-Magd, and H. Gilbert, Traffic management for B-ISDN services, *IEEE Net. Mag.* **6**(5), Sept. 1992.
18. A. E. Eckberg, B-ISDN/ATM traffic and congestion control, *IEEE Net. Mag.* **6**(5), Sept. 1992.
19. I. W. Habib and T. N. Saadawi, Controlling flow and avoiding congestion in broadband networks, *IEEE Commun. Mag.* **30**(10), Oct. 1991.
20. A. A. Lazar and G. Pacifici, Control of resources in broadband networks with quality of service guarantees, *IEEE Commun. Mag.* **30**(10), Oct. 1991.

21. M. Decina and T. Toniatti, On bandwidth allocation to bursty virtual connections in ATM networks, 1990 IEEE Int. Conf. Commun. Conf. Rec.
22. M. Decina T. Toniatti, P. Vaccari, and L. Verri, Bandwidth assignment and virtual call blocking in ATM networks, 1990 IEEE INFOCOM Proc.
23. A. Eckberg, D. Luan, and D. Lucatoni, An approach to controlling congestion in ATM networks, *Int. J. Digital Analogue Commun. Suppl.* **3,** April–June 1990.
24. G. Gallasi, G. Rigolio, and L. Verra, Resource management and dimensioning in ATM networks, *IEEE Net. Mag.* **4**(3), May 1990.
25. A. E. Eckberg, B. T. Doshi, and R. Foccolillo, Controlling congestion in B-ISDN/ATM: Issues and strategies, *IEEE Commun. Mag.* **29**(9), Sept. 1991.
26. *IEEE J. Selected Areas Commun.,* issue on Congestion Control in High Speed Packet Switched Networks, **SAC-9**(7), Sept. 1991.

7

Lightwave Networks

7.1. What Is a Lightwave Network?

The principles and techniques of broadband networks as presented thus far are based in large measure on a single enabling technology: VLSI. With VLSI, we can cost-effectively implement equipment capable of very sophisticated functionality. We can process the routing and protocol headers of information packets in real time, resolve contention for access rights to LANs and MANs, develop self-routing massively parallel space division packet switches, cast all types of telecommunication traffic into a standard, fixed-length cell format, and provide integrated bandwidth-on-demand to many users geographically dispersed over a very wide geography. Except for its use as the medium for high-speed point-to-point transmission links between pairs of electronic multiplexers, switches, or access modules, the technology of fiber optics is not fundamental to realization of the broadband vision.

In this chapter, we shall explore the potential for fiber optics and photonic technologies to emerge as key underpinnings of broadband networks, and we shall explore their use *not* for point-to-point transmission systems in an otherwise conventional broadband network context, but, rather, as the foundation for alternative broadband, multiuser, multimedia network architectures. The field of lightwave networks is quite new and is the subject of ongoing research worldwide. Because the field is still in its infancy, we shall take the liberty of defining terms, offering opinions, and even speculating a bit.

To date, most telecommunication applications of lightwave technology have been limited to point-to-point transmission systems, wherein signals are collected and multiplexed in remote concentrators (either reservation-based circuit-oriented or statistical multiplexers) and shipped over fiber-optic facilities to a switching office where the signals are demultiplexed and routed. Subsequent to routing, signals intended for a common remote destination (either another switching office or a remote demultiplexer for final distribution to their intended receivers) are remultiplexed onto a fiber-optic transmission system for delivery to that destination. The process of demultiplexing, routing, remultiplexing, and transmission may occur several times before signals are finally delivered to their destinations. The architecture of such a system is similar to that depicted in the B-ISDN Reference Architecture (Figure 5.2) in that direct, point-to-point fiber transmission systems are used to interconnect multiplexers and switches. From a *functional* perspective, the use of fiber as the medium for those transmission systems is not essential; twisted pair copper wiring, coaxial cable, or radio could just as well been used (such a statement, however, could not be made from a physical or practical perspective; see below). Despite the widespread use of fiber-optic links, the fundamental architectures of the telecommunication system (point-to-point transmission links interconnecting switching nodes and multiplexers) have not evolved at all from those of the prephotonic era.

From a physical level and practical perspective, optical fiber has clearly emerged as the medium of choice in telecommunication networks by virtue of its low loss and low dispersion, and because of the optical modulators/demodulators at the end of each link can economically operate at high data rates. Low loss and low dispersion, coupled with high-speed optical sources and detectors, permit the operation of fiber-optic transmission links at high data rates over long distances without requiring many expensive repeater stations. The figure of merit often used to compare fiber-optic transmission systems is the speed–distance product, that is, the product of the link data rate and unrepeatered distance over which a low bit error rate (less than 1×10^{-9}) can be maintained. Even without the use of optical amplifiers, fiber-optic systems can readily offer speed–distance products in excess of 100 Gbit/sec-km. Thus, link operation at a data rate of 1 Gbit/sec over an unrepeatered distance of 100 km is possible. However, because of an effect known as the electro-optic bottleneck, it is not feasible to run a link at 100 Gbit/sec over a distance of 1 km, although the speed–distance product for such a hypothetical link is again 100 Gbit/sec-km. Basically, the electro-optic bottleneck limits the peak rate of any optical link to the maximum bit rate at which practical optical transmitters and receivers can operate, which has nothing to do with either the geographical length of the link or the bandwidth of the medium. Practical links can run at peak rates of several gigabits per second; rates of several tens of gigabits per second have been reported in research laboratories. Although the electro-optic bottleneck is being slowly relaxed year after year, the analog bandwidth required by the maximum possible optical transmission or reception rate remains but a tiny fraction of the bandwidth of single-mode optical fiber itself. If we define the bandwidth of the medium by the low-loss window between the optical wavelengths of 1.1 and 1.6 μm where the loss is under 1 dB/km, then the bandwidth of the medium is approximately 100,000 GHz! Performance of a point-to-point optical link, however, is limited by two independent effects: its speed–distance product and the electro-optic bottleneck.

A lightwave network is not the use of fiber-optic transmission links in a conventional telecommunication architecture. Rather, a lightwave network consists of a geographically distributed shared medium to which are attached some number of geographically dispersed access stations, each of which provides service to an attached device through a user access port. A high-level, generic representation of a lightwave network appears in Figure 7.1. For our purposes, we shall assume that the medium is "passive," in the sense that no nonlinear or logical operations are performed within the medium itself: only linear operations (amplification, splitting, combining, wavelength-selective filtering) can be performed on the optical signals injected by the access stations onto the optical medium. Nonregenerative optical amplification is permitted within the medium (hence, "passive" is included in quotations). Here, however, the role of the optical amplifier is limited to linearly amplifying any applied optical signal (which may have incurred significant splitting loss and fiber attenuation) such as to maintain an output signal-to-noise ratio adequate for reliable detection of one of the access stations.

Signals on the medium exist in optical form throughout: there are no electronic elements located within the optical medium. All logical operations (signal routing, protocol processing, etc.) are performed within the access stations by electronic circuitry. The access stations also contain the electro-optic devices (semiconductor lasers, photodetectors) needed to convert signals to optical form and place them onto the medium, and to retrieve optical signals from the medium. Finally, the access stations accept user signals (which may arrive in either optical or electronic form) through the user interface, and electronically process, format, and prepare those signals for transfer over the optical medium. The access stations also prepare signals received from the medium for transfer back across the user interface.

Note that we are precluding the use of optical logic elements, both within the optical medium and within the access stations. We have done so because (1) optical logic elements which produce an optical output which is some logical

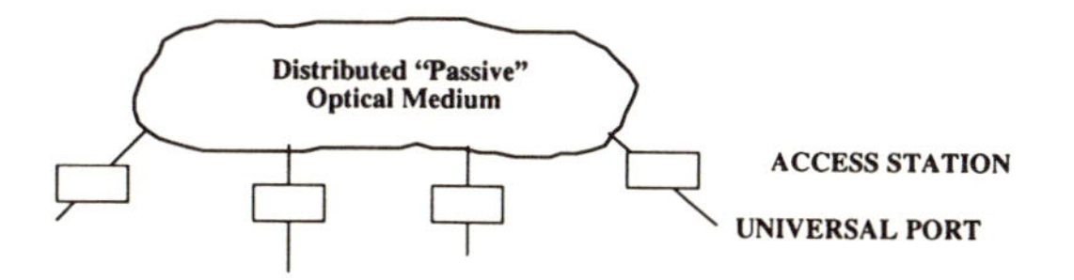

Figure 7.1. Generic model of a lightwave network.

function (AND, OR, etc.) of two or more optical input signals do not exist [except through the use of some intervening semiconductor electronics such as occurs within self electro-optic effect devices (SEEDs) or through such means as using photodetectors at the inputs, electronic logic gates internally, and laser diodes at the outputs; (2) optical logic which is dependent on some intervening semiconductor electronics may not offer any speed advantage relative to conventional electronic and electro-optic devices such as laser and photodetectors (i.e., the electro-optic bottleneck would still exist); and (3) any advantages of a hypothetical network using hypothetical optical logic, relative to the type of network implied by Figure 7.1, are yet to be discovered and, should any be found, may not be significant (clearly an opinion!).

As implied by Figure 7.1, the access stations communicate among themselves over the optical medium in a nonhierarchical fashion: there is no multiplexing and switching hierarchy such as appears in the B-ISDN Reference Architecture of Figure 5.2. Thus, the lightwave network is similar to a Local Area Network wherein geographically distributed users communicate over a shared channel and are connected to that channel via LAN access stations. However, three important differences are to be noted. First, the geographical distances associated with the lightwave network are potentially much larger than those applicable to LANs (thousands of kilometers rather than fractions of a kilometer). Second, whereas there exists but a single high-speed shared channel in the LAN environment, *many* such high-speed channels exist and are simultaneously in operation in the lightwave network environment. In fact, it is the intent of a lightwave network to enable the creation of many independent channels, each operating at a very high speed (perhaps approaching the rate set by the electro-optic bottleneck), and each somehow maintaining its own identity in a manner to be subsequently described, such that the amount of bandwidth actually consumed is commensurate with the bandwidth of the medium, rather than being commensurate with the bandwidth required of a single channel operating at a rate set by the electro-optic bottleneck. Each such channel would provide a direct, nonshared connection between precisely two access stations (one transmitting on that channel, one receiving from that channel; note that a media access protocol to enable channel sharing among a multitude of access stations is not required). Third, the number of access stations which may be interconnected can grow very large (hundreds of thousands or millions).

In this chapter, we will briefly present some basic background information concerning relevant underlying device technology. Then, we shall describe and study various architectural alternatives to the design and implementation of optical networks involving wavelength, time, and space division multiplexing of many independent high-speed channels onto the medium, along with selected combinations of these basic multiplexing techniques. Approaches which require so-called rapid optical agility, in which a packet switching capability is achieved by rapidly assigning (and reassigning) each of the many channels existing on the medium to a pair of access stations (one transmitting, one receiving) in response to packet arrival patterns will be contrasted with a more circuit-switching-oriented approach in which each channel is either permanently assigned to a pair of access stations, or is slowly reassignable to a different pair in response to changing real or virtual circuit patterns. Techniques to enable a self-routing packet switching service (such as ATM) on top of an inherently circuit-switched optical medium involving routing in the access stations will also be described.

As a prelude to further motivate the potential importance of lightwave networks and to help in comprehending some of the concepts involved, we will present some background and qualitatively describe one possible approach.

The enormous potential of optical networks has long been recognized and, over the years, a considerable body of knowledge and technology has been developed by researchers working in the field. Work in the early years tended to be focused almost exclusively on the physical and device layers (e.g., trying to stabilize the wavelength of semiconductor lasers so that a comb of independently modulated channels might be created on the physical medium to carry information via wavelength division multiplexing; trying to develop semiconductor lasers which can be rapidly retuned in wavelength to support fast packet switching in a dense WDM network). Later, as telecommunication systems researchers became attracted to the field, the focus began to slowly shift toward network-level architectures. It was then that the unique opportunities of optical networks (e.g., clear-channel circuit switching, physical/logical topological independence, "permanent plant," self-routing fast packet overlays, etc., to be described later) became recognized, along with the architectural constraints presented by device limitations (e.g., the electro-optic bottleneck). It became clear that the major issues involved in multiuser optical networks were vastly different from those of point-to-point optical transmission systems, and were even different from those encountered in the earliest attempts to use optical fiber as the shared, single-channel medium of a multiaccess network (i.e., DQDB; see Chapter 4). Issues involving signal detectability, power budget, fiber dispersion, coupling loss, and so forth, although still significant from a physical layer perspective, were joined by new and profound networking-related issues such as (1) user connectivity in a multichannel environment, wherein a great number of channels are created on a shared optical medium with each user station having only a limited number of optical transmitters and receivers and therefore having instantaneous access to only a limited number of such channels (each user cannot "hear" all activity, as required by simple single-channel architectures such as DQDB or CSMA/CD); (2) provisioning of fast, self-routing packet switching capability in such an environment; (3) determination of the best way (or, at least, a good way) to assign the channels among the access stations, along with algorithms which enable channels to be electronically reassigned so as to match the station-to-station connectivity with prevailing traffic patterns, network usage, and equipment failure/recovery (the last providing a high degree of fault-tolerance); and (4) traffic performance and fault management, including admission control, access control (policing), flow control, buffer and bandwidth management, and alternative routing. In short, the field has rapidly matured from one driven by device and physical-level accomplishments to one driven by network-level understanding, new network architectures, and new management/control strategies. The innovations of the earlier years produced a solid platform of physical-layer capabilities and architectural insights, upon which network-level innovations were developed to fulfill the potential of modular, wide-area, high-performing all-optical networks.

Each access station is equipped with a small number (say, P) of optical transmitters and receivers, where P is much smaller than the number of access stations N. The shared medium, in effect, provides a pool of $P \times N$ channels which can be assigned among the various transmitter–receiver pairs, creating a connectivity pattern among the access stations. It is important to note that the *connectivity* diagram among stations is independent of both the relative physical locations among the stations and (to a certain degree) the physical topology of the optical medium; it can, however, be electronically rearranged by reassigning the channels to different transmitter–receiver pairs in response to changing traffic patterns and equipment failure/recovery.

An N=8 node example of several types of passive physical topologies (bus, tree, star—to be further described later in this chapter) appears in Figure 7.2a. Each of these topologies superimposes the signals injected by each access station and broadcasts the superimposed signals back to all receivers. For this example, distinguishability among the signals is maintained by placing each on a separate wavelength of light. In Figure 7.2b is the resulting connectivity diagram for a particular assignment of 16 distinct wavelengths among the eight access stations, each such station having access to two transmit and two receive wavelengths. The connectivity diagram, being dependent only on the wavelength assignment among access stations, is independent of the physical topology of the network and is applicable to all three physical topologies shown in Figure 7.2a. The particular connectivity diagram shown is that of the recirculating perfect shuffle wherein transmitters of stations in the first column are connected to receivers of stations in the second column by a set of eight wavelengths assigned to the stations in a perfect shuffle pattern. The transmitters of stations in the second column are connected to receivers of stations in the first column by using a second set of eight wavelengths and a second perfect shuffle pattern. Assigning wavelengths among the stations in some other fashion would produce a different connectivity diagram.

For this introductory, qualitative example, we will further restrict attention

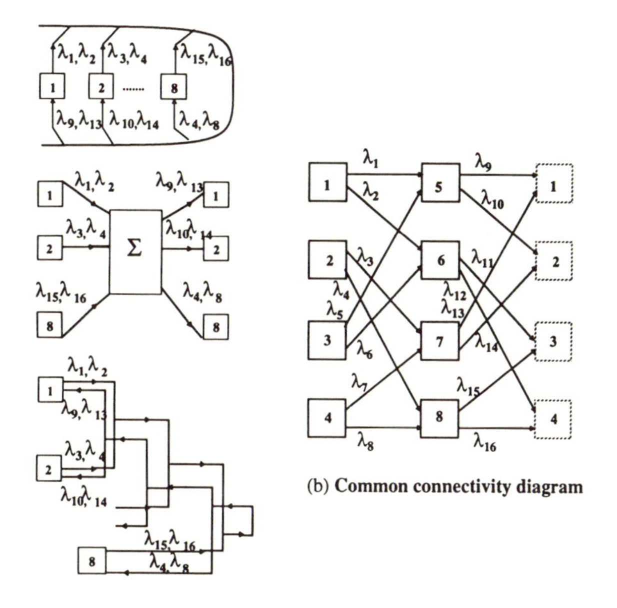

Figure 7.2. Several possible physical topologies, all producing a common connection diagram, with optical channels distinguished via wavelength division multiplexing.

to the case where the channel assignment among access stations is either fixed or slowly changeable. For the latter, wavelength agility is implied, meaning that each of the fixed number of transmitters and receivers associated with each access station can be tuned to any of the wavelengths in use. However, the tuning speed and setup procedures are slow: in effect, the optical medium provides a means of enabling circuit-switched, broadband (several gigahertz) analog connections among the access stations. The number of such connections which can be established by each access station is limited to P, the number of optical transmitters (receivers) deployed in each access station.

In this approach, the access stations serve three functions. First, the access stations present a "universal" interface to the user. Two such interfaces might be envisioned. The first is the "clear channel" interface in which a signal can be generated by a user in *any* format (analog or variable bit-rate digital) which is simply carried across the network by one of the optical channels to its intended receiving station. The second is a high-speed (multigigabit/second) Asynchronous Transfer Mode (ATM) format. Signals presented in this format can be delivered to any other user, and the number of ATM virtual connections which can be established by any user can be quite large (a user is by no means limited to P virtual connections if signals are presented in ATM format; more will be said of this momentarily).

The second function provided by the access stations is that of wavelength translation. Suppose station A wishes to establish a "clear channel" to station C. Unfortunately, all of station A's P wavelengths (one per transmitter) are already in use, with the exception of wavelength λ_3, and all of station C's receivers are in use, with the exception of the one which listens to wavelength λ_2. However, station B has an available receiver on wavelength λ_2 and an available transmitter on wavelength λ_3. In this case, channels from the medium would be assigned as follows: a channel on wavelength λ_3 would be created from station A to station B, and a second channel on wavelength λ_2 would be created from station B to station C. Within station B, the signal received from station A on wavelength λ_3 is electronically translated from wavelength λ_3 to wavelength λ_2 prior to being relayed on to station C, there by enabling the clear-channel connection from station A to C. An alternative, of course, would have been to provide a direct connection from A to C over the optical medium, but this is prevented by the fact that none of A's unassigned transmit wavelengths coincides with any of C's unassigned receive wavelengths. Since the number P of transmitters and receivers assigned to each station is small, the use of wavelength translation permits any station to access any other station, without requiring an update to the connection diagram. As we shall see, this is very important with regard to the provisioning of a packet switching service, and permits each access station to enjoy access to a number of virtual circuits far greater than its optical fan-out/fan-in P. In effect, full connectivity among the stations can be provided at the virtual circuit level. Clearly, for large N (which may range into the millions), we must insist that the number of transmitters/receivers *not be* equal to the number of stations, as would be needed to enable full connectivity at the optical channel level.

The third function to be implemented in the access station is that of store-and-forward ATM switching. As previously mentioned, each user can, if necessary, establish a virtual connection to every other user. To see how this is accomplished, we must return to the connection diagram among access stations. Suppose a direct connection exists between stations A and B, but not between stations A and C. However, as before, a direct connection does exist between B and C. We can, nonetheless, route ATM cells from A to C by first sending these cells to B. There, in electronics, the self-routing header (virtual circuit number) is read, and the cell is electronically switched to the outbound link from B to C. Since contention may arise for the use of the link from B to C, smoothing buffers are deployed in the ATM switch to permit cells to be delivered on a store-and-forward basis (such buffers must be present in any ATM switch). By generalization, we see that ATM cells can be routed from any user to any other user by means of relaying through intermediate access stations along the links of a suitably chosen path of the connection diagram. While accumulated queuing delay might seem to present an issue, upon reflection we see that we can easily control this by limiting the number of virtual connections assigned to each link of the connection diagram at call setup time such that the load presented to each link is less than the capacity of that link. The average number of ATM cells queued behind each link can therefore be maintained at a sufficiently small value to ensure some required end-to-end average latency or latency distribution. By using such store-and-forward techniques, a large, modularly growable distributed ATM switch can effectively be created over the optical medium; the electronics needed for the self-routing elements and buffers of a large centralized ATM switch are effectively decomposed and distributed among the geographically distributed access stations. Higher-level issues associated with traffic control and performance management of an ATM network, discussed in Chapter 6, are no different when the elements of a distributed ATM switch are interconnected by the channels of a shared optical medium.

The exemplary approach just described (a geographically distributed "passive" optical medium and high-functionality access stations) lets the optics do what optics does best (move vast amounts of information among widely dispersed ports), while letting electronics do what electronics does best (process protocols, route, statistical smooth via buffering). Among the benefits anticipated by this and other possible approaches to lightwave networks are:

1. Enormous aggregate capacity (easily growing to the tens or hundreds of gigabits per second if enough users demand service, although no user "sees" an access rate greater than that constrained by the electro-optic bottleneck, a rate which we shall assume to be in the range of several gigabits per second)
2. Modularity (adding new users involves merely adding more access stations and channels, and updating the logical connectivity diagram to include these new stations)
3. Permanent plant (once the optical medium, consisting of optical fibers, optical combiners, optical amplifiers, and optical switches, is deployed, it assumes the appearance of "permanence" in that new access stations can always be added and/or upgraded without requiring that the preexisting medium be retrofitted)
4. Multiple virtual network integration onto a common physical medium (the medium provides "clear channels" among stations; some can be used for an analog network, some for an ATM network, some for a voice network, some for a video network, etc.)

5. Integration of circuit/packet switching capability
6. Distributed ATM fabric (it is easier to build/deploy a large number of small ATM switch modules than it is to build/deploy a small number of large ATM switch modules, especially at port speeds greater than 1 Gbit/sec)
7. Enhanced reliability (failure of an access station affects only the user connected via that station; the failed station is then "bypassed" by changing the connection diagram, thereby preserving all other real and virtual connections)
8. Relative independence of the physical topology from the connectivity diagram (the connectivity diagram is easily modified in response to changing traffic patterns, service requests, and equipment failure/restoral, without requiring a physical change to the optical medium)

7.2. Essentials of Lightwave Technology

The significance of optical fiber as the medium for point-to-point transmission systems and/or lightwave networks is illustrated in Figure 7.3, which shows the attenuation of selected media as a function of frequency (note that the abscissa is a highly compressed log scale). Twenty-two-gauge twisted pair copper wiring is a relatively lossy medium. Even at low frequencies compatible with analog telephone signals, the loss is about 4 dB/km; at higher frequencies which might

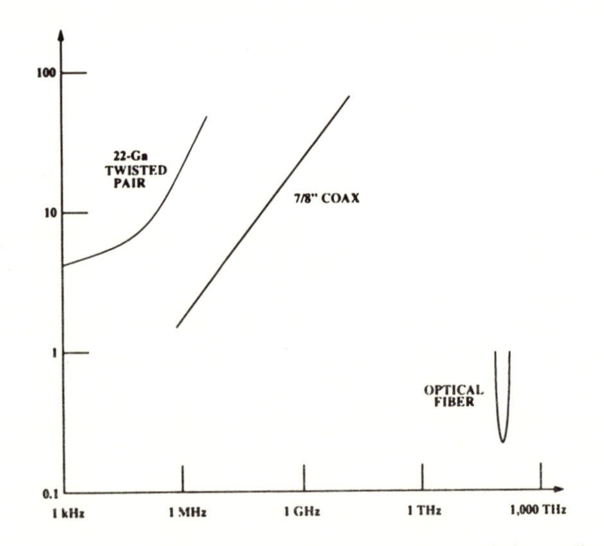

Figure 7.3. Attenuation characteristics of various transmission media.

correspond to digital signals at a data rate of several megabits per second, the loss increases to 30–40 dB/km. Furthermore, the medium is highly dispersive to such signals; those spectral components in the kilohertz range would be attenuated much less than those components in the megahertz range, causing severe signal distortion and intersymbol interference. Coaxial cable is capable of operation at higher frequencies and over a broader spectrum. Its loss is about 1 dB/km at 1 MHz, rising to about 20 dB/km at 1 GHz. By comparison, optical fiber has a minimum loss of only 0.25 dB/km, and the amount of optical spectrum over which the loss is less than 1 dB/km is about 100,000 GHz wide. Not only is the loss very low, but the loss variation over the spectral range corresponding to a digital optical signal modulated at the maximum rate permitted by the electro-optic bottleneck (a few tens of gigahertz) is negligible throughout the low-loss band. Thus, the dispersion caused by gain variation is almost totally absent.

A more substantial contribution to optical dispersion is caused by the difference in the velocity at which different wavelengths propagate longitudinally along an optical fiber, and a much more substantial contribution arises from the fact that if the diameter of the fiber is too large, then the optical signal will be carried in different transversal modes as it propagates longitudinally along the fiber, each with its own longitudinal velocity. Optical fiber operates on the principle of total internal reflection. As shown in Figure 7.4, the fiber consists of a core of glass, surrounded by glass cladding of lower refractive index. Rays of light originating within the core which strike the core–cladding interface at an angle greater than some critical angle will be totally reflected back into the core. As the ray is reflected back into the core, it strikes the opposite core–cladding bounding and is again reflected back into the core, causing the ray to be guided longitudinally along the fiber. As the ray propagates down the fiber, reflecting continuously between core–cladding boundaries on opposite sides, constructive and

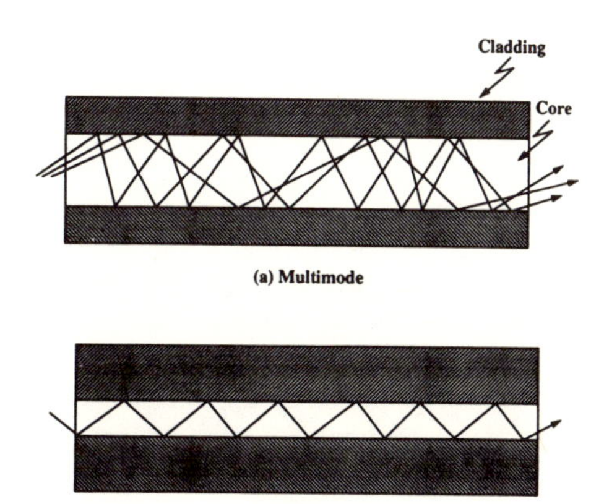

Figure 7.4. Multimode and single-mode transmission.

destructive interference will arise; those wavelengths at which the interference is constructive define the operating modes of the optical waveguide. If the core diameter is very large relative to the wavelength of the incident light, the optical waveguide will be capable of supporting many such modes, each with its own angle of incidence. The actual path length traversed by a ray associated with a mode having a smaller angle of incidence will be longer than that of a ray associated with a mode having larger angle of incidence. If the optical energy of a single pulse of light is split into several such modes, then a different delay will be encountered by each component as it propagates along the fiber. If the length of fiber is sufficiently long, then a single incident pulse of light may emerge as several distinct, nonoverlapping pulses. The effect of such multimode dispersion is to limit the data rate of the link to a value such that the bit period is large relative to the delay spread among the various propagating modes, and the speed–distance product of transmission systems using multimode fiber is limited to about 3 Gbit/sec-km. While this may be adequate for relatively low-speed links operating over short distances, long-distance links and wide-area optical networks require much higher speed–distance products. Fortunately, it is possible to fabricate fiber for which the core diameter is sufficiently narrow so that only a single propagating mode is supported (that is, the relationship between core diameter and wavelength is such that, for a particular wavelength, only one angle of incidence exists which allow constructive interference among multiple-reflected waves). Although dispersion for such single-mode fiber is not entirely eliminated since each frequency within the spectrum of a modulated signal propagates at a slightly different velocity, dispersion is *much* lower than for multimode fiber, and speed–distance products in excess of 100 Gbit/sec-km are readily achieved. Although the wavelength of minimum loss for single-mode fiber (1.5 μm) is different from the wavelength of minimum dispersion (1.3 μm) where the delay variation across the spectrum of a modulated signal is smallest, this is not a significant factor with regard to architectures and capabilities of lightwave networks.

A simplified diagram of a lightwave transmission system appears in Figure 7.5. It consists of a light source (typically a semiconductor laser for a high-quality single-mode system, or a light-emitting diode for a lower-speed, multimode system), the job of which is to convert an electronic signal into an optical signal. If the electronic signal is in binary format, then the simplest, but not the only, way in which the information can be carried is by converting each logical “one” into a pulse of light; each logical zero causes no light to be produced. The optical signal produced by the light source is then coupled into the core of the optical fiber (or, stated more correctly, a fraction of the light produced by the source is coupled into the fiber since not all of the light can be focused into the narrow core, especially if single-mode fiber is used). At the other end of the fiber (which may span a distance of tens or hundreds of kilometers), optical power not lost to attenuation and/or other loss mechanisms which might be encountered, shines on a photo-

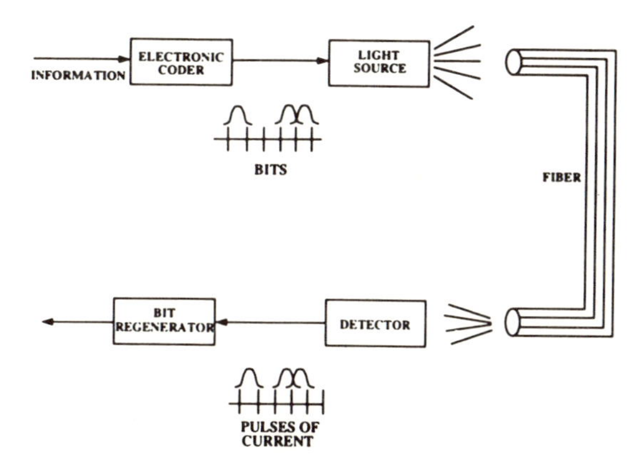

Figure 7.5. Simplified diagram of an optical transmission system.

detector, the job of which is to convert the received optical signal back to an electrical signal. Ideally, if the dispersion produced by the fiber is slight, then the electrical signal re-created by the photodetector should be a scaled replica of that generated at the transmitter. If the signal is digital, the final step is to regenerate the signal (that is, to decide in each symbol interval which digit was most likely to have been transmitted). This is a decision performed in the electronic domain; the sole function of optics in the above system is to serve as the carrier of the information-bearing signal over an extended geography. For a properly designed digital link, as determined by the linearity of the light source, the amount of optical power coupled into the fiber, the degree of signal attenuation and dispersion, and the conversion efficiency of the photodetector, the resulting bit error rate (BER) produced by the electronic regenerator should be small.

7.3. Direct Detection and the Quantum Limit*

The quantum limit tells us the absolutely minimum amount of optical power needed at the photodetector in a faithful reproduction of the transmitted signal to ensure some specified BER, and derivation of the quantum limit is instructive. We assume on–off binary optical signaling, with "ones" and "zeros" equally likely to occur. As shown in Figure 7.6, the bit period is T seconds, the signaling is rectangular, and no distortion is produced within the fiber. The conversion efficiency of the photodetector is 100% (each arriving photon in a logical "one" produces one electron). The "photo counter" is again ideal, counting the number of electrons generated, N, in each T-second period. The optical power at the receiver corresponding to a logical "one" is P_R.

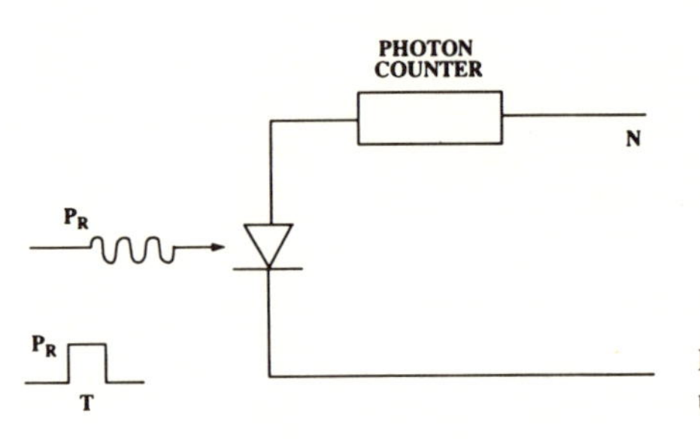

Figure 7.6. Model of the ideal optical receiver used to compute the quantum limit.

The number of photons contained in a T-second pulse of light power level P_R is a Poisson distributed random variable with average $E(N)$ equal to the energy in the pulse divided by the energy contained in each photon. Thus, if a logical zero is sent, no light is present within that T-second interval and electrons will be neither produced nor counted. If a logical "one" is sent, the number of electrons produced is Poisson-distributed. The probability distributions for N, conditional on logical "0" or "1" being present, are therefore

$$P(N = 0 | \text{"0"}) = 1 \tag{7.3.1}$$

$$P(N = k | \text{"0"}) = 0, k, = 1, 2, \ldots \tag{7.3.2}$$

$$P(N = k | \text{"1"}) = e^{-\alpha P_R T} \frac{(\alpha P_R T)^k}{k!} \tag{7.3.3}$$

$$\langle N \rangle = \alpha P_R T = P_R T / h\vartheta \tag{7.3.4}$$

where h is Planck's constant and ϑ is the frequency of the light.

Since $P(N = 0 | \text{"0"}) > P(N = 0 | \text{"1"})$, and, for $k > 0$, $P(N = k | \text{"0"}) < P(N = k | \text{"1"})$, the optimum receiver will decide:

$$\begin{aligned} &\text{"0" if } N = 0 \\ &\text{"1" if } N > 0 \end{aligned} \tag{7.3.5}$$

Now, if a logical "0" is sent, then no electrons are produced, and using the decision rule (7.3.5), the receiver will always decide "0", i.e., it will never make a mistake if a "0" is sent. Similarly, if a logical "one" is sent and one or more photons are counted, the receiver will always correctly decide that a "one" was sent. The only way that an error can occur would be for a "one" to be transmitted and no photons counted. From (7.3.3) and (7.3.4),

$$P(N = 0 \mid \text{"1"}) = e^{-\langle N \rangle} \tag{7.3.6}$$

and since "ones" are sent half the time, the BER or probability of bit error, *P(E)*, is simply

$$P(E) = \tfrac{1}{2}P(N = 0 \mid \text{"1"}) = \tfrac{1}{2}e^{-\langle N \rangle} \tag{7.3.7}$$

The quantum limit follows directly from (7.3.7). If a BER = 1×10^{-9} is sought, then (7.3.7) tells us immediately that $\langle N \rangle = 20$. From this and (7.3.4), we can compute the required received optical power level corresponding to a logical "one," P_R, for any bit time T. Since the bit time T is inversely proportional to the signaling rate (link speed) we conclude that the minimum required received power increases directly with link speed. Also, since we need, on average, $\langle N \rangle = 20$ photons when a logical "one" is sent, and "ones" are present only half the time, the required received power averaged over "ones" and "zeros" is $(1/2)P_R$. Thus, the average number of photons needed per bit, averaged over "ones" and "zeros," is $(1/2)\langle N \rangle = 10$. This is the quantum limit: on average, the receiver must be presented with at least 10 photons per bit to ensure a BER = 1×10^{-9}.

For a variety of reasons, performance of a practical optical receiver is substantially inferior to that predicted by the quantum limit: the counter is really an integrating filter, the fiber introduces dispersion, and, most importantly, the photocurrent produced is so weak as to require the presence of an electronic amplifier before any bit decision can be made. Such an amplifier introduces noise, the intensity of which is typically much greater than the variability in photon count resulting from the weak signal of a quantum-limited receiver. To enable reliable detection, sufficient power must be present in the electronic signal produced by the photodetector so that a suitably high signal-to-noise ratio appears at the output of the amplifier. Thus, greater optical power must shine on the photodetector. Depending on the quality of the amplifier (lower noise equating to higher quality) and the data rate of the signal, the actual optical power level required at the photodetector may be between 10 and 20 dB greater than that predicted by the quantum limit (in general, the degradation is greater at higher bit rates, approaching 20 dB for data rates as high as several gigabits per second).

7.4. Wavelength Division Multiplexing

As previously mentioned, the electro-optic bottleneck prevents any one optical transmitter or receiver from accessing a fiber link at a rate commensurate with the bandwidth of the medium. Wavelength division multiplexing is a technique which allows multiple independent transmitters and receivers to collectively access a common fiber link at a rate greater than that set by the electro-optic bottleneck. The basic approach is shown in Figure 7.7. Here, each of three high-speed information sources modulates one of three laser diode optical transmitters, each diode producing light at a different wavelength. As shown, the modulation bandwidth of each source is smaller than the spacing between the optical wavelengths λ_1, λ_2, and λ_3 such that the modulated spectrum of the three signals do not overlap. The three spectrally nonoverlapping optically modulated signals are combined in an optical wavelength division multiplexer, a passive structure formed entirely out of glass and free space (more will be said of this later); for now, it is important to note only that the optical multiplexer is entirely passive and operates independently of the nature of the modulating signals, i.e., neither detection nor regeneration of the signals is required, analog and digital signals could be equally well wavelength multiplexed, etc. The combined signal is transmitted along a length of optical fiber, and the identities of the signals are maintained by virtue of their being carried on different wavelengths with non-overlapping spectra. At the receiving end of the link, the combined signal is separated into its three spectral components (one at λ_1, one at λ_2, and one at λ_3) by a wavelength demultiplexer, another all-optical passive device. Each of the spectrally separated signals is now detected by its own dedicated receiver, and we note that, effectively, the link has been operated at a rate three times greater than the data rate of any single optical transmitter or receiver. In principle, the number of such independent signals which can be wavelength multiplexed onto a common fiber is quite large and the spectra of the three signals appearing in Figure 7.7 could be replaced by a comb of such spectra containing many such signals.

The optical demultiplexer of Figure 7.7 uses a diffraction grating (a specially fabricated piece of glass with accurately cut facets) to separate the arriving light into its spectral components. A diffraction grating operates very much like an optical prism, a cylinder of glass with triangular cross section. Light shining incident on one of the triangular faces of the prism (perpendicular to the axis of

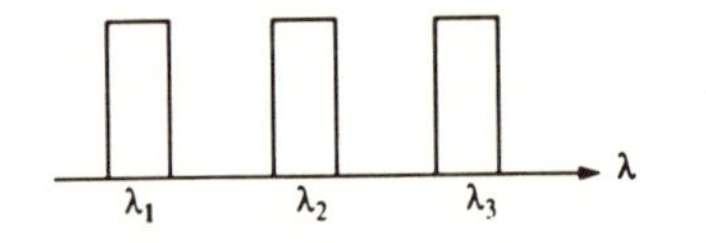

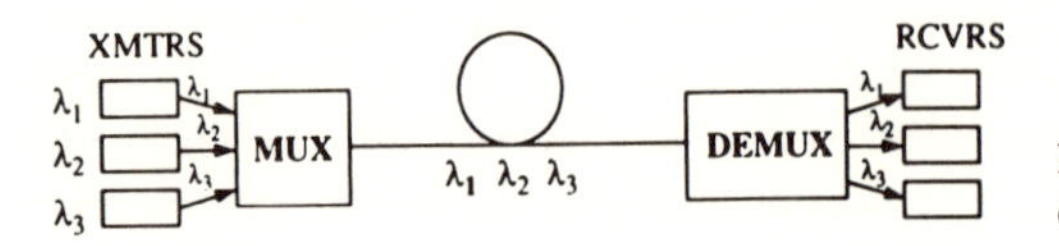

Figure 7.7. Principles of wavelength division multiplexing.

the cylinder) will be separated into its spectral components. Because of the difference in refractive index between air and glass, a ray of light which enters the glass nonperpendicularly to one of the triangular facets will be bent toward a line perpendicular to that facet at the point of incidence. Further bending occurs as the light leaves the glass after traversing the prism. The amount of bending which occurs is dependent on the wavelength of the light. (We have all observed such an effect in everyday life: sunshine or artificial white light containing many spectral components is decomposed by a prism into the colors of the rainbow.) Each color or wavelength of the light incident on the diffraction grating will exit at a different angle. Thus, each wavelength present in the incident light leaves the diffraction grating in a direction specific to that wavelength. A diffraction grating is a reciprocal device: if several wavelengths of light are incident on such a device, each arriving at a specific angle unique to that wavelength, then the output will be a single beam of light containing the superposition of all incident wavelengths. Thus, the wavelength multiplexer of Figure 7.7 is simply another diffraction grating.

The spectral resolution of a diffraction grating determines the minimum allowable separation between the various wavelengths in use such that the combined signal can be cleanly separated into its individual spectral components. In general, light of a specific wavelength leaves a diffraction grating not in one rigidly specific direction but in a cone-forming narrow band of directions. If the wavelengths present in the arriving signal are spaced too closely, then their "cones of departure" will partially overlap, and a portion of the power present in any one wavelength will shine on the receiver intended for an adjacent wavelength, thereby interfering with the detectability of that signal and causing a higher BER than would otherwise occur.

The spectral resolution of a diffraction grating is one measure of the quality of the grating and generally increases as the geometric dimensions of the grating grow larger. A high-quality grating can cleanly separate two lines in the low-loss optical window if their wavelengths are separated by about 0.5 nm (that is, the cross talk between the two signals will be less than 30 dB if the lines are separated by 0.5 nm). At wavelengths corresponding to the low-loss window, a separation of 0.5 nm equates to a bandwidth separation of about 60 GHz. Thus, a pair of 1 Gbit/sec wavelength-multiplexed optical signals, each with a modulation bandwidth of about 1 GHz, will require a frequency separation of about 60 GHz to prevent undesirable cross talk in the wavelength demultiplexer. A comb of N such signals therefore requires about $(60 \times N)$ GHz of bandwidth to convey N gigabits per second of information, and the optical spectral utilization efficiency is less than 2%. Fortunately, the low-loss optical spectrum is so wide that the spectral utilization efficiency of a wavelength-multiplexed lightwave network may not be of primary concern. Furthermore, increasing the signaling rate of each wavelength will improve spectral utilization efficiency because a greater proportion of the required 60-GHz separation between signals will be used, but this rate is ultimately limited to the electro-optic bottleneck. Finally, coherent detection offers the potential to substantially reduce the required spectral separation among the signals of a WDM system because the separation is done in the electronic domain using microwave filters, rather than in the passive optical domain.

7.5. Principles of Coherent Detection*

A coherent receiver operates much like a radio receiver, and we use similar terminology (signal frequency instead of wavelength, etc.). A block diagram of a coherent detector is shown in Figure 7.8. The incident signal is assumed to be an optical carrier at optical frequency ω_R, modulated by a single information source. For simplicity, we assume that the modulation format is either on–off signaling or phase shift keying (in the latter, the information is carried in the phase of the optical carrier which changes every T-second symbol interval). If phase shift keying is used, the incident power level is a continuous P_R watts. If on-off signaling is used, the incident power level during each T-second symbol interval is either P_R or 0. For the symbol interval shown, the phase of the incident signal is ϕ. The signal amplitude is proportional to the square root of its power. To explain the principles of coherent detection, it is necessary that the various optical signals be expressed in terms of the amplitude and phases of their respective sinusoidal fields, as shown in Figure 7.8.

Prior to the photodetector, a strong, unmodulated optical signal at optical frequency ω_L and power level $P_L \gg P_R$ is added to the incident information-bearing signal. This unmodulated optical tone is called the local oscillator. Thus, we may represent the optical signal, u, incident on the photodetector as:

$$u = \sqrt{P_L} \cos \omega_L t + \sqrt{P_R} \cos (\omega_R t + \phi) \tag{7.5.1}$$

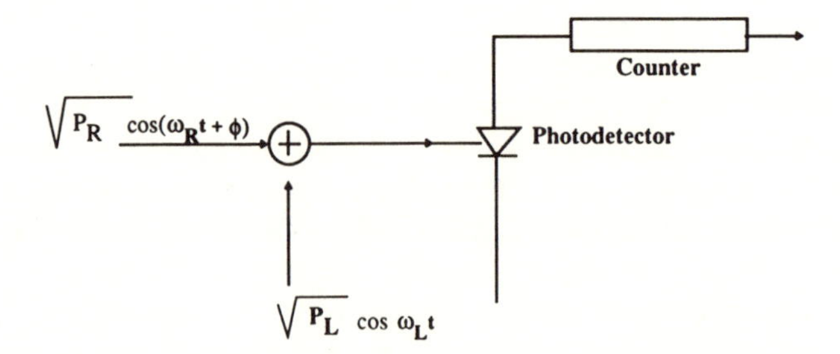

Figure 7.8. Model of the optical receiver used to study coherent detection.

As before, production of photoelectrons by the photodetector is a Poisson process with instantaneous rate proportional to the instantaneous power in u. To find the instantaneous power in u, we must express u in terms of a single sinusoid with, possibly, a time-varying amplitude. This is easily done. Recognizing that

$$\omega_R = \omega_L + \omega_R - \omega_L \tag{7.5.2}$$

we can use the trigonometric formulas for the cosine of the sum of two angles to write:

$$\cos(\omega_R t + \phi) = \cos(\omega_L t + \omega_R t - \omega_L t + \phi) \tag{7.5.3}$$

$$= \cos(\omega_R t - \omega_L t + \phi)\cos\omega_L t - \sin(\omega_R t - \omega_L t + \phi)\sin\omega_L t \tag{7.5.4}$$

Thus,

$$u = [\sqrt{P_L} + \sqrt{P_R}\cos(\omega_L t - \omega_R t + \phi)]\cos\omega_L t - [\sqrt{P_R}\sin(\omega_L - \omega_R + \phi)]\sin\omega_L t \tag{7.5.5}$$

The instantaneous power in u is therefore:

$$P_u = [\sqrt{P_L} + \sqrt{P_R}\cos(\omega_L t - \omega_R t + \phi)]^2 + [\sqrt{P_R}\sin(\omega_L - \omega_R + \phi)]^2 \tag{7.5.6}$$

$$P_u = P_L + P_R + 2\sqrt{P_L P_R}\cos(\omega_L t - \omega_R t + \phi) \tag{7.5.7}$$

We note that the instantaneous power in u consists of a constant term plus a time-varying term which varies sinusoidally at the difference frequency between the two applied optical sinusoids. We further note that the phase of the incident signal, ϕ, is preserved in (7.5.7). This difference frequency, $\omega_L - \omega_R$, is known as the intermediate frequency (IF).

The counter in Figure 7.8 counts the number of photoelectrons produced in consecutive periods of time Δ which is long compared with the period of an optical oscillation but short compared with the period of the IF oscillation, that is, $\Delta \gg 1/\omega_L$, $\Delta \gg 1/\omega_R$, $\Delta \ll 1/(\omega_L - \omega_R)$. The average number of photoelectrons produced in one Δ-second interval is given by

$$E(N) = \alpha\Delta[P_L + P_R + 2\sqrt{P_L P_R}\cos(\omega_L t - \omega_R t + \phi)] \tag{7.5.8}$$

where $\alpha = 1/h\vartheta$ [see (7.3.4)]. Since $P_L \gg P_R$, we interpret these results as follows.

The mean number of photoelectrons produced in each Δ-second interval has a constant component $\alpha\Delta(P_L + P_R)$ plus a signal-dependent component $2\alpha\Delta\, P_L\, P_R \cos(\omega_L t - \omega_R t + \phi)$. The actual number of photoelectrons produced is a Poisson-distributed random variable, but since P_L is so large, the Poisson distribution has a shape which is approximately Gaussian with mean given by $E(N)$ and variance also given by $E(N)$ (the mean of a Poisson random variable is equal to its variance). Again, since $P_L \gg P_R$, we may safely approximate the variance of the Gaussian to be a constant equal to $\alpha\Delta P_L$. Thus, the number of photoelectrons produced is approximated by a Gaussian random variable R with mean

$$E(R) = \alpha\Delta(P_L + P_R) + 2\alpha\Delta\sqrt{P_L P_R}\cos(\omega_L t - \omega_R t + \phi) \qquad (7.5.9)$$

and variance

$$\mathrm{Var}(R) = \alpha\Delta P_L \qquad (7.5.10)$$

The large constant term (7.5.10) carries no signal-bearing information, and we may safely remove it by subtracting from R a constant equal to $\alpha\Delta(P_L + P_R)$, thereby producing a new Gaussian random variable r with:

$$E(r) = 2\alpha\Delta\sqrt{P_L P_R}\cos(\omega_L t - \omega_R t + \phi) \qquad (7.5.11)$$

$$\mathrm{Var}(r) = \alpha\Delta P_L \qquad (7.5.12)$$

We now recognize that r has a signal component $s = E(r)$ and a noise component n which is Gaussian, zero-mean, and with variance $\mathrm{Var}(r)$:

$$r = S + n \qquad (7.5.13)$$

where

$$S = 2\alpha\Delta\sqrt{P_L P_R}\cos(\omega_L t - \omega_R t + \phi) \qquad (7.5.14)$$

$$\mathrm{Var}(n) = \alpha\Delta P_L \qquad (7.5.15)$$

Several possibilities now arise. Suppose $\omega_L = \omega_R$. Then, we may increase Δ to the full symbol duration T to obtain:

$$S = 2\alpha T\sqrt{P_L P_R}\cos\phi \qquad (7.5.16)$$

$$\mathrm{Var}(n) = \alpha TPL \qquad (7.5.17)$$

Suppose further that Binary Phase Shift Keying (BPSK) is the modulation format, that is, a logical "zero" corresponds to $\phi = 0$ and a logical "one" corresponds to $\phi = 180°$. Let "zeros" and "ones" be equally likely to occur. Then,

$$S = \begin{cases} 2\alpha T\sqrt{P_L P_R}, & \text{if "0" is sent} \\ -2\alpha T\sqrt{P_L P_R}, & \text{if "1" is sent} \end{cases} \tag{7.5.18}$$

We are now faced with a classical communications scenario: detect one of two possible outcomes in the presence of Gaussian noise. From (7.5.18), it is apparent that the optimum detector will inspect r and decide "zero" if $r > 0$, and "one" otherwise. Suppose "zero" is sent. Then,

$$r = 2\alpha T\sqrt{P_L P_R} + n \tag{7.5.19}$$

and the probability of error is given by

$$P(\varepsilon) = Q\left[\frac{2\alpha T\sqrt{P_L P_R}}{\sqrt{\mathrm{Var}(n)}}\right] = Q\left[\frac{2\alpha T\sqrt{P_L P_R}}{\sqrt{\alpha T P_L}}\right] = Q\left[2\sqrt{\alpha T P_R}\right] \tag{7.5.20}$$

where

$$Q(x) = \int_x^\infty \frac{1}{\sqrt{2\pi}}\, e^{-t^2/2} dt \tag{7.5.21}$$

From symmetry, $P(E)$, given by (7.5.20), is also the probability of error if a "one" is sent, and is therefore the unconditional probability of error. Furthermore, using the inequality

$$Q(x) \le e^{-x/2} \tag{7.5.22}$$

we obtain the result:

$$P(E) \le e^{-2\alpha T P_R} = e^{-2\langle N\rangle} \tag{7.5.23}$$

where $\langle N\rangle$ is the expected number of photons contained in the T-second incident symbol. Comparison of (7.5.23) with (7.3.6) shows that the homodyne coherent receiver ($\omega_L = \omega_R$) with BPSK modulation achieves a BER performance that is actually 3 dB superior to that of the quantum limit. Since a local oscillator with large power is involved, the signals appearing in the coherent receiver are stronger than those appearing in a direct detection receiver and are less affected by any noise which may be added by subsequent electronic amplification. Thus, in principle, a coherent receiver can actually achieve performance comparable to that predicted by the quantum limit, implying that the same BER can be produced by a coherent receiver as would be produced by a direct detection receiver, while requiring 10 to 20 dB less incident signal power.

If on–off signaling is used with a homodyne receiver, and $\phi = 0$ when a logical "one" is sent, then

$$S = \begin{cases} 2\alpha T\sqrt{P_R P_L}, & \text{if "1" is sent} \\ 0, & \text{if "0" is sent} \end{cases} \tag{7.5.24}$$

The optimum receiver will then decide

$$\begin{array}{c} \text{"one" if } r > \alpha T\sqrt{P_L P_R} \\ \text{"zero" otherwise} \end{array} \tag{7.5.25}$$

producing

$$P(E) = Q\left[\frac{\alpha T\sqrt{P_R P_L}}{\sqrt{\mathrm{Var}(n)}}\right] = Q\left[\frac{\alpha T\sqrt{P_R P_L}}{\sqrt{\alpha T P_L}}\right]$$

$$= Q[\sqrt{\alpha P_R T}] \leq e^{-\alpha P_R T/2} = e^{-\langle N\rangle/2} \tag{7.5.26}$$

Thus, again, performance resembling that of the quantum limit is produced, but performance is 3 dB inferior.

As a final example, we shall consider BPSK signaling, but with a heterodyne receiver (ω_L different from ω_R). In this case, we regard the consecutive "counts" produced every Δ-second interval as a time-varying function, and "mix" or multiply r with a microwave signal at frequency $\omega_M = \omega_L - \omega_R$ prior to summing the counts over the entire T-second symbol interval (this is identical to the processing that is done in a conventional digital radio receiver; the coherent optical detector has converted the problem of deciding the logical level of an arriving optical signal which contains a random number of photons into that of the much more familiar problem of deciding the logical level of a microwave signal embedded in additive Gaussian noise). Assume that $T = K\Delta$, where K is a large integer. Then, after mixing and summing, we obtain the result:

$$y = \sum_{k=1}^{K} r_k \cos(\omega_L t - \omega_R t) \tag{7.5.27}$$

$$\cong \int_0^T r(t) \cos(\omega_L t - \omega_R t)dt \tag{7.5.28}$$

Substituting (7.5.13) and (7.5.14), we obtain:

$$y \cong \int_0^T 2\alpha\sqrt{P_L P_R}\cos(\omega_L t - \omega_R t + \phi)\cos(\omega_L t - \omega_R t)dt + \int_0^T n(t)\cos(\omega_L t - \omega_R t)dt \tag{7.5.29}$$

where $E[n(t)] = 0$, Var $[n(t)] = \alpha P_L$, and $E[n(t)\ n(\tau)] = 0$, t not equal to τ. Thus,

$$y \cong \int_0^T \alpha\sqrt{P_L P_R}\,[\cos\phi + \cos(2\omega_L t - 2\omega_R t + \phi)]\,dt + N \tag{7.5.30}$$

where

$$N = \int_0^T n(t)\cos[\omega_L t - \omega_R t]\,dt \tag{7.5.31}$$

Now, assuming that $(\omega_L - \omega_R)\,T \gg 1$,

$$y = S + N \tag{7.5.32}$$

where $S = \alpha T\, P_L P_R \cos\phi$, and N is zero-mean and Gaussian with variance:

$$\begin{aligned}\text{Var}(N) &= \iint E[n(t)n(\tau)]\cos(\omega_L t - \omega_R t)\cos(\omega_L \tau - \omega_R \tau)\,dt d\tau \\ &= \int_0^T \alpha P_L \cos^2(\omega_L t - \omega_R t)dt \\ &= \tfrac{1}{2}\int_0^T \alpha P_L\,[1 + \cos(2\omega_L t - 2\omega_R t)]dt \\ &\cong \tfrac{1}{2}\alpha T P_L\end{aligned} \tag{7.5.33}$$

Since BPSK modulation is employed,

$$S = \begin{cases} \alpha T\sqrt{P_R P_L}, & \text{if "0" is sent} \\ -\alpha T\sqrt{P_R P_L}, & \text{if "1" is sent}\end{cases} \tag{7.5.34}$$

Thus, the optimum detector chooses "zero" if $R > 0$ and chooses "one" otherwise. The resulting BER is given by

$$\begin{aligned}P(\varepsilon) &= Q\left[\frac{\alpha T\sqrt{P_L P_R}}{\sqrt{\text{Var}(n)}}\right] = Q\left[\frac{\alpha T\sqrt{P_L P_R}}{\sqrt{\frac{1}{2}\alpha T P_L}}\right] \\ &= Q[\sqrt{2\alpha T P_R}] \le e^{-\alpha T P_R}\end{aligned} \tag{7.5.35}$$

Thus, we see that performance of a coherent heterodyne receiver with BPSK modulation is exactly that predicted by the quantum limit.

We see that, for all three cases considered (homodyne BPSK, homodyne on–off signaling, heterodyne BPSK), performance approximating that of the quantum limit is obtained, and that the same BER can be achieved with 10 to 20 dB less signal power than that required by a practical direct detection receiver with performance degraded from that of the quantum limit because of the noise added by the requisite electronic amplifier. This improved signal detectability may be significant for point-to-point transmission links. Also of potential interest is the ability to use complex multilevel signaling formats such as Quadrature Amplitude Modulation, familiar from digital microwave radio, to send several bits of information per signaling interval T. However, for lightwave networks, the real interest in coherent reception arises from its potential to support wavelength multiplexing with much closer channel spacing than that permitted by the spectral resolution of a diffraction grating.

7.6. A Coherent Lightwave System

Consider the systems shown in Figure 7.9. Here, each of three sources of information independently modulates a laser at a unique wavelength. These three modulated signals are superimposed in an optical combiner. This is yet another passive glass optical component which might be constructed out of optical waveguide couplers. A 3-dB coupler is a passive optical device with two inputs and two outputs, which can exactly split the power received from either input among the two outputs. If a signal appears on each input, then each output will contain the linear superposition of the two input signals, each attenuated by 3 dB in power. Arrays of such couplers can be used to superimpose N input signals onto each of N outputs; if N signals are superimposed, then each signal (as it appears on any one of the outputs) will be attenuated in power by the factor $1/N$. Such an N-input, N-output device is known as a star coupler. In Figure 7.9, only one output of the star coupler is used, and the superimposed signals appearing at this output are carried over an optical fiber link to a remote receiver where another star coupler splits the combined signals into three waveguides, each with one-third the power of the incident combined signals. Each of the three scaled replicas of the three signals is then added to a strong local oscillator prior to being applied to a photodetector and microwave signals at the intermediate or difference frequencies between the optical local oscillator and each of the three optical signals will appear at the output of each photodetector. Thus, subsequent to each photodetector, the signals appear as frequency-multiplexed microwave signals. If we now "mix" these signals with a microwave frequency corresponding to the intermediate frequency of only one of them, and sum or integrate the product for T seconds,

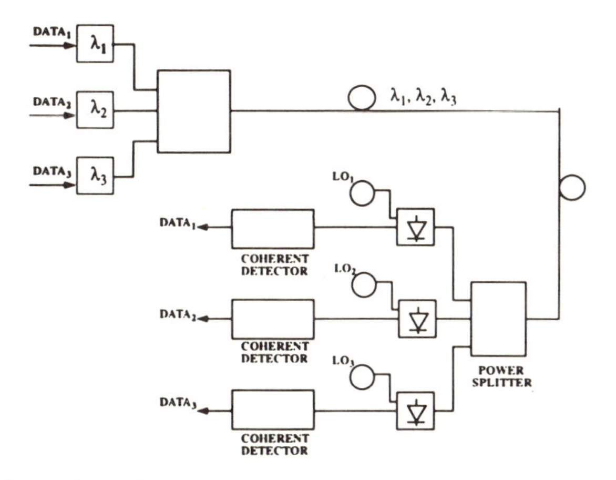

Figure 7.9. A coherent lightwave system which can also be used as a circuit switch.

then all products will integrate to zero except for that one with intermediate frequency equal to the frequency of the microwave mixing signal; we assume here that for all other signals, $(\omega_{IF} - \omega_M)\,T \gg 1$. Using such microwave techniques, the required optical spacing between the wavelength-multiplexed signals can be reduced to about six times the modulation linewidth. Thus, 1 Gb/sec signals, each requiring about 1 GHz of optical spectrum, can be placed onto optical channels separated by 6 GHz, instead of the 60-GHz separation required by a diffraction grating type of wavelength demultiplexer.

Returning again to Figure 7.9, we note that if the same microwave mixing frequency is used in all three coherent detectors, but the local oscillator frequencies of the three are different, then in each receiver only that signal whose optical frequency differs from that of the local oscillator by an amount equal to the mixing microwave frequency will appear subsequent to the low-pass filter. Each of the three coherent receivers is therefore used to detect a different one of the three optical signals; the signal detected by each is dependent on the frequency of its optical local oscillator. Furthermore, if the optical local oscillator is wavelength agile, then the signal detected by each receiver can be chosen by tuning the optical local oscillator such that the difference between its frequency and that of the desired signal is precisely equal to the frequency of the mixing microwave source. Thus, a coherent system much as shown in Figure 7.9 can be used as the basis for a switch: if N signals of different wavelengths are combined, then each receiver can choose which signal to accept by appropriately choosing the frequency of its optical local oscillator.

7.7. Optical Devices

Having covered some basic principles of optical communications, we now turn our attention to some of the devices which may be found in an optical communication system. Semiconductor lasers, the optical sources needed for high-quality transmission systems or lightwave networks, are essentially resonant cavity-type devices such as shown in Figure 7.10a. Current passed through a semiconductor junction causes a population inversion in which the electron population in the conduction band exceeds that in the valence band. Under these conditions, the energy contained in a photon passing through the junction is more likely to stimulate an election to drop from the conduction band to the valence band, thereby producing an additional photon, than it is to cause a valence electron to jump to the conduction band, which would cause the absorption of the initiating photon. Thus, as light passes through such a population-inverted junction, it grows in intensity. The walls of the laser cavity are mirrored to cause most of the amplified light to be reflected back into the cavity where it undergoes further amplification. If the round-trip gain through the cavity exceeds the losses through the ends of the cavity, optical frequency oscillations will build up. The frequencies at which the oscillations develop are determined by the dimensions of the cavity: as the growing fields are continually reflected from the ends of the laser, oscillations will develop at those frequencies where the superposition of the multiple reflected signals add constructively. In general, oscillations will occur at each frequency where the length of the cavity is an integral multiple of a half-wavelength, provided that the round-trip gain of the laser is positive at that frequency (stimulated emission will occur only over a band of wavelengths roughly 50 nm

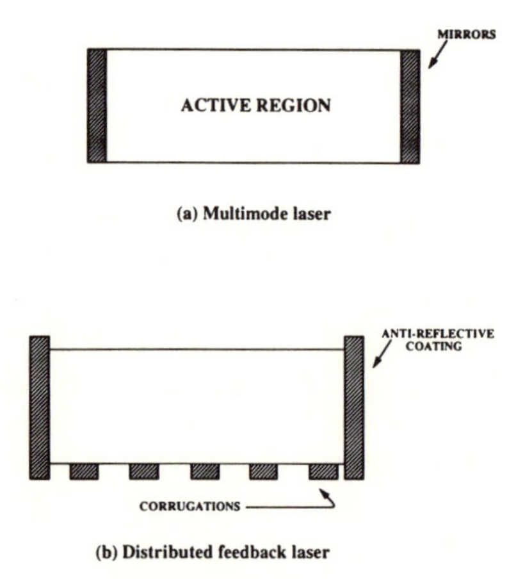

Figure 7.10. Resonant cavity of a semiconductor laser.

wide centered around a wavelength corresponding to the energy band-gap between the conduction and valence bands). The length of the cavity typically will support several oscillating modes, or frequencies, and the resulting multimode laser produces power at many discrete wavelengths within its 50-nm band.

The distributed feedback laser (DFB) shown in Figure 7.10b replaces the cavity edge mirrors by a series of longitudinal corrugations. A portion of the light passing longitudinally through the laser, in both directions, will be reflected by each corrugation. By properly choosing the spacing of the corrugations, constructive interference among the multiple reflections will occur at only one wavelength, and light at a single optical wavelength will be produced.

The amount of optical power developed by both the multimode and DFB laser is about 10 mW, and of this, about 1 mW can be successfully captured by the narrow core of a single-mode fiber. Before the advent of the optical amplifier, only a very limited portion of this 1 mW of optical power successfully coupled into the fiber was available at the receiver for the purpose of signal detection, and optical link margins were correspondingly limited. For example, assuming a quantum-limited receiver (10 photons/bit) and a data rate of 1 Gbit/sec, we would need a received optical power level of

$$\begin{aligned} P_R &= 10 \text{ photons/bit} \times (h\vartheta) \text{ watt-sec/photon} \times 10^9 \text{ bit/sec} \\ &= 10^{10}\, h\vartheta \text{ watts} \end{aligned} \qquad (7.7.1)$$

The energy per photon at a wavelength of 1.3 μm is about 1.3×10^{-19} joule. Thus, the required received power is 1.3 nW. If the laser transmitter produces 1 mW of power, then about 58 dB of loss can be tolerated between the transmitter and the receiver. A practical receiver with performance 20 dB inferior to that of the ideal quantum-limited receiver could therefore tolerate a link loss of only 38 dB.

There are several techniques whereby the wavelength of a semiconductor laser may be made tunable. These involve thermal tuning, injection current tuning, mechanical tuning, and acousto-optic tuning. By heating or cooling a laser, wavelength tuning of about 0.1 nm/°C can be achieved, but the tuning range is small (about 2 nm or 200 GHz total) and the tuning speed is slow (thermal time constants corresponding to several milliseconds). Use of thermal tuning is typically limited to wavelength stabilization in conjunction with a wavelength-sensing feedback loop. Injection current tuning can achieve very rapid tuning (time constants measured in the several nanosecond range) over a maximum range of 10 to 15 nm (about 1 THz bandwidth). As the injection current changes, so does the refractive index of the laser, corresponding to a shift in the optical length of the cavity. Practical difficulties involve long-term thermal drift of the wavelength as the junction temperature changes in response to changes in the injection current. Mechanical tuning may be used in conjunction with external cavity lasers wherein the laser cavity (the region between the end reflective mirrors) consists of an active region containing the semiconductor laser junction plus an external, free-space region, the length of which can be electromechanically adjusted, thereby changing the optical frequency of oscillation. Such a mechanically tunable laser has a wide tuning range, essentially extending over the entire 50-nm active band of the junction. This corresponds to a tuning range of about 5 THz, but the tuning time constants are long (milliseconds). Acousto-optic tuning also involves an external cavity with broad tuning range (50 nm), but the time constants are significantly shorter (several microseconds). These lasers are not continuously tunable, however, and permit simultaneous oscillations in several modes of the external cavity.

A semiconductor optical amplifier is essentially a laser diode (Figure 7.11) in which the end mirrors have been carefully removed by the deposition of antireflective coatings. Light which enters at one end will be amplified as it propagates through the active region, and will exit from the other end with a larger intensity. The end-to-end gain of such a device may be quite high (greater than 30 dB), implying the need for strong antireflective coatings; if as little as 0.1% of the power leaving the cavity is reflected back into the cavity, the high gain can cause oscillations to occur. Also, as with all types of amplifying devices (optical, electronic, acoustical, etc.), gain saturation will occur, that is, the device exhibits a linear, small signal gain only if the applied signal is small; as either the strength of the applied signal or the small signal gain of the device increases, the output level will eventually saturate at some maximum value. Since a semiconductor laser amplifier is essentially a laser diode with antireflective coatings, the maximum optical power to be drawn from such a device is limited to about 10 mW. Of greater concern than saturated power level is intermodulation distortion. The amplifier is essentially a nonlinear device: its gain is a diminishing function of applied power level. If two or more wavelength-multiplexed signals are present in the input, this nonlinear gain compression will cause some of the power in each wavelength to be converted into power at other wavelengths, interfering with the signals present at those wavelengths. To minimize such nonlinear intermodulation distortion, the amplifier should be operated in its small-signal regime.

Another type of optical amplifier is the Erbium-Doped Fiber Amplifier. This

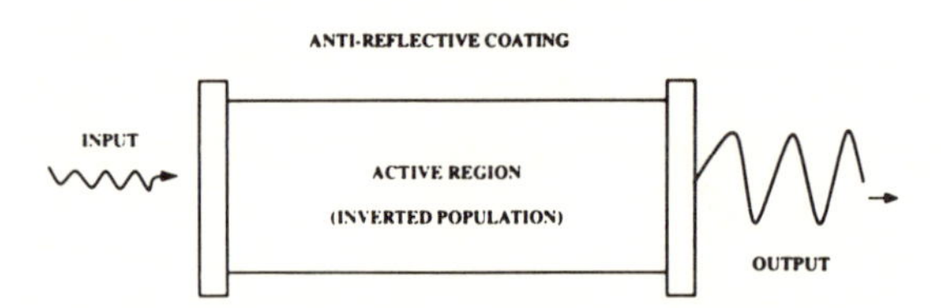

Figure 7.11. Diagram of a semiconductor optical amplifier.

consists of a special type of optical fiber having the property that light of sufficient intensity present in the 0.9-μm wavelength range will cause a population inversion to occur with a band gap corresponding roughly to 1.5 μm. Thus, light traveling down the fiber at 0.9 μm serves as the "optical pump," providing an active medium to amplify signals at 1.5 μm propagating down the same fiber. The saturated power obtainable from such an amplifier can be higher than that developed by a semiconductor amplifier because a different coupling mechanism involving continuous interaction along the length of the erbium-doped fiber section is involved. Also, the time constants associated with gain saturation are long (milliseconds), implying that rapid, small-signal intensity variations will not cause significant intermodulation distortion. The so-called fiber-to-fiber gain of such an amplifier is about 20 dB.

The advent of the optical amplifier is a very important milestone in the development of lightwave transmission systems and networks since the loss of signal arising from attenuation, coupling, splitting, etc. can now be partially offset. By using amplifiers, the rigid link loss margin corresponding to the spread between the maximum power produced by a laser and the minimum needed for detectability by a receiver can be considerably relaxed. However, it should *not* be concluded that a link margin of A dB can be increased to $(A + G)$ dB by the inclusion of G dB of optical gain. Like all types of amplifiers, the optical amplifier introduces noise of its own (in the case of optical amplifiers, the noise is caused by spontaneous emission as electrons spontaneously drop from the conduction band to the valence band, emitting a photon in the process). If an optical amplifier is followed by an optical filter (to be described subsequently), then the power of the spontaneous noise appearing at the filter's output is given by:

$$P_{\mathrm{N}} = 2h\vartheta(G - 1)B \tag{7.7.2}$$

where G is the power gain of the amplifier and B is the bandwidth of the optical filter. If several amplifiers are cascaded in a chain along a link, then a subsequent amplifier, in addition to introducing noise of its own, also amplifies the noise of prior amplifiers in the chain. Since signal detectability is ultimately limited by noise which corrupts the detection process, introduction of an optical amplifier of gain G dB increases the link margin by an amount smaller than G dB; the actual link margin improvement is dependent on the number of amplifiers in a chain, the signal power level appearing at the input of each amplifier in the chain, the attenuation between consecutive amplifiers in the chain, and other such factors.

Another optical device important for multiwavelength lightwave networks is the optical filter, a wavelength-sensitive device capable of selecting one signal out of several that have been wavelength multiplexed together. These filters typically involve cavity-like resonators (much like microwave filters), the lengths of which are adjusted such that, after entering the cavity from one side and after multiple reflections from the cavity edges, only light of a particular narrow band of wavelengths leaves from the other side. Essentially, the multiple reflections in this narrow band cause constructive reinforcement in the direction of propagation through the cavity, and the multiple reflections at all other wavelengths effectively cause the cavity to reflect all such other wavelengths back in a direction opposite to that of incidence. Several types of tunable optical filters exist, and are characterized on the basis of their tuning speed, tuning range, and passband width. The ideal filter, of course, would tune rapidly over a broad range, and would be highly selective in choosing which signal to pass (narrow passband). The spectral width of the passband defines the resolution of the filter. Waveguide Mach-Zehnder filters have a resolution of about 0.04 nm (about 5 GHz), but are limited in tuning range to about 5 nm (about 1 THz) and having tuning time constants measured in milliseconds. An electro-optic tunable filter has a resolution of about 0.6 nm (100 GHz), a tuning range of 16 nm (3 THz), and tunes very rapidly (time constant of several nanoseconds). The acousto-optic tunable filter has the poorest resolution (1 nm, or about 200 GHz), but has a *very* broad tuning range (400 nm, the entire low-loss optical band) and a reasonably fast tuning time constant (about 10 μsec). Thus, the acousto-optic tunable filter may be a particularly useful filter for use in a noncoherent dense wavelength-multiplexed lightwave network.

Another device of potential significance for optical networks is the waveguide switch. This is a four-part device (two input, two output) shown in Figure 7.12. It consists of a lithium niobate substrate onto which titanium has been diffused to create two waveguides. Lithium niobate is an electro-optically active material: its refractive index can be altered in response to an applied electrical voltage impressed across the two electrodes shown. Underneath the electrodes, the two titanium-diffused waveguides are brought into close proximity, and significant coupling between signals applied at the input of the two waveguides occurs. The amount of the coupling is dependent on the effective length of the interaction region, and for certain lengths, all of the power applied to one waveguide will be coupled into the other, and vice versa. At other effective lengths, the power applied to either waveguide will remain entirely within that waveguide. At yet other effective lengths, a portion of the incident power applied to the first waveguide will be coupled into the second, with the balance of the applied power remaining in the first. Under these conditions, the same fraction of the power applied to the second waveguide will be coupled into the first, with the balance remaining in the second. By changing the voltage applied across the electrodes, the refractive index can be made to change, thereby changing the propagation velocity of the optical signals and causing a shift in the effective length of the interactive region. Thus, the apparatus shown in Figure 7.12 is essentially that of an electrically controlled optical switch or power splitter. Application of a specific voltage can cause a fraction α of the power applied to waveguide A to be coupled into waveguide B, with the balance of $1-\alpha$ remaining in waveguide A. Under these

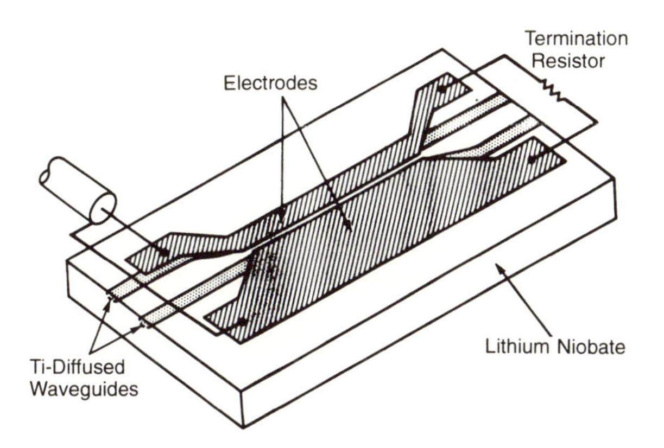

Figure 7.12. An electronically controlled waveguide switch.

conditions, the same fraction of the power applied to waveguide B is coupled into waveguide A, with the balance remaining in waveguide B. Thus:

$$P(\text{out, A}) = (1 - \alpha)\, P(\text{in, A}) + \alpha P(\text{in, B}) \tag{7.7.3}$$

$$P(\text{out, B}) = \alpha P(\text{in, A}) + (1 - \alpha)\, P(\text{in, B}) \tag{7.7.4}$$

where α can be varied between 0 and 1 by changing the applied voltage. Such a device can therefore be used as a 2 × 2 switch, a variable power divider/combiner, and even as an external modulator if only one input is active. The change of state of such a switch can be effected in about 1 nsec.

An Acousto-Optic Tunable Filter (AOTF) is another four-part device (two input, two output) in which the coupling between the two inputs is responsive to the intensity and frequency of an acoustic wave impressed across its interaction region. A very important distinction between an AOTF and a lithium niobate switch is that the coupling is wavelength-sensitive. To effect a certain coupling between applied signals at wavelength λ_A, an acoustic signal of a particular frequency (in the range of about 100 MHz) must be applied. To *simultaneously effect a different degree of coupling* between two signals each at a different wavelength λ_B, a second acoustic signal at an appropriate frequency can also be applied. Thus, the AOTF behaves like several independent lithium niobate devices all running in parallel, each responsive to a different wavelength of the incident light. If signals of wavelength $\lambda_A, \lambda_B, \ldots, \lambda_D$ are simultaneously applied, then the coupling coefficients $\alpha_A, \alpha_B, \ldots, \alpha_D$ corresponding to each can be independently chosen. The resolution of the AOTF is about 1.5 nm (signals separated in wavelength by more than 1.5 nm or about 300 GHz can be operated on independently), and the time to effect a change of state is on the order of several tens of microseconds.

7.8. Broadcast-and-Select Lightwave Packet Networks

Using some of the optical technology described in the previous section, an optical network along the lines of that shown in Figure 7.13 is readily envisioned. This network uses an optical star coupler such that the power of each of N inputs signals is equally split among the N outputs, and all of the power-divided input signals are superimposed on each output. Each access station contains a laser diode which transmits at a wavelength unique to that station, and an optical receiver which can be rapidly tuned to select one of the signals superimposed by the star coupler. All transmitters and receivers operate at the same data rate, which may or may not be the fastest allowed by the electro-optic bottleneck.

As drawn, this scheme requires that each station must be preinformed of each arriving packet so that it can tune its receiver to accept the wavelength of the corresponding sender. In general, a given station must be able to accept packets time-sequentially produced by different sources and, to accept a packet from station B immediately subsequent to receiving a packet from station A, that station's receiver must rapidly retune from A's wavelength to B's wavelength.

Several issues must now be addressed. First, what is the required tuning speed? Since a station cannot accept a packet during the transient when its receiver is being tuned to a different wavelength, then, to maintain high throughput to each receiver, the tuning transient must be small relative to the packet size. Thus, if very short packets are involved, the tuning speed must be very fast. For example, if

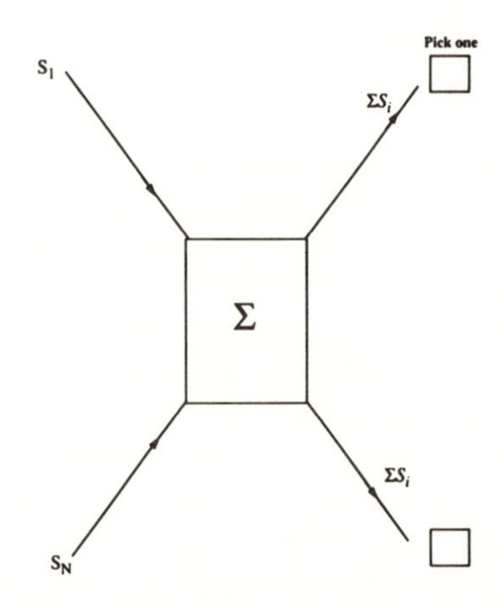

Figure 7.13. A broadcast-and-select lightwave network.

53-byte ATM cells are involved and the data rate of each optical channel is 1 Gbit/sec, then the cell size is 424 nsec, and the switching transient should be no greater than a few nanoseconds. Tunable lasers using injection-current tuning can tune this fast, but can be used to implement a wavelength-selective filter only in conjection with coherent detection; here, the rapidly tunable laser would serve as the local oscillator, allowing one signal to be selected out of the sea of signals superimposed by the star coupler. Furthermore, the tuning range is limited to about 10 to 15 nm, and if the channels are spaced as closely as every 10 GHz, then the number of access stations would be limited to somewhere in the vicinity of 200. Finally, while implementation of a coherent receiver with a resolution of 10 GHz is certainly feasible, stabilizing the wavelengths of such a large number of geographically dispersed transmitting lasers to permit such dense spacing could be very difficult. If the transmitters are spaced on 10-GHz centers and the modulation bandwidth is a few gigahertz, then the permissible wavelength drift for each laser should be not greater than about 1 GHz of spectrum. Near 1.3 μm, this implies a drift of no more than about 0.01 nm, which translates into a required long-term stability of 0.01 nm/1.3 μm, or about 1 part in 10^5.

To permit use of tunable filters (thereby avoiding coherent detection but substantially increasing the spacing among the transmitting wavelengths), slower tuning speed must be accepted, implying that packets of lengths considerably longer than ATM cells must be used. Suppose AOTFs are used to implement wavelength-selective receivers. The tuning transient is on the order of several tens of microseconds, implying the need to ship packets of average duration equal to several hundred microseconds. Suppose an average packet size of around 100 μsec is chosen and the data rate remains 1 Gbit/sec. Then, the average-size packet will contain around 10,000 bits (about 1 kbyte). Furthermore, the resolution of an AOTF is about 1.5 nm and the tuning range is 400 nm. Thus, if we were to use the entire optical band, about 250 stations could, in principle, be supported in an application environment favoring long packets, such as long file transfer among host computers. Such long packets would not, however, favor multimedia communication which involves shipment of short packets, long packets, and continuous bit rate traffic having various real-time delay requirements.

A second issue involves signaling. How much information is required to tell each receiver which channel to tune to? How will signaling be implemented? If ATM cells are sent at a data rate of 1 Gbit/sec (approximate length = 0.5 μsec per earlier computation) and N = 1000 stations, then 10 bits of signaling information must be sent to each receiver to advise it as to which channel it should tune to, and the required signaling data rate is therefore 10 bits to each of 1000 stations every 0.5 μsec:

$$\begin{aligned} R_{\text{signaling}} &= (10 \text{ bits/station} \times 1000 \text{ stations})/(0.5\ \mu\text{sec}) \\ &= 20 \text{ Gbit/sec!} \end{aligned} \tag{7.8.1}$$

Worse yet, this required signaling rate scales as $N\log N$, since $\log N$ bits are required to select one of N transmitters. Thus, an $N = 100{,}000$ user network would require a signaling rate of $R_{\text{signaling}} = 3400$ Gbit/sec! Again, we see that to maintain a reasonable signaling rate, such a network would favor a modest number of access stations along with long packets (receivers must be updated only at the packet rate). Thus, an $N = 250$ station example with access rate of 1 Gbit/sec and average packet size of 1 *K*byte (reasonable from the perspective of timing speed and range as discussed earlier) requires a signaling rate of only

$$\begin{aligned} R_{\text{signaling}} &= (8 \text{ bits/station} \times 250 \text{ stations})/(100\ \mu\text{sec}) \\ &= 20 \text{ Mbit/sec} \end{aligned} \tag{7.8.2}$$

A third issue involves the need for a scheduler or controller. Suppose two stations have packets intended for the same receiver. Since the receiver can only accept one packet at a time, which access station should transmit? Furthermore, in each time slot, we can choose transmit/receive pairs in such a way that the maximum amount of information is transferred? The situation is exactly analogous to one which we have previously encountered: modified input queuing with sorting at the input, followed by a nonblocking router (see Section 3.9 and, in particular, Figure 3.9). Let each of the packets arriving at the input to each transmitting access station be sorted by destination as in Figure 3.9. Then, each time slot, a centralized controller schedules the queued cells in such a way that each access station having any cells to transmit is asked to transmit at most one cell, the chosen cells are each intended for a distinct receiver, and the maximum possible number of cells (always, of course, less than or equal to N) are transferred every time slot: maximum length diagonals of cell matrix $\mathbf{C} = \{C_{i,j}\}$ [see (3.9.8)] must be chosen by the controller each time slot.

With such a scheduler, the distributed lightwave network of Figure 7.13 can be used to replace the nonblocking routes of Figure 3.9 since within any time slot, any unused output can accept a cell from any input, independent of the pattern of input-to-output connections already established for that time slot; that is, there are no blocked input–output combinations. The need to signal queued-packet information from each of several geographically distributed access stations to the central scheduler implies a geographically small network. The need for geographical compactness is further amplified by the need to maintain tight synchronization. Each ATM cell must arrive at the star coupler *precisely* at the right time to fill the time slot allotted to that cell by the scheduler since all cells assigned to the same time slot (but appearing, of course, on different wavelengths) were chosen to originate at different sources and are intended for different receivers. The need for synchronization further implies that each station must know its propagation delay from the center of the network (the star coupler) since, after receiving its time sequence of assignments, each transmitter (receiver) must locally know when to transmit (receive) in order to place a packet onto (receive a packet from) a specific time slot as defined at the star coupler.

For example, suppose the network controller is scheduling cells for the *k*th time slot, where time slots are defined at the output of the star coupler (it is here that time slot boundaries must be time aligned since, from here, the superimposed signals traveling to a common receiver incur the same propagation delay, and receiver is hopping among channels synchronously with time slot boundaries in accordance with instructions received from the controller.) Suppose station No. 1 is located 10 miles away from the star coupler. Then, to strike the *k*th time slot with its assigned cell, station No. 1 must transmit that cell about 105 μsec earlier (the speed of light in fiber is about ½ foot/nsec). Furthermore, to receive a cell from time slot *k*, station 1 must tune its receiver to the assigned channel 105 μsec later than the point in time when time slot *k* occurs at the star. Finally, considering the propagation delays involved, it is apparent that, each time slot, the network controller computes an assignment or schedule to be effected some *fixed number of time slots in the future.* The delay from computation time to scheduled time is needed to allow the scheduling information to be conveyed to the transmitters and receivers, and to allow the transmitters to begin to transmit their assigned cells sufficiently early that, after their respective propagation delays, the cells arrive at the star in their assigned time slots.

In a variant of the broadcast-and-select architecture of Figure 7.13, we can reverse the roles of the access station transmitters and receivers. Rather than permanently assigning a unique wavelength to each transmitter and sequentially instructing each receiver to tune to the appropriate wavelength, we can permanently assign to each receiver a unique wavelength, and, for each packet to be transferred, ask each transmitter to tune its wavelength to that of the intended receiver. This approach always requires tunable lasers at the transmitter. Fast tuning speed can be achieved if injection current tuning is used, but to avoid the aforementioned stabilization problem the lines should be spaced by about 100 GHz, implying only about 20 lines possible within the 15-nm tuning range. Since each receiver permanently "listens" to only one wavelength, signaling, control, and synchronization are not needed; to send a packet to a given receiver, a transmitter simply tunes its laser diode to that receiver's wavelength. Then, provided that no other transmitter is attempting to access that receiver at the same time, that transmission will be successful. However, if two or more transmitters attempt to access the same receiver at the same time, then a collision will occur on that receiver's wavelength, and the information content of all packets involved in the collision will be destroyed. In operation, such a system will behave much like *N* random-access Aloha systems running in parallel, and the maximum utilization efficiency of such a system is 18% (that is, if at each source, each packet is equally likely to be bound for any receiver, and if all links operate at data rate *R*, then each transmitter and each receiver will enjoy a maximum sustainable data rate of 18% × R).

To see the analogy to Aloha, consider a time-slotted system (in general, such a system would seek to avoid rigidly defined time slots such that tight time synchronization is not required). Suppose that the system is overloaded, that is, each transmitter always has N supplies of fixed-length packets, one supply intended for each of the N receivers. Each time slot, each transmitter randomly selects a receiver and transmits a packet to that receiver. Now, consider a particular receiver. The probability that the receiver receives a noncolliding packet in a given time slot, P_s, is equal to the probability that precisely 1 out of N transmitters has selected that receiver in that time slot. Since the number of transmitters selecting that particular receiver is a binomially distributed random variable,

$$P_s = Np(1 - p)^{N-1} \tag{7.8.3}$$

where p is the probability that a given transmitter selects that particular receiver. Since each transmitter selects at random any of N receivers,

$$P_s = N(1/N)(1 - 1/N)^{N-1} = (1 - 1/N)^{N-1} \tag{7.8.4}$$

and, for large N,

$$\lim_{N\to\infty} P_s = 1/e \cong 36\% \tag{7.8.5}$$

For unslotted access, a simple extension of the analysis for unslotted Aloha would yield a maximum throughput of 18% per receiver.

In reality, the maximum throughput for such a select-and-transmit scheme might be considerably less than 18%. Unlike Aloha, wherein each transmitter can listen to the single shared channel and determine whether or not a collision has occurred, in the multichannel system, the only way that a transmitter can determine that its packet was successfully received is for the receiver to generate an acknowledgment back to the transmitter for every successfully received packet. If the transmitter does not receive an acknowledgment within some predetermined time of sending a packet, it will assume that a collision occurred and will retransmit that packet. The need to send acknowledgments detracts from the overall efficiency of information transfer since the channels must carry acknowledgments as well as user traffic. To keep the acknowledgment overhead low, it would be desirable for the length of the acknowledgment to be short relative to the length of the typical packet whose receipt is being acknowledged. For example, if the length of the acknowledgment is equal to that of the typical packet, then half of the traffic on the network is acknowledgment traffic, and the maximum utilization efficiency would be limited to 9% (this is an upper bound; the acknowledgement of a successfully transmitted packet might itself suffer a collision, thereby causing a second transmission of that already-sent packet). Again, we see that long packet lengths are to be preferred.

To summarize, constraints imposed by the need for (1) rapid agility, (2) wavelength stability, (3) signaling and control, (4) accurate measure of time delay from the access stations to the core, and/or (5) acknowledgments in a random access environment, imply that simple broadcast-and-select packet lightwave networks (or their dual, select-and-transmit networks) are best suited for applications involving a relatively small number of access stations (under, say, 200) which are geographically separated by relatively short distances (local or metropolitan area service regions, not wide area networks), and which can conveniently use long packet lengths (1 Kbyte or longer, such as needed to transfer long data or image files).

7.9. Multihop Lightwave Packet Networks

As we have just concluded, the conceptually simple broadcast-and-select architecture for lightwave packet networks is limited to applications involving a small number of long-file data users dispersed over a relatively small geography. These limitations were caused by the need for rapidly agile optical elements (laser diodes and/or optical filters), signaling rates that grow rapidly with the number of access stations, elaborate real-time scheduling algorithms, and the need for tight time and wavelength synchronization. An alternative approach is needed to enable lightwave packet networks which (1) can be dramatically scaled up in both permissible number of access stations and range of geographical coverage and (2) can support multimedia, bandwidth-on-demand traffic which is best formatted into small, fixed-length cells. The multihop architecture for lightwave packet networks represents one such approach.

The basic approach behind the multihop architecture is illustrated in Figure 7.14, which shows an eight-node multihop network. Here, the nodes or access stations are positioned linearly along a bus although, as will become apparent, the physical architecture can be that of a bus, rooted tree, physical star, or any combinations thereof (see, for example, Figure 7.2).

As shown in Figure 7.14, each access station is equipped with two optical transmitters and two optical receivers; in general, each of N users is equipped with p transmitters and p receivers. Each transmitter operates at a wavelength unique to that transmitter (for example, station No. 1's two transmitters operate on wavelengths λ_1 and λ_2, respectively; no other transmitter operates on λ_1 or λ_2). The signals produced by the various transmitters are combined on the shared medium

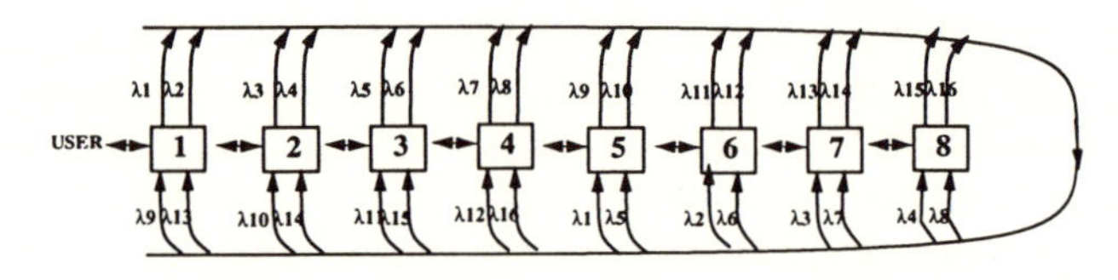

Figure 7.14. An exemplary 8-node multihop network.

by means of wavelength multiplexing. The signals produced by each access station are coupled onto the "transmit" optical bus using passive waveguide directional couplers, and propagate along the upper bus toward the right (head end). There, the combined signals are looped back and propagate to the left on the lower "broadcast" bus. Again using waveguide couplers, a portion of each signal is coupled to each station. Using optical filters, each access station then accepts the two wavelengths intended for local reception. Each receiver accepts a wavelength unique to that receiver (for example, station No. 1's two receivers accept wavelengths λ_9 and λ_{13}, respectively; no other receiver accepts λ_9 or λ_{13}). Each access station is equipped with a bidirectional user port. Except for the optical transmitters and receivers, all other functions of the access stations are implemented in electronics.

Operation is as follows. Suppose User No. 1 generates an ATM cell which is destined for User No. 5. That cell arrives at access station No. 1 through User No. 1's port and, in electronics, the access station reads the cell header to determine its destination. Knowing that station No. 5 has a receiver which accepts wavelength λ_1, station No. 1 will route that cell to its transmitter operating on λ_1. The receiver at station No. 5 which accepts λ_1 simply converts the optical signal arriving on λ_1 into an electronic signal. Now, when the cell from station No. 1 arrives at station No. 5, the cell header is read and, since that cell is now at its destination, access station No. 5 will route that cell across the interface to User No. 5.

Next, suppose that User No. 1 generates a cell destined for User No. 2. Now we have a problem. Since station No. 1 can only transmit on λ_1 and λ_2, and station No. 2 can only accept λ_{10} and λ_{14}, it is not possible to directly transfer the cell from station No. 1 to station No. 2. Instead, station No. 1 will route the cell to its transmitter which operates at λ_1. Since only station No. 5 accepts λ_1, that cell will be transferred to station No. 5. Here, upon reading the cell header, station No. 5 recognizes that the arriving cell is intended for User No. 2, and routes that cell to its transmitter which operates on λ_{10}. Since station No. 2 has a receiver which accepts λ_{10}, the cell will arrive at its intended destination after a second pass over the optical medium (hence, the name "multihop"). As will subsequently be explained, the assignment of wavelengths to transmitters and receivers is done in such a way that a cell originating at any access station can be delivered to any other access station after a suitable number of "hops."

The linear bus topology of Figure 7.14 was chosen for illustrative convenience; as mentioned above, other physical topologies can be used. In fact, for large *N*, a different physical topology might be preferred since the cumulative optical insertion loss incurred as the signals propagate past the directional couplers on the transmit and broadcast buses can become excessive. As will become apparent, it is not even necessary that all transmit signals be combined and broadcasted to each receiver; all that is really needed is a physical architecture that permits each receiver to access the wavelength to which it is assigned.

Since each transmitter is assigned to a unique wavelength and each receiver accepts from a unique wavelength, then assignment of a wavelength to the transmitter of one access station and the receiver of another effectively creates a dedicated one-way channel existing at that wavelength between the two access stations. In Figure 7.14, the wavelengths were assigned to the access stations in such a way as to provide the dedicated one-way connections shown in Figure 7.15. A diagram such as that appearing in Figure 7.15 is known as a connection graph. For multihop lightwave networks, the connection graph bears no relationship to the physical topology of the network. We see that, from a connectivity perspective, the eight stations of Figure 7.14 are arranged in two columns of four stations each. The four stations in the first column are connected to the four stations in the second column via a perfect shuffle diagram using eight wavelengths; the four stations in the second column are reconnected back to those of the first column via a perfect shuffle diagram using a second set of eight different wavelengths. Such a connection diagram is known as a recirculating shuffle.

The connection diagram of Figure 7.15 contains two columns with four stations each. In general, recirculating shuffle diagrams can be constructed which

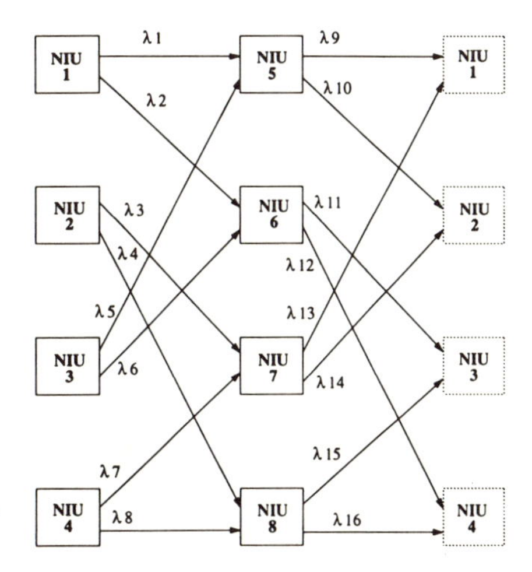

Figure 7.15. Recirculating perfect shuffle connection graph for an 8-node multihop network.

contain k columns with 2^k stations each. Thus, the number of access stations which can be interconnected by a recirculating shuffle diagram is given by

$$N = k2^k, \qquad k = 2, 3, \ldots \tag{7.9.1}$$

where k is the number of columns in the diagram. If each access station is equipped with p transmitters and p receivers, then recirculating shuffle diagrams with k columns can be defined wherein the number of access stations per column is equal to p^k, and (7.9.1) generalizes to

$$N = kp^k, \qquad k = 2, 3, \ldots \quad p = 2, 3, \ldots \tag{7.9.2}$$

Examples of $k = 3$, $p=2$ ($N = 24$) and $k = 2$, $p = 3$ ($N = 18$) recirculating shuffle diagrams appear in Figures 7.16 and 7.17, respectively. For arbitrary k and p, the number of wavelengths needed, W, is given by

$$W = kp^{k+1}, \qquad k = 2, 3, \ldots \qquad \text{and } p = 2, 3, \ldots \tag{7.9.3}$$

since each of $N = kp^k$ access stations needs p transmitters and p receivers.

Techniques for implementing multihop networks with a number of access stations *not* conforming with the values permitted by (7.9.2) will be discussed in Section 7.13.

For fixed p, the number of wavelength-multiplexed channels grows linearly with N and, for large N, it may become difficult to multiplex such a large number

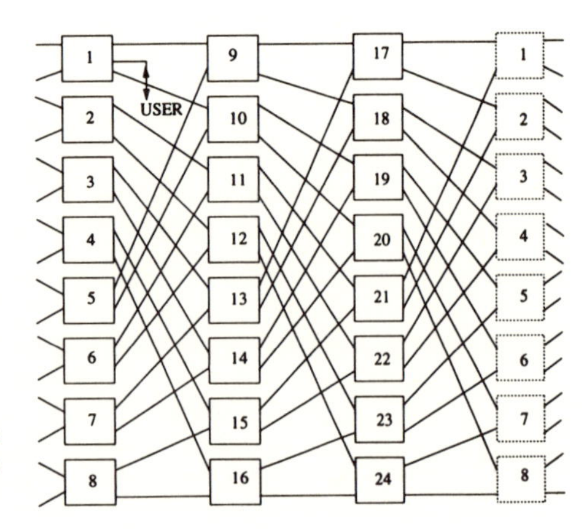

Figure 7.16. Recirculating perfect shuffle connection graph for a $k = 3$, $p = 2$, $N = 24$ multihop network.

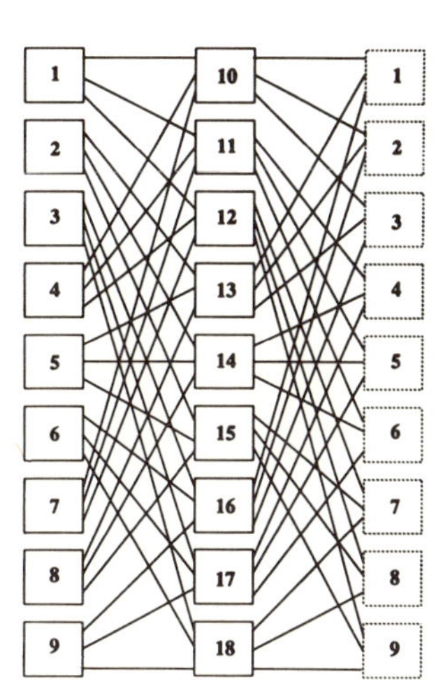

Figure 7.17. Recirculating perfect shuffle connection graph for a $k = 2$, $p = 3$, $N = 18$ multihop network.

of wavelengths onto a common medium. It should be apparent from the connection diagrams of Figures 7.15–7.17, however, that although the creation of a multihop network with a large number of access stations requires a large number of one-way channels, it is *not* necessary that each of these channels consists of a unique wavelength multiplexed onto a common medium with all other wavelengths. For example, referring to Figure 7.14 and 7.15, it is possible to replace the single fiber bus of Figure 7.14 with two fiber buses running in parallel, as shown in Figure 7.18. Suppose the transmitters of each station in column 1 of Figure 7.15 are each assigned to a unique wavelength which are all multiplexed onto the transmit side of the first bus, and the receivers of each station in column 2 are connected to the broadcast side of the first bus where each accepts a unique wavelength appearing on that bus. Similarly, the transmitters of each station in column 2 are each assigned to a unique wavelength which are all multiplexed onto the transmit side of the second bus, and the receivers of each station in column 1 are connected to the broadcast side of the second bus where each accepts a unique wavelength appearing on that bus. In this fashion, the connectivity diagram of Figure 7.15 is maintained, but since physically distinct buses are involved, it is possible to use the same set of wavelengths on both buses, and the physical topology of Figure 7.18 results. Many other physical arrangements are possible which permit a limited set of wavelengths to be reused on physically disjoint optical media while maintaining the connectivity diagram of a recirculating shuffle among a large number of access stations. Large multihop networks need a large number of channels to create the large number of directed links needed by the connectivity graph, but this does not necessarily imply that a correspondingly

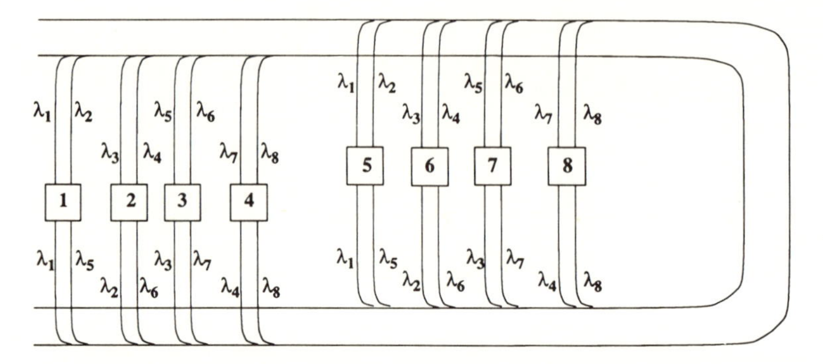

Figure 7.18. A simple example of multihop network with wavelength reuse.

large number of wavelengths be used. If it should become necessary to reuse wavelengths as described above, however, then the flexibility to arbitrarily change the connection graph by simply tuning each receiver filter to accept a different channel may be compromised, since a receiver connected to a given medium cannot accept a channel created by a transmitter attached to a different medium. Fortunately, it is possible to regain this flexibility if lithium niobate optical switches and/or wavelength-selective AOTFs are deployed within the medium. The need to rearrange the connection graph of a multihop network is discussed in Section 7.13, along with methods for selecting a good connectivity graph, and the use of slowly updated optical switches is described in Section 7.16.

A block diagram of the multihop access station appears in Figure 7.19, drawn for $p = 2$ optical transmitters and receivers. On the input side, the access station contains its two optical receivers, each of which accepts its unique channel from the optical medium (coherent channels are shown; channels could just as well have been chosen by means of optical filters). The electronically detected signals (consisting of ATM cells) along with ATM cells injected via the user input comprise the input signals to a small, output-queued, fully connected 3×3 electronic ATM switch. Each input to the switch is broadcast on its respective bus, and each cell is sent to its correct switch output by three banks of channel filters, each bank responsible for selecting cells intended for one of the switch outputs. Each channel filter is a β element which interrogates cell headers in order to correctly deliver each cell to its intended output. Each β element is provided with a memory lookup routing table. Each ATM cell header interrogated by each β element serves as an input to the corresponding routing table, which replies with a binary decision: pass the cell to the corresponding output or ignore the cell. The routing tables contain the identity of each virtual circuit which "belongs" to the corresponding output; at call setup time, the call processor uses the connection

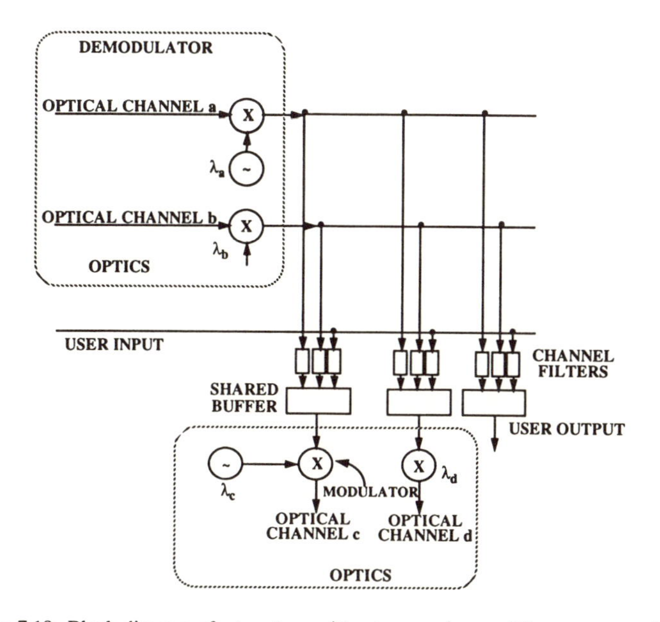

Figure 7.19. Block diagram of a two-transmitter, two-receiver multihop access station.

diagram to select a path from source to destination, and informs the appropriate routing tables located within the access stations along that path. Thus, it is not necessary that each access station have knowledge of the connection diagram; routing instructions to each access station along the path of a newly created virtual circuit are provided to that access station at call setup time.

The cells which are accepted by the channel filters associated with a given output then enter a shared buffer serving that output. As in any self-routing ATM switch, such buffers are needed to resolve contention: if cells should simultaneously arrive on two or more switch inputs, all bound to the same output, then those cells which cannot be immediately delivered to the output are stored for deferred delivery. Two of the switch outputs are sent to the access station's transmitters from whence the corresponding ATM cells are sent onto the correct channels of the optical medium. The third switch output contains ATM cells which have arrived at their destination access station; these are delivered to the corresponding user for subsequently reassembly.

Generalization to an arbitrary number of transmitters and receivers is straightforward: an access station with p transmitters and p receivers requires a $(p + 1) \times (p + 1)$ ATM switch. In fact, the traffic of more than one user can be multiplexed through each access station: if m users are multiplexed through each station, then each station will require a $(p + m) \times (p + m)$ ATM switch.

7.10. Capacity of a Multihop Network—Analysis*

Each channel of the recirculating shuffle diagram is assumed to operate at the same data rate S (say, 1 Gbit/sec). The total amount of traffic that can be carried by the network will now be found; we shall refer to this as the *aggregate capacity*. To determine the aggregate capacity, we shall make the uniform traffic assumption: an ATM cell which originates at any given access station is equally likely to be destined for any of the N stations, including the originating station. Under this condition, routing instructions can be issued such that all links of the perfectly symmetrical recirculating shuffle connection diagram will carry the same amount of newly generated traffic and the same amount of relay traffic. Furthermore, the number of hops taken by a representative ATM cell in reaching its destination is a random variable whose probability distribution is the same for all starting nodes. Let E(hops) be the expected number of hops taken by a representative cell to reach its destination. The total bit rate in use is equal to $p \times N \times S$ (the number of channels per station, times the number of stations, times the data rate per channel), but each packet appears, on average, on E(hops) links. The aggregate capacity C_T of the N-node multihop network is therefore:

$$C_T = \frac{PNS}{E\{\text{hops}\}} \tag{7.10.1}$$

We shall find E(hops) with the help of the $k = 3$, $p = 2$, $N = 24$ network shown in Figure 7.16. Suppose a new ATM cell is generated at station No. 1. In one hop, that cell can reach two stations in the second column (Nos. 9 and 13). In two hops, that cell can reach four stations in the third column (Nos. 17, 18, 19, 20). In three hops, that cell can reach any of the eight stations in the first column (Nos. 1–8). In four hops, that cell can reach any of the six remaining stations in the second column (Nos. 11–16) not already reached after one hop. In five hops, that cell can reach any of the four remaining stations in the third column (Nos. 21–24) not already reached after two hops. Each station is reachable within a maximum of five hops. Since the cell is equally likely to be destined for any of the 24 stations, it follows that, for this particular case,

$$\begin{aligned} E\{\text{hops}\} &= \frac{1}{24} \sum_{j=1}^{S} j \times (\text{number of stations reached after } j \text{ hops}) \\ &= \frac{1}{24} [(1 \times 2) + (2 \times 4) + (3 \times 8) + (4 \times 6) + (5 \times 4)] \\ &= 3.25 \end{aligned} \tag{7.10.2}$$

In general, for arbitrary k and $p = 2$, we see that 2^j access stations can be reached after j hops, $j = 1, 2, \ldots, k$. Also, after $(k + j)$ hops, the remaining $(2^k - 2^j)$ stations in column $j + k$ not reachable after j hops can be reached, $j = 1, 2, \ldots, k - 1$. The maximum number of hops needed to reach any destination is k-1. Again, since all $N = k2^k$ destinations are equally likely,

$$E\{\text{hops}\} = \frac{1}{k2^k}\left[\sum_{j=1}^{k} j2^j + \sum_{j=1}^{k-1}(k + j)(2^k - 2^j)\right] \tag{7.10.3}$$

The summations in (7.10.3) are readily expressed in closed form, yielding

$$E\{\text{hops}\} = \frac{1}{k2^k}\left[(3k^2 - 3k)\, 2^{k-1} + 2k\right] \tag{7.10.4}$$

Finally, for $p = 2$ transmitters, 2 receivers per access station, we can substitute (7.9.1) and (7.10.4) into (7.10.1) to obtain an expression for the aggregate capacity, valid for any multihop network with uniform traffic and $N = k2^k$ access stations:

$$C_T = \left[\frac{k2^{2k+1}}{3(k - 1)\, 2^{k-1} + 2}\right] S \tag{7.10.5}$$

For arbitrary p and $N = kp^k$ stations, this result generalizes to:

$$C_T = \left[\frac{2(p - 1)kp^{2k+1}}{(p - 1)(3k - 1)p^k + 2p(1 - p^{k-1})}\right] S \tag{7.10.6}$$

Finally, by dividing the total aggregate capacity by the number of access stations $N = kp^k$, we obtain the sustainable throughput per user, C_u:

$$C_u = \frac{S}{k2^k}\left[\frac{k2^{2k+1}}{3(k - 1)2^{k-1} + 2}\right] = \frac{4S}{3(k - 1) + (1/2^{k-2})}, \qquad p = 2 \tag{7.10.7}$$

$$C_u = \left[\frac{2p(p - 1)}{(p - 1)(3k - 1) - 2 + 2/p^{k-1})}\right] S, \qquad p \geq 3 \tag{7.10.8}$$

For large N (corresponding to large k), we note from (7.10.5) that

$$C_T \approx \frac{2^k S}{3} = \frac{k2^k S}{3k} = \frac{NS}{3k}, \qquad k \text{ large} \tag{7.10.9}$$

Also, from (7.9.1),

$$\log_2 N = \log_2 k + k \cong k, \qquad k \text{ large} \tag{7.10.10}$$

Thus, for large N,

$$C_T \cong \frac{NS}{3\log_2 N} \tag{7.10.11}$$

In (7.10.11), we note that the numerator represents the fact that as N increases, the number of optical channels actually in use scales linearly with N. The denominator is an approximation for E (hops) when N is large.

Similarly, from (7.10.7), we note that as the number of stations increases, the sustainable throughput per user diminishes, again reflecting the fact that as users are added, each brings its pro rata two more optical channels into use, but the expected number of hops increases. For large N (large k).

$$C_u \cong \frac{4S}{3k} \cong \frac{4S}{3\log_2 N}, \qquad k, N \text{ large} \tag{7.10.12}$$

7.11. Capacity of a Multihop Network—Results

The aggregate capacity which can be produced by a multihop network and shared among its users is tabulated in Table 7.1 (the aggregate capacity is also known as the throughput). Here, we assume that traffic patterns are uniform (a virtual circuit created by a given access station is equally likely to be terminated at any access station); as will be shown in Section 7.13, this is a pessimistic assumption in that real-world nonuniformities can be exploited to increase the overall capacity. Also, these results assume that each access station is equipped with two optical transmitters and two optical receivers [the throughput is, therefore, given by (7.10.7)]. Finally, we assume that each user port and each optical link operates at a data rate of $S = 1$ Gbit/sec; no user can access at a rate greater than $S = 1$ Gbit/sec. All capacity results (as well as the user access rate) scale linearly with S.

In Table 7.1, k is the number of columns in the recirculating shuffle connectivity graph, $N = k2^k$ is the resulting number of access stations, and the required number of optical channels is $2 \times N$. We see that, as we add more stations, the aggregate capacity grows without bound; as more stations are added, more optical channels are brought into use and the capacity increases. However, the capacity does not increase linearly with the number of channels brought into use; as we add more stations, the number of columns in the connectivity graph increases, along with the expected number of hops [which grows, approximately, logarithmically

Table 7.1. Aggregate Capacity of A $p = 2$ Multihop Network

k	N	No. of channels	Capacity (Gb/sec)
2	8	16	8
3	24	48	14.8
4	64	128	27.7
5	160	320	52.8
6	384	768	101.9
7	896	1792	198.8
8	2048	4096	389.8
9	4608	9216	767.7

Based on $p = 2$ transmitters, 2 receivers per user.
No. of regenerative users = $k2^k$, k = 2, 3, 4, ...
No. of channels = $k2^{k+1}$.

with N; see (7.10.9)–(7.10.11)]. The aggregate capacity therefore increases monotonically with N, but at a rate which is slower than linear.

Consider the $k = 7$ line of Table 7.1. This corresponds to a network containing close to 900 access stations, and the total aggregate capacity is nearly 200 Gbit/sec! Each access station can transfer ATM cells at a peak rate of 1 Gbit/sec, and all of the ATM electronics operate at a transfer rate of 1 Gbit/sec. If all stations are busy and the traffic pattern is uniform, then the sustainable transfer rate per station is about 200 Mbit/sec.

In Figure 7.20, we compare the performance of a multihop packet network against that of a more conventional single-channel bus topology packet network using a fiber-optic medium (for example, DQDB, wherein all access stations contend for a single high-speed channel). We assume that the peak data rate for each access station is set by the electro-optic bottleneck (for convenience, let's call

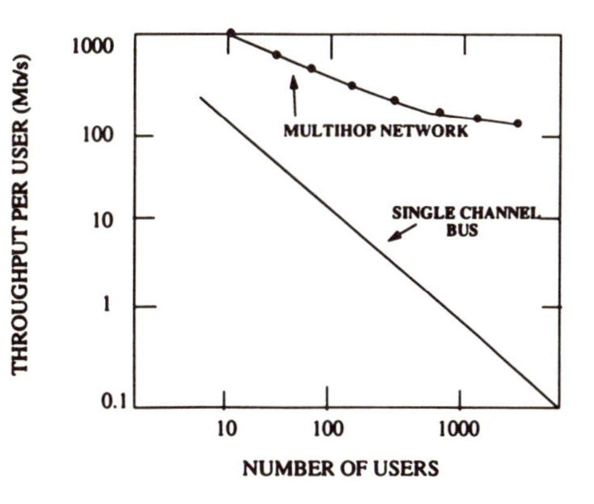

Figure 7.20. A comparison of the sustainable user throughput for a $p = 2$ multihop network and a conventional single-shared-channel network.

this 1 Gbit/sec). Plotted is the sustainable throughput per access station as a function of the number of access stations. If all N access stations contend for a single high-speed channel, then the maximum sustainable data rate per station is (1 Gbit/sec)/N, assuming that the bus access mechanism is perfect capture, i.e. , no time is lost to collisions. By contrast, by recognizing that the fiber medium can support many independent wavelength-multiplexed physical channels, we can do much better. The physical topologies of the networks compared in Figure 7.20 might be identical (both might, for example, be linear buses). However, by permanently assigning the access stations to different wavelengths and interconnecting the access stations in a recirculating shuffle pattern, we can greatly increase the sustainable throughput per user as shown.

Similar results are presented in Figure 7.21. Here, however, we assume that for the multihop networks, each access station is equipped with p transmitters and p receivers, $p = 2, 3, 4, 5, 6, 7, 8$, each operating at 1 Gbit/sec, and that p users are multiplexed through each access station. Thus, each access station contains a $(2p \times 2p)$ self-routing ATM switch. In effect, we have deployed 1 transmitter/receiver pair for each user multiplexed through an access station. The traffic pattern among users is again assumed to be uniform. The abscissa in Figure 7.21, N_u, is the actual number of users, *not* the number of access stations. We note that, for a fixed number of users, the sustainable throughput per user increases as more users are multiplexed through a diminishing number of access stations. The explanation is simple. For fixed N_u, as p increases, the number of access stations required to support a fixed number of users diminishes inversely with p. At the same time, the aggregate capacity of the network increases since the total number of channels in use, N_u, is fixed but the connectivity of each access station is enhanced (greater fan-out/fan-in), implying that E(hops) is smaller. A larger pool of aggregate capacity shared among a constant number of users produces a greater sustainable throughput per user. By way of specific example, for $N_u = 2000$, if two users are

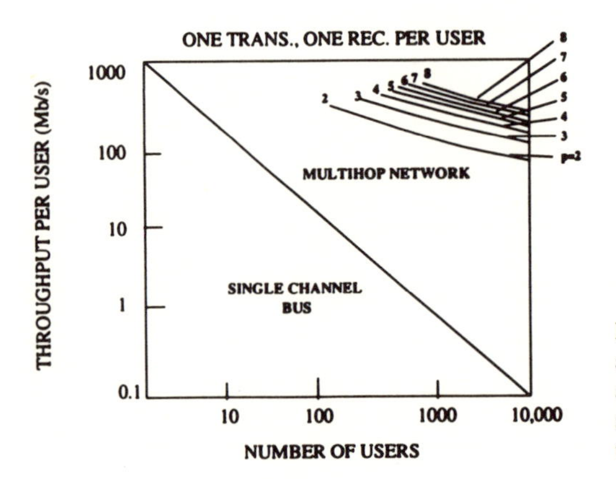

Figure 7.21. A comparison of sustainable throughput per user when p users are multiplexed through a common access station equipped with p optical transmitters and p optical receivers.

multiplexed through each access station, then each user enjoys a sustainable throughput of about 100 Mbit/sec; if eight users are multiplexed through each access station, then each user enjoys a sustainable throughput of about 400 Mbit/sec.

7.12. Delay and Lost Cell Performance of Multihop Networks*

Since the multihop scheme involves merging of traffic streams at each access station to produce each output stream, congestion can occur whenever the instantaneous rate at which messages arrive exceeds the maximum rate at which messages may be discharged. On average, the arrival rate to any access station (counting new and relayed messages) must be less than the rate at which messages are discharged. However, because of statistical fluctuations, queues will appear in the output buffers.

We shall now find an approximate expression for the mean queuing delay experienced by a representative packet in traversing the network. From this, the mean total delay can be found simply by adding the propagation delay, appropriately modified by the expected number of times that a representative message must propagate around the network. Since propagation delay is related to the geographical extent of the network's service region and to the locations of the individual access stations, no attempt will be made to encompass propagation delay into the analysis. Since, as in any real network, the amount of buffering provided will be limited, we also estimate, for a fixed-size buffer, the statistical probability that the buffer will overflow, resulting in lost data.

To simplify the analysis, we shall assume that, rather than being of fixed length, the packets which are routed by the multihop network are of variable length with a common exponential probability distribution having mean value T seconds. (We note that there is nothing in the multihop architecture which precludes variable-length packets.) For simplicity, we will concentrate on systems with $p = 2$ transmitters and receivers per access station, and one user is served by each access station. Consider the three arrivals to an access station. Assume that these arrivals are Poisson, and that messages are characterized by an exponentially distributed message length. We now focus on one access station output. With probability P_1, a message arriving from an optical channel is destined for that access station, and with probability $(1 - P_1)/2$, an arriving message must be routed to a given one of the access station's outputs (for each message, both optical outputs are equally likely to be selected). Also, a message generated by the user attached to that access station is equally likely to go to one of the two optical output channels. Finally, the message arrival rate from an end-user must equal the discharge rate to an end-user. From these considerations, we conclude that λ is the rate at which an end-user generates messages and if arrivals from an optical channel are Poisson with rate α, then the overall arrival rate to one of the optical output buffers is also Poisson with arrival rate λ^* given by:

$$\lambda^* = \frac{\lambda}{2} + \frac{2\alpha - \lambda}{2} = \alpha \tag{7.12.1}$$

Let T be the average time taken to discharge a message at the channel speed S. Then, the system will be stable provided $\alpha T < 1$.

We model the queuing system associated with a given access station output port as an M/M/1 queue. If the arrivals from an optical channel are indeed Poisson, then, as shown above, the arrivals to a particular access station output are also Poisson, and the departures from an access station output are again Poisson. Since all access stations are identical, the traffic is assumed to be symmetric, and the access stations are interconnected in a recirculating perfect-shuffle pattern with one output feeding the next input, the assumption of Poisson arrivals is at least plausible.

Under these assumptions, the mean delay, D_1, experienced by a packet at one such output is given by [see (4.3.32)]:

$$D_1 = T \frac{\alpha T}{1 - \alpha T} \tag{7.12.2}$$

From this, the mean delay in multihopping through the network is simply the sum of the delays experienced for each hop. Note that, in (7.12.2), we have ignored the actual transmission time of the packet since we do not require receipt of the entire packet at any access station before we can begin to transmit that packet onto an empty output link. The delay is therefore queuing delay only, not queuing delay plus transmission delay. Since the random variable corresponding to the number of hops taken by a representative packet is independent of the delay experienced in passing through one output buffer, the total mean queuing delay D is given by:

$$D = D_1 E\{\text{hops}\} \tag{7.12.13}$$

Finally, we note that the total volume of messages appearing on the optical links must equal the total volume of messages generated by the users multiplied by the expected number of optical links that each message appears on. Since, for this example, each access station is equipped with two optical transceivers,

$$N\lambda E\{\text{hops}\} = 2N\alpha \tag{7.12.4}$$

Thus, substituting (7.12.4) into (7.12.3), we conclude that the average end-to-end queuing delay experienced by a representative message is given by:

$$D = TE\{\text{hops}\} \frac{\lambda TE\{\text{hops}\}/2}{1 - \lambda TE\{\text{hops}\}/2} \tag{7.12.5}$$

Results are plotted in Figure 7.22. The offered load in these curves is that offered directly by the end-users, and is given in terms of the number of channels which are effectively used. For example, from Table 7.1, we see that the maximum aggregate traffic which may be offered by an eight access station network ($k = 2$) is eight channels; at this offered load, the system becomes unstable, and the queuing delay goes to infinity.

In Figure 7.22, the mean delay is expressed in terms of the average number of message transmission times T. We see from these curves that, as the number of users and permissible offered load increase, more delay is encountered since the average number of hops taken by a representative message increases. However, for all cases considered, the queuing delay is less than 10 message transmission times if the offered load is less than about 80% of the aggregate capacity.

We evaluate the lost packet performance by again invoking the M/M/1 approximation. Assuming infinite buffers at either of the two optical channel outputs for any access station, the probability P_j that there are exactly j packets awaiting transmission is given by [see (4.3.30)]:

$$P_j = (1 - \alpha T)(\alpha T)^j \tag{7.12.6}$$

Then, for a fixed-length buffer of length B messages, a good approximation to the lost packets probability P_L is given by

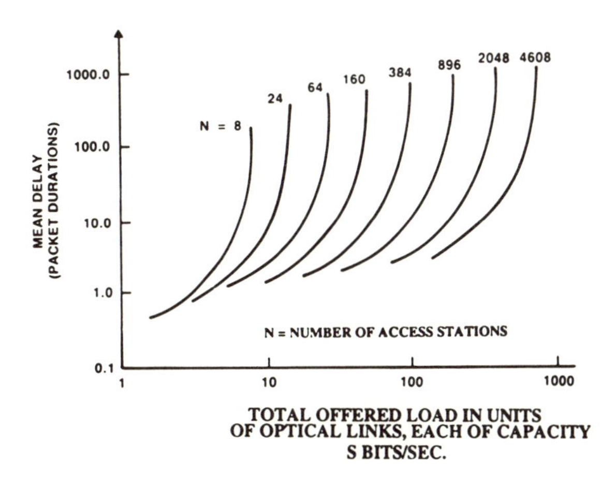

Figure 7.22. Queuing delay for a $p = 2$ multihop network with uniform traffic.

$$P_L = \sum_{j=B}^{\infty} P_j = (\lambda T)^B \tag{7.12.7}$$

Results are summarized in Figure 7.23. Offered load has been normalized by the aggregate capacity for the particular $N = 4608$ configuration. For a buffer size equal to 128 messages, the lost packet rate can be held at less than 1 in 10^6 if the offered load is no greater than 90% of the achievable capacity.

We have seen that multihop represents a fresh new approach to achieve concurrency in lightwave networks, and that aggregate capacities far in excess of the electro-optically limited access speed of each end-user are readily achievable. Among the advantages are the ability to tap a sizable portion of the enormous optical band without requiring rapidly agile optical components, and the ability to sustain average user throughputs which, although lower, are nonetheless comparable to the access speed of the electronics (in this regard, vastly superior performance is provided relative to the much lower capacity single-shared channel network). The approach is modular and reliable, requires no centralized controller or user coordination, and is compatible with bus, star, and tree topologies.

7.13. Rearrangeable Lightwave Packet Networks

The basic concept behind multihop lightwave packet networks is very simple. To enable the scalable growth of lightwave networks capable of supporting many access stations distributed over a wide-area geography, it is necessary to avoid the need for rapidly wavelength-agile (packet-by-packet) optical components and the concurrent need for elaborate signaling and control mechanisms.

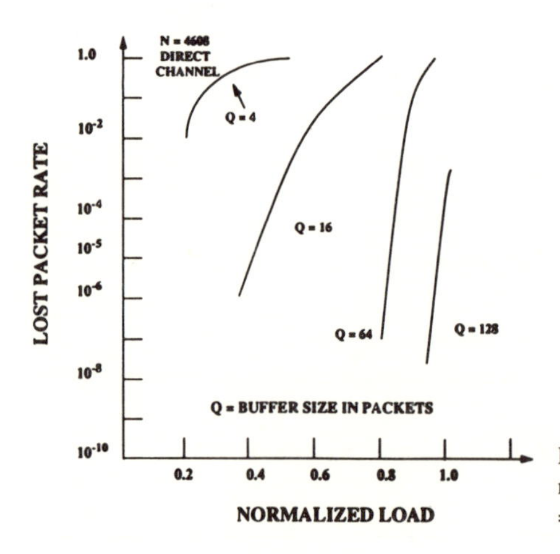

Figure 7.23. Lost cell performance of a $p = 2$ multihop network with uniform traffic and $N = 4608$ access stations.

One way to accomplish this consists of permanently assigning each of many wavelength-multiplexed channels to the transmitter of one access station and the receiver of a second access station. Since the number p of transmitters/receivers deployed in each access station is much smaller than N (the number of networked access stations), each station does not enjoy a dedicated link to every other station and, to allow virtual circuits to exist between every pair of stations, it becomes necessary for the access stations to serve as regenerative repeaters, accepting cells on one wavelength and repeating them on another wavelength. Thus, for each virtual circuit, a path from source to destination is found from the connectivity diagram, involving several intermediate access stations which serve as relay stations.

The recirculating shuffle connectivity diagram discussed in the previous section is a good choice when the traffic patterns are uniform. However, it does not fully exploit the potential of lightwave technology in that it assumes a permanently fixed connectivity diagram. However, by introducing optical components which are slowly tunable to any wavelength over a broad spatial range, it is possible to rearranged the connection diagram, without altering the physical topology of the network, simply by commanding the receivers of some access stations to tune to different wavelengths, thereby creating a link from a different access station. Alternatively, the transmitters might be tunable and the receivers stationary. In either case, the connectivity diagram can be updated, not on a packet-by-packet basis, but in response to events which occur on a time scale which is long relative to the characteristic tuning time.

Some possible advantages of rearranging the connectivity diagram are:

1. To better match the connectivity diagram to prevailing, time-varying traffic patterns. Consider for a moment the eight-node recirculating diagram of Figure 7.15. Suppose that station No. 1 has a great number of high-volume virtual connections established to station No. 7. Unfortunately, as drawn, the path from station No. 1 to station No. 7 is the longest path in the network (three hops). Since the volume of traffic between 1 and 7 is so high, it would clearly be advantageous to assign one channel for a direct connection from station No. 1 to station No. 7. This might be done, for example, by having the first of station No. 7's receivers retune to wavelength λ_1. At the same time, one of station 5's receivers (the one that had been receiving λ_1) would retune to accept wavelength λ_3. By rearranging the connection diagram to exploit nonuniformities in the traffic patterns, the amount of relayed traffic can be reduced relative to the uniform traffic case, and both greater aggregate capacity and greater throughput per user will be enjoyed.
2. To improve reliability. Although the recirculating shuffle diagram is inherently reliable, improvement can be realized by rearranging the con-

nectivity diagram in response to equipment failures. Referring again to Figure 7.15, suppose that station No. 6 should fail. We note that station No. 6 is a relay point for virtual connections from station No. 1 to stations 3 and 4, and from station 3 to stations 3 and 4. Fortunately, for $p = 2$ transmitters/receivers, there is always at least one other path from each source to each destination. For example, messages from station No. 1 could be sent to station No. 3 by means of stations 5, 2, and 8. Thus, failure of an access station does not prohibit virtual connections among any pair of surviving stations (of course, the user connected to the failed station would be without service). We note, however, for the example given above, that the alternate path is considerably longer than the original path, and that the links associated with the alternate path may already be carrying full traffic loads. Although connectivity can be maintained around the failed station, the quality of service to existing virtual connections may be unacceptably degraded. To maintain a higher quality of service to virtual connections not originating or terminating at the failed station, it would be desirable to rearrange the connection diagram among survivors in such a way as to avoid congestion bottlenecks. Ideally, the new connection diagram would seek to minimize the largest load found on any link. At a minimum, if node 6 should fail, the transmitters which had been connected to 6, and the receivers which had been accepting from 6, should be reassigned. For example, station 3's receiver which had been accepting λ_{11} could retune to λ_2, thereby creating a link from station 1 to station 3; simultaneously, station 4's receiver which had been accepting λ_{12} could retune to λ_6, thereby creating a link from station 3 to station 4.

3. To permit modular growth. For a given p, the recirculating shuffle connectivity diagram will accommodate only a limited, specific number of access stations $N = kp^k$, $k = 2, 3, \ldots$. Thus, recirculating shuffle connectivity diagrams are not modular; it is not possible to add one more access station to the network. However, by exploiting the rearrangeability property, it is *always* possible to add one more station, even for very large networks. Each new station brings p new wavelengths, along with p receivers. To add one station to an existing network, it is merely necessary for each of p receivers from the existing network (preferably from distinct access stations) to retune to a different one of the new station's transmit wavelengths. Concurrently, each of the new station's receivers would tune to a different one of the wavelengths to which the now retuned receivers from the existing network had been tuned. A more sophisticated approach might be to add the p new wavelengths to the existing channel pool and reassign the enlarged set of channels among the enlarged set of access stations such that the new connection diagram is well matched to the new traffic pattern.

4. To better match the connection diagram to the physical separation between access stations in such a way as to reduce the end-to-end propagation delay. In a wide-area network, the effect of overall propagation delay on quality of service cannot be ignored. A poor connection diagram might be one in which a virtual channel from New York to Philadelphia requires a virtual path first to Los Angeles, then to Miami, then to San Francisco, before arriving at Philadelphia. Incidentally, it is the optical distance between stations over the medium, rather than actual geographical separation, that determines the propagation delay incurred along direct links between two stations.

The generic connection diagram problem can be stated as follows. Given traffic descriptors of all one-way virtual channels existing between every pair of access stations (one sending, one receiving), and given the set of optical distances between every pair of stations, find a "good" connectivity graph. What is meant by "good"? Several possibilities exist. A "good" graph might be one which (1) allows the quality-of-service objectives to be met for each virtual channel, or (2) does the "best" job within some allowable computation time, or (3) satisfies some other objective function. Problems of this type are typically very difficult if not impossible to solve rigorously, and good heuristics are often sought.

One possible quality-of-service metric that we may seek to "optimize" is the virtual circuit blocking probability. If we are not concerned with message latency, and a suitable low lost-packet rate can be maintained by deploying very large buffers within the access stations, then a sufficient set of traffic descriptors consists of the average traffic load presented between each source–destination pair. This can be expressed as an $N \times N$ traffic intensity matrix:

$$\mathbf{T} = \begin{bmatrix} t_{1,1} & t_{1,2} & \cdots & t_{1,N} \\ t_{2,1} & t_{2,2} & \cdots & t_{2,N} \\ \vdots & & & \\ t_{N,1} & t_{N,2} & \cdots & t_{N,N} \end{bmatrix} \tag{7.13.1}$$

where element $t_{i,j}$ represents the average traffic presented by all virtual connections originating at station i and destined for station j. Since delay is assumed to be of no concern and the lost packet rate can be made arbitrarily small by increasing the size of the buffers, then, to satisfy the traffic intensity matrix $\mathbf{T}$, we need to find a connectivity graph and a set of routes for each source–destination pair such that the average traffic flow on every optical link is less than the link capacity. Suppose that the network is accommodating a given traffic matrix T_1. In response to a newly requested virtual connection with known average traffic, a new traffic matrix T_2 is generated by adding the average traffic to be presented by the new request to the appropriate element of T_1. If a connectivity and set of routes can be found such that, when T_2 is applied, the resulting average traffic flow on each link is less than the link capacity, then the new request can be admitted and the network appropriately reconfigured. If not, then the new request must be blocked and the network remains unchanged.

To make such a determination, the call controller would first seek to fit the new request into the existing connection graph. What are the shortest paths from source to destination? Does one of those paths have adequate reserve capacity to satisfy the average traffic of the new request? If so, accept the connection and assign the corresponding virtual path. If not, then seek a nonminimum length path. If none can be found, attempt a reconfiguration. Only if a suitable connectivity graph could not be found would the requested connection be blocked. Ultimately, before blocking a request, the call processor would need to answer the following question: If, in response to the new traffic, a connectivity graph and set of routes for each source–destination pair is found which minimizes the largest flow on any optical link, does that largest flow exceed the link capacity? If so, then the request must be blocked; the call processor now knows that no connectivity graph can accommodate the new request.

7.14. Problem Definition and Reconfiguration Heuristic*

Mathematically, to determine whether or not a new request for a virtual connection must be blocked, the call processor must solve the following problem:

Given

The new traffic matrix $\mathbf{T} = \{t_{i,j}\}$, $i = 1, 2, \ldots, N$; $j = 1, 2, \ldots, N$

Find

a. The number of directed optical links $Z_{i,j}$ between station i and station j, and
b. The amount of traffic $F_{i,j}$ (u,v) flowing on any optical link between station i and station j corresponding to traffic originating at station u and destined for station v (remember, this is a multihop network and traffic originating at u and terminating at v will flow over one or more optical links)

Such that

The largest flow on any link is minimized.

If the largest flow on any link exceeds the capacity of that link, then the new request must be blocked.

Assuming that $Z_{i,j}$ optical links exist from station i to station j, then the amount of traffic $F_{i,j}$ flowing on each such link from i to j can be expressed as:

$$F_{i,j} = \sum_{u=1}^{N} \sum_{v=1}^{N} f_{i,j}(u, v)/Z_{i,j} \tag{7.14.1}$$

In (7.14.1), we have summed the contributions to the flow on the optical link from i to j over all virtual circuit source–destination pairs having traffic flowing on that link, and divided by the number of directed links from i to j to get the amount of traffic on each such link. In (7.14.1) we assume that $F_{i,j}$ (u,v) = 0 if $Z_{i,j} = 0$ (i.e. , traffic can only flow over optical links which exist!). The problem now becomes one of identifying the optical link (i,j) having the greatest flow $F_{i,j}$, and choosing the $F_{i,j}$ (u,v) and the $Z_{i,j}$'s which minimize this maximum flow:

$$\text{Minimize} \left\{ \max_{i,j} \left[F_{i,j} = \sum_{u=1}^{N} \sum_{v=1}^{N} f_{i,j}(u, v)/Z_{i,j} \right] \right\} \tag{7.14.2}$$

Certain constraints apply. For each "species" of traffic (a "species" corresponds to a particular source–destination pair), the traffic flowing into any station i must equal the traffic flowing out of that station, adjusted for the traffic which is generated or terminated in that station:

$$\sum_{j} f_{i,j}(u, v) - \sum_{j} f_{i,j}(u, v) = \begin{cases} tu, v, & u = i \\ -tu, v, & v = i \\ 0, & \text{otherwise} \end{cases} \tag{7.14.3}$$

The total traffic flowing from station i station j must be less than the total capacity of all directed links between i and j:

$$\sum_{u=1}^{N} \sum_{v=1}^{N} f_{i,j}(u, v) \le Z_{i,j}\, S, \qquad i = 1, 2, \ldots, N; j = 1, 2, \ldots, N \tag{7.14.4}$$

where S is the link capacity. The number of directed links out of any station i must be equal to the number of transmitters p. Similarly, the number of directed links into any station j must be equal to the number of receivers p:

$$\sum_{j=1}^{N} Z_{i,j} \le p, \qquad i = 1, 2, \ldots, N \tag{7.14.5}$$

$$\sum_{i=1}^{N} Z_{i,j} \le p, \qquad j = 1, 2, \ldots, N \tag{7.14.6}$$

$$Z_{i,j} \ge 0, \qquad i = 1, 2, \ldots, N; j = 1, 2, \ldots, N \tag{7.14.7}$$

$Z_{i,j}$ must be an integer

$$f_{i,j}(u, v) \ge 0 \tag{7.14.8}$$

Let the solution to (7.14.2), subject to constraints (7.14.4)–(7.14.8), be denoted by F_M. Then, if F_M does not exceed the link capacity S, the newly requested connection can be accommodated. If $F_M > S$, then the newly requested connection must be blocked. Note that the solution to (7.14.2) also yields the resulting connectivity graph, source–destination routes, and flow for each route since the various $Z_{i,j}$'s and $F_{i,j}$ (u,v)'s are found as part of the solution.

The mathematical problem which we have just defined is a mixed integer–linear programming problem in that both integer variables (the $Z_{i,j}$'s) and continuous variables [the $F_{i,j}$ (u,v)'s] are involved. Having carefully defined the problem, we can now state that problems of this type have no known analytical solution. Thus, we must resort to heuristics, that is, techniques which permit *a* solution (not necessarily the best, but hopefully a reasonably good solution) to be found. For example, suppose we try to separate the problem into two components: first, find a "good" connectivity graph, then flow the traffic (find routes for each source–destination pair) such that the maximum flow on any link is minimized. A not-unreasonable approach for defining a good connectivity graph might be to seek the graph which accepts the maximum-possible amount of single-hop traffic. In principle, this is conceptually simple. Starting with the traffic matrix T (7.13.1), select exactly p elements in every row such that the sum of elements selected is as large as possible, subject to the constraint that exactly p elements are chosen in each column. (Note that row i contains only traffic elements corresponding to virtual connections which originate in station i; column j contains only traffic elements corresponding to virtual connections which terminate in column j.) If we find the p largest elements in any row and assign a directed optical link to the corresponding p source–destination pairs, then, for the access station which corresponds to that row, we will have maximized the amount of its traffic which is delivered in one hop. However, we cannot perform such a selection independently for each row since we would then have no assurance that each terminating access station was chosen exactly p times (each station has only p receivers). Thus, when selecting elements, we must be sure not only that each row has exactly p chosen elements but also that each column contains exactly p chosen elements. Then, if element $t_{i,j}$ is chosen, a directed optical link should be established from station i to station j. By choosing elements which sum to the largest possible value, the amount of traffic which is carried in one hop is maximized, and the connectivity graph will be optimized or matched with regard to one-hop traffic.

Now, we can formally state the problem to be solved by this heuristic (the same variables are used as before).

Given

The traffic matrix $T = \{t_{i,j}\}$

Find

The number of directed connections $Z_{i,j}$ from station i to station j such that the one-hop traffic F_0 is maximized:

$$F_0 = \text{maximum} \left\{ \sum_{i=1}^{N} \sum_{j=1}^{N} t_{i,j} Z_{i,j} \right\} \tag{7.14.9}$$

subject to the following constraints:

$$\sum_i Z_{i,j} = p, \qquad j = 1, 2, \ldots, N \tag{7.14.10}$$

$$\sum_j Z_{i,j} = p, \qquad i = 1, 2, \ldots, N \tag{7.14.11}$$

$$Z_{i,j} = 0 \text{ or } 1, \qquad i = 1, 2, \ldots, N; j = 1, 2, \ldots, N \tag{7.14.12}$$

Fortunately, this problem is similar in structure to a well-known transportation problem which always yields (0,1) integer solutions. The integer constraints can therefore be dropped and the problem solved using conventional linear programming techniques.

With a connectivity graph now at hand, we can solve the second problem:

Find

The amount of traffic $F_{i,j}(u,v)$ on link (i,j) corresponding to virtual circuits which originate at station u and terminate at station v, such that the maximum flow on any link is minimized.

This corresponds to finding $F_M = \min(F_{i,j})$ in (7.14.2), subject to constraints (7.14.3) and (7.14.8). The important difference here is that the $Z_{i,j}$'s are now known, not variables to be found. This problem has now been cast in another well-known form, that of a multicommodity flow problem, also solvable by linear programming techniques.

A specific numerical example of this heuristic will now be offered. Consider an eight-station network with the applied traffic matrix shown in Figure 7.24a. Note the ringlike nature of the traffic: station 1 has a great deal of traffic for station 2; station 2 has a great deal of traffic for station 3; etc. We assume two transmitters and two receivers per access station. Application of the heuristic produces the connection diagram shown in Figure 7.24b. Note that a backbone ring corresponding to the ringlike nature of the traffic has been produced, along with several chords which make good use of the remaining transmitters and receivers. Also shown on each link is the resulting traffic flow.

Another heuristic which may be applied involves a novel procedure known as simulated annealing. To explain this iterative approach, we assume $p = 2$ transmitters/receivers per access station. The approach is initialized with an arbi-

0	10	0.9	1	0.8	1	1.1	0.8
0.9	0	11	0.9	1	1	0.8	1
1.1	0.9	0	8	0.8	0.9	1	0.9
1.1	1	1	0	9	0.9	0.9	1
0.8	1	0.8	0.9	0	10	1.1	0.9
0.9	0.9	1	1	0.9	0	12	1.1
0.8	1.1	1.1	1	0.8	0.9	0	9
11	0.9	0.8	0.7	1	1.1	0.8	0

(a) Traffic matrix

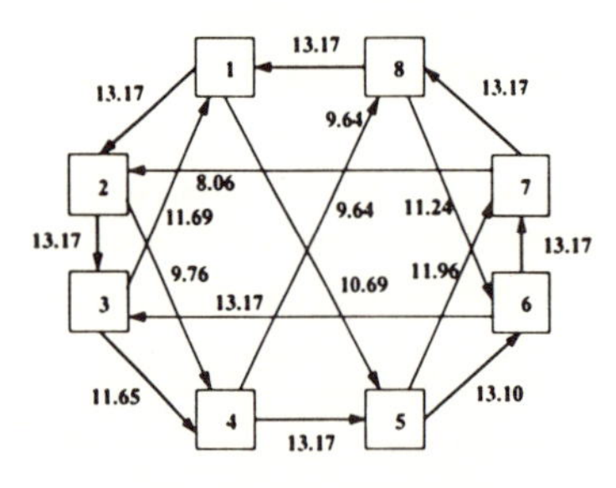

(b) Connection graph / traffic flow

Figure 7.24. Resulting connection graph for an 8-node multihop network with a specific type of traffic non-uniformity.

trary connectivity graph having two directed links to and from each station. For this connection graph, the flow problem is solved. Next, two links are chosen at random and a so-called *operation branch* is performed. To execute a branch exchange operation, each of the two receivers which terminate the two chosen links are tuned to the wavelength formerly accepted by the other. The flow problem is then solved for the new connection graph. If the largest flow on any link, F_M, is reduced, use this connection graph for the next iteration. If F_M is increased, then with probability P_1, use this as the new connection graph for the next iteration; with probability $(1 - P_1)$, return to the original connection graph. In either case (F_M reduced, F_M increased) we are presented with a connection graph and are ready to perform another branch exchange operation on two randomly chosen links. Again, the branch exchange is executed, and the flow problem is solved for the resulting connection graph. Now, however, if F_M is increased, we retain this new configuration for the next iteration with probability $P_2 < P_1$. It turns out (for reasons that are not fully understood) that if this procedure is iterated many times, with the probability of retaining a new, inferior connection graph (F_M larger) slowly reduced on subsequent iterations, a not-unreasonable connection diagram results. Simulated annealing has also been applied to find "good" connection diagrams for a given applied traffic pattern when total delay (the sum of the queuing and propagation components) is the objective function to be "minimized."

7.15. Traffic-Handling Capability of a Rearrangeable Multihop Network

We shall now evaluate the traffic-handling performance of a rearrangeable lightwave packet network when used as a distributed ATM switch. We shall do this by randomly generating virtual circuit traffic matrices and finding the probability that such a representative matrix cannot be accommodated. The "rejection ratio" will then be compared against that of an ideal centralized ATM switch with output queuing.

To determine the rejection ratio, the traffic statistics of all virtual circuits will be assumed to be the same, and a given matrix will be accepted if a connection graph and flow assignment can be found such that the average traffic on each link is less than the link capacity (large buffers are used, and delay is not an issue). For a given traffic matrix, the connection diagram which maximizes the maximum flow will then be found and the resulting largest flow compared against the link capacity. We assume that the capacity of each link (either optical or user input/output) is a factor of g times greater than the average traffic presented by a virtual circuit.

To generate an $N \times N$ traffic matrix, we shall assume that each element is drawn from a Poisson probability distribution:

$$P[t_{i,j} = k] = e^{-\Lambda}(\Lambda)^k/k!, \qquad i = 1, 2, \ldots, N; j = 1, 2, \ldots, N; k = 0, 1, 2, \ldots \tag{7.15.1}$$

where Λ is a parameter representing the expected number of one-way virtual connections between station i and j. Note that this is a constant, independent of i and j. Thus, the expectation of the traffic matrix is uniformly distributed, that is, on average, there are as many virtual connections between a given pair of stations as there are between any other pair. However, a specific matrix with elements drawn from the probability distribution (7.15.1) will generally exhibit nonuniformly distributed traffic, and it is this nonuniformity that the network connectivity diagram will adapt to.

The ideal, centralized switch will reject a given matrix if and only if one or both of the following conditions exist:

1. The number of virtual circuits generated by any user exceeds g (if the sum of the elements in any row exceeds g, then the user corresponding to that

row is attempting to initiate more virtual circuits than the capacity of the user input port can accept).

2. The number of virtual circuits terminating at any user exceeds g (if the sum of the elements in any column exceeds g, then the user corresponding to that column is attempting to accept more virtual circuits than the capacity of the user output port can accept).

In addition to the occurrence of either of these two conditions, the distributed ATM switch will reject a traffic matrix if

3. No connection diagram/flow assignment can be found for which the number of virtual circuits on each optical link is less than or equal to g (the number of full or partial virtual circuits on each link must be less than the link capacity). In attempting to minimize the maximum flow on any link, we will allow the call controller to bifurcate a virtual circuit by dividing that virtual circuit's traffic among several paths, if necessary, to minimize the maximum flow. In practice, of course, this could not occur, but for our study this is a good, convenient approximation, since the number of virtual circuits on each path is large and it is only the fractional parts of these large numbers that correspond to bifurcated traffic. If we were to insist on only whole numbers of virtual circuits assigned to each link, then the flow parameters $f_{i,j}\ (u,v)$ would need to be restricted to whole numbers, and a simple linear programming problem would become a complex integer programming problem. Since, in most cases, the traffic matrix can be accepted (that is, there is spare capacity on each link since we are interested only in low rejection ratios), the call processor could usually, in practice, adjust the computed flow parameters such that no bifurcation occurs.

Thus, the rejection ratio of the distribution switch must be greater than that of the centralized switch, since in addition to conditions (1) and (2), condition (3) must also apply.

Mathematically, the above three conditions can be expressed as:

$$\sum_{j=1}^{N} t_{i,j} \leq g, \qquad i, = 1, 2, \ldots, N \tag{7.15.2}$$

$$\sum_{i=1}^{N} t_{i,j} \leq g, \qquad j, = 1, 2, \ldots, N \tag{7.15.3}$$

$$F_{\mathrm{M}} \leq S/g \tag{7.15.4}$$

where F_{M} is the solution to (7.14.2) and S is the link speed.

A comparison of the centralized and distributed switches appears in Figure 7.25 drawn for the case $g = 10$ (each virtual circuit consumes, on average, 10%

of a link's capacity). Shown is the rejection ratio for each switch as a function of the normalized offered load per input or output link $\rho = N\,\Lambda/g$ (we see that ρ is simply the average number of virtual circuits established either from or to any user, normalized by the number of virtual connections that can be handled by an I/O link), These results were obtained by numerical simulation. For a given offered load, many traffic matrices were generated at random, and the fraction which could not be accommodated was normalized by the number of random matrices generated. Results shown are for a 24-station network, with $p = 2$ transmitters/receivers per station. We see that when $g = 10$ (again, each link has adequate capacity to serve 10 virtual circuits), there is virtually no difference between the rejection ratios of the two switches over the range of offered loads considered. Thus, the traffic nonuniformities were successfully exploited to permit rearrangement of the distributed lightwave network via the heuristic presented earlier in this section. Essentially, for this case, rejection is caused by saturation of the input/output links, and the internal optical links retain excess capacity. However, for g = 100 (Figure 7.26, each virtual circuit consumes 1% of a link's capacity), we see that, to reproduce the rejection ratio of the centralized switch, it is necessary that each optical link run at twice the speed of each I/O link ($g = 100$ for the I/O links; $g = 200$ for the optical links). Without optical link speedup, saturation occurs on the optical links before occurring on the I/O links. If the speedup factor for the optical links is reduced to 1.5 ($g = 150$ on the optical links), a departure from the blocking performance of the centralized switch is noted. What has happened is that, for $g = 100$, each of the randomly generated traffic matrices begins to appear uniform; the links can handle so many virtual connections that the law of large

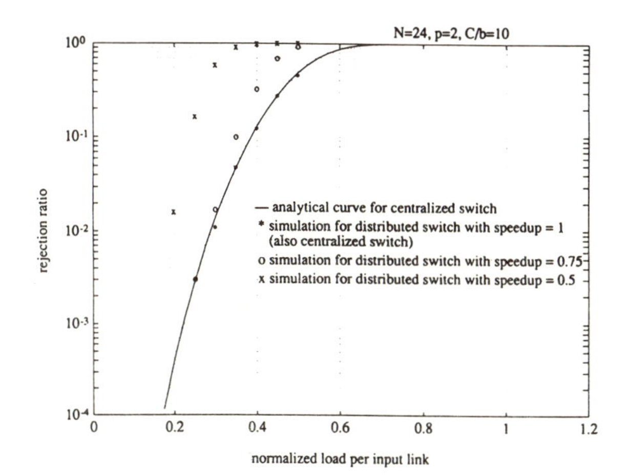

Figure 7.25. A rejection-ratio comparison of a centralized ATM switch and a multihop network for 24 stations and $p = 2$; each virtual circuit consumes 10% of a link's bandwidth.

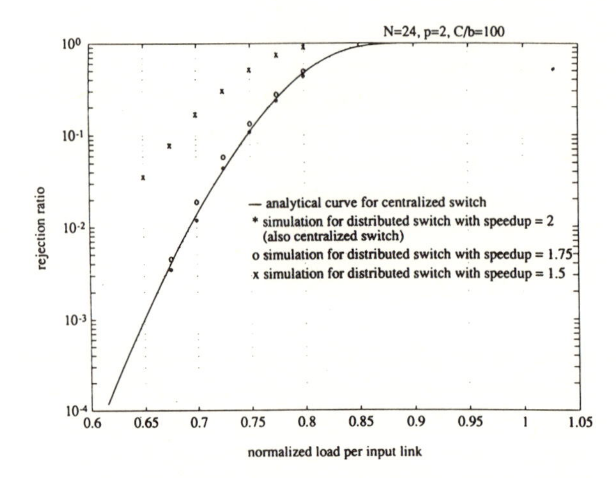

Figure 7.26. A rejection-ratio comparison of a centralized ATM switch and a multihop network for $N = 24$ stations and $p = 2$; each virtual circuit consumes 1% of a link's bandwidth.

numbers takes effect, and the number of one-way virtual circuits established between any pair of stations begins to approach its mean value. Since, for this exercise, these mean values are all the same, the traffic matrices begin to appear somewhat uniform, and rearrangeability cannot be exploited. If, however, the mean number of virtual circuits between pairs of station was *not* the same for all possible pairs, as would realistically be expected in practice, then exploitable nonuniformities would appear, and less (or no) optical link speedup would be needed.

Thus, rearrangeability is a property of lightwave packet networks which, under the right set of circumstances, can be used to support a modular, reliable, fully decentralized ATM service. By using a shared optical medium, the potential advantages described earlier in this chapter can be enjoyed. For example, it might be possible to rapidly reconfigure the network to deliver infrequently occurring long data files (e.g. , high-resolution, uncompressed images) by means of one-hop direct connections, thereby leaving only the somewhat more homogeneous, highly bursty traffic to be carried by multihop ATM, an opportunity which might mitigate some of the traffic control/performance management issues described in Chapter 6.

7.16. Rearrangeable Multihop Networks with Wavelength Reuse

The techniques described thus far in this section are predicated on arbitrary rearrangeability: at the optical level, *any* desired connectivity can be achieved by appropriately assigning channels from a large available pool, and the connection diagram is totally independent of the physical topology of the medium. However, as discussed earlier in this chapter, arbitrary rearrangeability may be physically infeasible. For example, there may be several physically disjoint optical media to permit the creation of a large pool of channels, without a commensurately large number of wavelengths, by means of wavelength reuse among the disjoint optical media. Fortunately, by judiciously deploying lithium niobate or AOTFs to interconnect the various media, it is often possible to maintain sufficient flexibility to permit any desired connectivity graph.

For example, consider the network shown in Figure 7.27a. Here, there are three optical media, and wavelengths λ_1 through λ_K are reused on all three. On each medium, all wavelengths except λ_a can be used to arbitrarily interconnect the transmitter of any station attached to that medium to the receiver of any other station also connected to that medium (as usual, each access station contains p transmitters/receivers). However, λ_a is used to interconnect the transmitter of a station on one medium to the receiver of a station on a different medium. As shown, wavelength λ_a is brought out of each medium on an optical fiber and is switched by a 3 × 3 lithium niobate rearrangeably nonblocking switch fabric

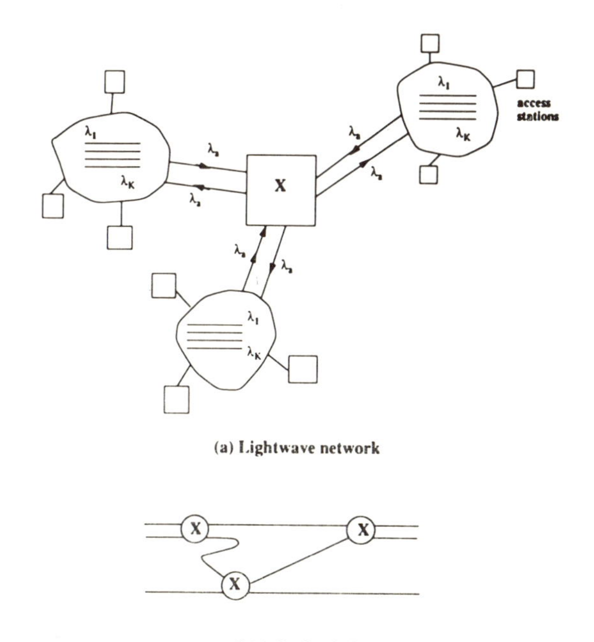

(a) Lightwave network

(b) A 3 x 3 switch

Figure 7.27. An example of wavelength reuse employing a space-division optical switch.

appropriately constructed from three 2 × 2 switches (see Figure 7.27b). Thus, since within one medium, any transmitter and any receiver can be assigned to wavelength λ_a, it becomes possible to interconnect any one transmitter on any medium to any one receiver on a different medium; the signal created on wavelength λ_a on any medium can be injected into any other medium (of course, it is not permissible for two transmitters assigned to λ_a on different media to access the same receiver on the third medium; the space division lithium niobate switch maintains the required spatial separation between wavelength λ_a signals created on different media). In principle, a rearrangeably nonblocking switch fabric of arbitrary dimensionality, Q, can be constructed from 2 × 2 piece-parts, and can be used to interconnect a common wavelength on Q disjoint media. In practice, Q will be limited by the insertion losses and cross talk of the 2 × 2 elements, although optical amplifiers can be used to greatly offset the latter. Some number R of such switches might be deployed, each operating at a single wavelength, to independently interconnect each of R wavelengths among the Q media. Thus, by using such switches, "almost arbitrary" connection diagrams can be established among the access stations connected to disjoint media. The number R of such switches will, however, limit the degree of flexibility, with smaller R causing greater restrictions. If the users are assigned to media in such a way that users which predominantly communicate along themselves are assigned to a common medium, with limited intermedia traffic (strong communities of interest), then even a small value of R will effectively permit realization of any desired connectivity diagram which reflects prevailing traffic patterns. In any event, once the desired connectivity graph is found, the optical switches are then set in the correct state to realize that diagram via remote electronic command.

The use of AOTFs can facilitate the interconnection of R wavelengths among Q media without requiring R distinct physical switches. A Q-input Q-output AOTF fabric constructed from 2 × 2 AOTF elements, each capable of handling R wavelengths, can independently switch those wavelengths, thereby replacing R lithium niobate switches. For example, a single 4 × 4 AOTF with $R = 3$ can interconnect three wavelengths on each of four media, as shown in Figure 7.28a. By cascading media interconnected via AOTFs (each capable of independently switching R wavelengths) as shown in Figure 7.28b, very large "almost arbitrarily" rearrangeable networks can be implemented. Note that, in Figure 7.28b, it is possible for a transmitter on medium 1 using λ_a to connect with a receiver on medium 5 also using λ_a by routing λ_a from medium 1 to medium 5. For even greater flexibility, in realizing any desired connectivity diagram, it is possible to interconnect a transmitter using λ_a on medium 1 to a receiver using λ_b on medium 5 if we use an access station on, for example, medium 2 to (1) accept that transmitter's signal on λ_a, (2) electronically regenerate, and (3) relay that signal on λ_b to medium 5. Thus, intermediate access stations can simultaneously serve as

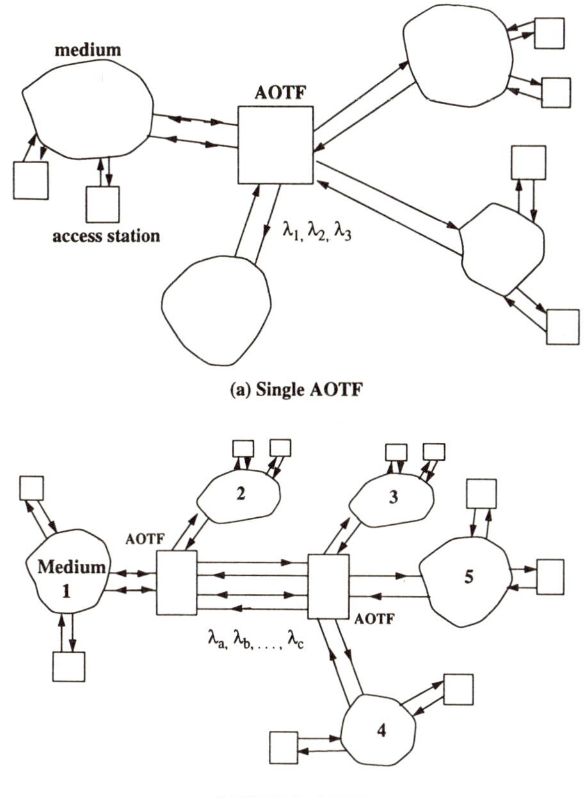

Figure 7.28. Example of wavelength reuse employing Acousto-Optic Tunable Filters.

user ports, multihop ATM repeaters, and wavelength translators. Values of R as large as 16 might realistically be deployed in practice. Not only do techniques such as these permit large, almost totally rearrangeable networks but, also, they permit such networks to be realized with a limited number of wavelengths. Although, in principle, a great number of wavelengths could be multiplexed onto the medium (the bandwidth of the medium is on the order of 50 THz; the modulation linewidth of each channel is only on the order of several gigahertz), such an approach would require the development and refinement of dense WDM techniques. Problems associated with such techniques involve (1) stabilization of the individual transmitting semiconductor lasers, geographically dispersed within the access stations; (2) gain variation in the passband of the optical amplifiers needed to maintain adequate signal-to-noise ratio after WDM channels have been combined and distributed; (3) bandlimitedness of the optical amplifiers (the medium may have bandwidth in excess of 50 THz, but the optical amplifiers do not, and the total amount of usable bandwidth would therefore be limited to that of the amplifier's passband); (4) linearity, gain saturation, and intermodulation distortion of the optical amplifiers in a dense WDM environment; and (5) inability to tune transmitters and receivers over a wide optical range as would be needed to produce the desired connection graph. By using a combination of wavelength and space division multiplexing, a large number of channels can be created by reusing a limited number of wavelengths. Since none of the optical fibers, which comprise a major element of the medium, carry more than this limited number of wavelengths, the physical level problems associated with dense WDM are thereby avoided, without compromising the overall network capacity. Furthermore, by deploying wavelength-selective optical switches, we can once again assign channels to produce some desired connection diagrams among access stations.

As before, the medium (media) is "passive" in the sense that no logical operations on the information-bearing signals are performed within the medium *per se.* Although amplifiers are to be deployed, their sole function is to linearly amplify wavelength-multiplexed signals, thereby maintaining adequate signal-to-noise ratios. Moreover, although wavelength-selective optical switches are present, these too are linear: their sole function is to effect the desired connection pattern among access stations, and they are controlled by electronic signals generated by a network configuration manager which is responsive to requests for "clear channel" connections among the stations and to equipment status (failures/recoveries).

In effect, from a network of perspective, the optical medium provides a means of enabling circuit-switched, broadband (several gigahertz) analog connections among the access stations. The number of such connections which can be established by each access station is limited to p, and by appropriately assigning channels to the transmitters and receivers of the access stations, an arbitrary connection graph can be achieved. Electronic self-routing switches can then be deployed within the access stations to relay information among the channels, thereby enabling a self-routing packet switching service with full connectivity among the access stations at the virtual circuit level.

7.17. Conclusion

The novel architectures and unique opportunities of lightwave packet networks raise many intriguing issues and possibilities concerning the relationships among the various elements comprising a distributed telecommunication/computing environment. Given the existence of a passive optical telecommunication infrastructure, can we further develop a well-matched set of higher-layer protocols which exploit the lightwave network? How will these impact the design of a well-matched computer bus within the devices attached via the network user ports? How might application code be optimized in this environment? The advent of lightwave packet networks may permit questions such as these to be addressed in an integrated systems fashion, and may enable end-user applications that can be supported only at great difficulty by means of conventional techniques. While no guarantees can be made, the potential of lightwave networks cannot be ignored.

7.18. Problems

7.1. An on–off optical transmission link uses a 10-mW semiconductor laser. The laser-to-fiber coupling loss is 10 dB, and there are five optical splices, each involving 1-dB insertion loss. The link is 100 km long, the fiber loss is 0.3 dB/km, and the data rate is 100 Mbit/sec. (a) Find the required receiver sensitivity, that is, the required energy per bit such that a BER = 10^{-9} is produced. (b) What is the permissible degradation of this receiver, in decibels, relative to the quantum limit (assume that the wavelength is 1.3 μm, yielding an energy per photon of 1.3×10^{-9} joule)?

7.2. Performance of a particular homodyne receiver with binary phase shift keying is 5 dB inferior to the ideal. The receiver is used as part of a transmission link which has 30 dB of coupling loss (which includes laser-to-fiber loss). The laser produces 10 mW of power, the data rate is 1 Gbit/sec, and the energy per photon is 1.3×10^{-9} joule. Assuming that the fiber loss is 0.3 dB/km, compute the maximum range of the system, in kilometers.

7.3. Every T seconds, a binary source produces a symbol, equally likely to be "0" or "1." A logical "0" is modulated onto an optical carrier as a $T/2$-second pulse of light having a power level of $P = 2h\upsilon/T$; a logical "1" is carried as a T-second pulse of light of the same intensity. The receiver consists of an ideal counter which observes the number of photoelectrons produced over the T-second source interval, followed by decision circuitry. (a) Find the decision rule which minimizes the probability of making a bit error. (b) Find the probability of error conditioned upon a "0" being sent. (c) Find the probability of error conditioned upon a "1" being sent.

7.4. A coherent homodyne system is used to carry the information produced by a four-level source. The symbol interval is T seconds, and the modulation format is quadrature phase shift keying. The signal power availability at the receiver is A^2, and the local oscillator power is B^2 (B^2 is much greater than A^2). Assuming that the source produces equally likely outcomes, find an expression for the probability of error, in terms of the Q-function.

7.5. Each of three binary sources is assigned its own optical carrier at frequencies W_1, W_2, and W_3, respectively, on–off modulation is used, the pulse duration is T, and the three signals are combined in a wavelength division multiplexer. At the receiver, the power level of each "on" signal is A^2. A coherent receiver is used to detect the signal at optical frequency W_2. The power level of the local oscillator is B^2, much greater than A^2 and its frequency is W_L (which does not coincide with any of the three signal frequencies). A microwave filter, installed after the photodetector, removes all spectral components except the one (if any) at frequency $W_2 - W_L$. Find an expression for the probability of error for the signal at frequency W_2 in terms of the Q-function.

7.6. An optical network uses a star coupler to wavelength multiplex and broadcast the signals produced by each of four users. All users produce fixed-length information packets of equal size and all transmissions are synchronized to arrive at the star within time-aligned slots. The receivers are wavelength agile, and each must be commanded each time slot to its correct wavelength. The number of packets originating at user i and destined for user j may be represented by a 4×4 matrix $\mathbf{T} = \{t_{i,j}\}$. Suppose that

$$\mathbf{T} = \begin{bmatrix} 0 & 3 & 4 & 2 \\ 8 & 0 & 3 & 1 \\ 4 & 8 & 0 & 5 \\ 3 & 6 & 5 & 0 \end{bmatrix}$$

and that all packets in **T** must be completely scheduled and delivered before any new arrivals. Neglecting the signaling associated with the scheduling of packets, (a) find the minimum number of time slots needed to schedule the packets of **T**, and (b) produce a possible nonconflicting schedule which uses the minimum number of time slots.

7.7. Consider a time-slotted multiwavelength broadcast star optical LAN containing N access stations. Each receiver is assigned a unique wavelength, and all transmitters are fully tunable on successive time slots. A given receiver is accessed with a fixed-length packet by tuning the transmitter's wavelength to that of the intended receiver and shipping the packet within one time slot. All transmissions are synchronized so that slot boundaries are time-aligned at the star. The transmitters do not coordinate their transmissions among themselves and, each time slot, any transmitter may seek to access any receiver. If two or more transmitters attempt to access the same receiver during the same time slot, a collision occurs and all packets on that wavelength are destroyed. Each packet newly arriving at a transmitter is equally likely to be destined for any receiver, and the packet arrival rate is sufficiently high that each transmitter always has a copious supply of packets awaiting transmission to each receiver. Derive an expression for the total LAN throughput, that is, find the percentage of time slots which carry packets not having suffered a collision. Assume that the data rate on each wavelength is a common S bits/per second.

7.8. A broadcast-and-select lightwave network uses a star coupler to interconnect 32 microprocessors. Each receiver is assigned a unique wavelength and the transmitters are fully tunable. Fixed-length packets are used and all transmissions are synchronized so that the slot boundaries are time-aligned at the star. The data rate of each optical link is 100 Mbit/sec. A centralized scheduler is used to resolve receiver contention. Each transmitter has a separate bidirectional signaling link to the controller, by means of which it notifies the controller of the destination of every newly arriving packet, and through which it receives a receiver assignment for every time slot. The controller can schedule a receiver for every transmitter once every 100 μsec. (a) What is the minimum slot time? (b) What is the corresponding packet size, in bits? (c) What is the required data rate on each signaling link?

7.9. Draw the block diagram of the access station for a multihop network. Each station has p transmitters and p receivers, and multiplexes the traffic from M users. The data rates of all optical links and user access links are the same.

7.10. Consider a recirculating perfect shuffle optical network with uniform traffic (each packet supplied by a given user is equally likely to be destined for any access station, including that to which the user having generated the packet is connected). Let the number of access stations be N and the number of transmitter–receiver pairs per station be p. Using semilog paper, plot the expected number of hops and $\log_p N$ as a function of N for $p = 2, 3, 4, 6, 8, 12, 16$. Do not consider values of N greater than 10^8. Estimate the range over which $\log_p N$ is a reasonable approximation for $E\{\text{hops}\}$.

7.11. Referring to Problem 7.10, show that, for N suitably large and any value of p, $E\{\text{hops}\}$ is approximately equal to $\log_p N$.

7.12. Derive Eq. (7.10.6).

7.13. Consider an $N = kp^k$ station recirculating shuffle multihop lightwave network where p is the number of transmitter–receiver pairs for each station and k is the number of columns in the graph. A packet generated at any access station is equally bound for any station except that it is never bound for its originating station. Find an expression for the expected number of hops taken by a representative packet.

7.14. Consider an $N = kp^k$ station recirculating shuffle multihop lightwave network where p is the number of transmitter–receiver pairs per station and k is the number of columns in the connection graph. A packet generated at any access station is equally bound for any access station including the one at which it was generated. Packet arrivals at each station are Poisson, with rate λ, and the length of each packet in bits is an exponentially distributed random variable with mean value B (packets are sufficiently long that the length granularity arising from the integer values for the number of bits may be safely ignored). The data rate of each optical link and each user access link is S bits/per second. The aggregate data arrival rate, normalized by the link speed, is therefore $\lambda BN/S$. (a) Find and plot an expression for the average delivery delay in units of mean packet time B/S as a function of $\lambda BN/S$ for $k = 2, 3, \ldots, 8$ and $p = 2, 3, 4, 6, 8, 12$. For this part, you may assume that all buffers contained within the access stations are capable of holding an arbitrarily large number of packets. (b) Let each access station contain a fully connected packet switch with output buffering and one shared buffer per output, and let Q be the maximum number of packets which may be stored in any one buffer. Find and plot the packet loss probability as a function of $\lambda BN/S$ for $Q = 2$, 4, 16, 64, 256 and the following combinations of other parameters: ($k = 2, p = 3$), ($k = 2, p = 12$), ($k = 4, p = 3$), ($k = 4, p = 8$), ($k = 7, p = 3$), ($k = 7, p = 8$).

7.15. Referring to Problem 7.14a, suppose the physical topology is that is that of a star and that all access station-to-star coupler links are of equal length with propagation delay 100 B/S. Plot the total mean delay (queuing plus propagation) for the same value of k and p.

7.16. Referring to Problem 7.14, suppose that the network is virtual connection oriented, and that the average traffic intensity per virtual connection is $S/1000$ (the data rate per VC is sufficiently small relative to the link speed that the number of VCs in progress is approximately the same between any pair of stations). (a) Under the conditions of Problem 7.14a, and for the same combinations of k and p, graphically find the maximum number of VCs that may be admitted such that the mean delay is less than 100 B/S. (b) If the admission controller limits the number of admitted calls in accordance with part a, what fraction of the maximum possible throughput is actually used in each case?

7.17. A multihop lightwave network is used to interconnect $N = 6$ access stations. Station-to-station traffic patterns are nonuniform and are in accordance with the following traffic matrix:

$$\mathbf{T} = \begin{bmatrix} 0 & A & 0 & 0 & 0 & A \\ A & 0 & 0 & A & 0 & 0 \\ 0 & 0 & 0 & 0 & A & A \\ 0 & A & 0 & 0 & A & 0 \\ 0 & 0 & A & A & 0 & 0 \\ A & 0 & A & 0 & 0 & 0 \end{bmatrix}$$

where A is some arbitrary unit of traffic. Each station has two transmitters and two receivers. Subject to these constraints on the number of transmitters and receivers per station, any logical connection graph can be realized. (a) Draw the logical connection pattern which maximizes the permissible value of A. (b) Assume that the units of A are virtual connections, packet arrivals to each virtual connection are Poisson with rate λ, and packet durations are exponentially distributed random variables, each with mean T. The data rate of all optical and access links is S bits/per second, and $\lambda T = 1/50$. Each access station contains a fully connected packet switch with output queuing, and each buffer can store an arbitrarily large number of packets. Find the maximum value of A permitted by the admission controller such that the mean delay is less than 100 T.

References

1. T. Li and R. A Linke, Multigigabit-per-second lightwave systems research for long-haul applications, *IEEE Commun. Mag.* **26**(4), April 1988.
2. P. S. Henry, Lightwave primer, *IEEE J. Quantum Electron.* **QE-21**(12), Dec. 1985.
3. J. C. Palais, *Fiber Optic Communications,* 3rd ed., Prentice–Hall, Englewood Cliffs, N.J., 1992.
4. P. K. Cheo, *Fiber Optics Devices and Systems,* Prentice–Hall, Englewood Cliffs, N.J., 1992.
5. B. E. A. Saleh and M. Teich, *Fundamentals of Photonics,* Wiley, New York, 1991.
6. R. A. Linke, Optical heterodyne communication systems, *IEEE Commun. Mag.* **27**(10), Oct. 1989.
7. H. Kobrinski and K.-W. Cheung, Wavelength tunable optical filters: Applications and technologies, *IEEE Commun. Mag.* **27**(10), Oct. 1989.
8. T. P. Lee and C.-E. Zah, Wavelength tunable and single frequency semiconductor lasers for photonic communications networks, *IEEE Commun. Mag.* **27**(10), Oct. 1989.
9. A. S. Acampora and M. J. Karol, An overview of lightwave packet networks, *IEEE Net. Mag.* **3**(1), Jan. 1989.
10. A. S. Acampora, A multichannel multihop local lightwave network, 1987 IEEE GLOBECOM Conf. Rec., Tokyo.
11. A. S. Acampora, M. J. Karol, and M. G. Hluchyj, Terabit lightwave networks: The multihop approach, *AT&T Tech. J.* Nov./Dec. 1987.
12. M. G. Hluchyj and M. J. Karol, Shufflenet: An application of generalized perfect shuffles to multihop lightwave networks, 1988 IEEE INFOCOM Conf. Rec., New Orleans.
13. C. A. Brackett, Dense wavelength division multiplexing networks: Principles and applications, *IEEE J. Selected Areas Commun.* **8**(6), Aug. 1990.
14. P. S. Henry, High capacity lightwave local area networks, *IEEE Commun. Mag.* **27**(10), Oct. 1989.
15. P. E. Green, *Fiber Optic Communication Networks,* Prentice–Hall, Englewood Cliffs, N.J., 1992.
16. P. E. Green, An all-optical computer network: Lessons learned, *IEEE Net. Mag.* **6**(2), March 1992.
17. B. Mukherjee, WDM-based local lightwave networks, Part 1: Single-hop systems, *IEEE Net. Mag.* **6**(3), May 1992.
18. B. Mukherjee, WDM-based local lightwave networks, Part 2: Multihop systems, *IEEE Net. Mag.* **6**(4), July 1992.
19. P. E. Green, The future of fiber-optic computer networks, *IEEE Comput. Mag.* **24**(9), Sept. 1991.
20. M. S. Goodman *et al.,* The LAMBDANET multiwavelength network: Architecture, applications, and demonstrations, *IEEE J. Selected Areas Commun.* **SAC-8**(6), Aug. 1990.
21. P. R. Prucnal, M. A. Santoro, and T. R. Fan, Spread-spectrum fiber optic local area networks using optical processing, *IEEE J. Lightwave Technol.* **LT-4,** May 1986.
22. P. R. Prucnal, M. A. Santoro, and S. K. Sehgal, Ultrafast all-optical synchronous multiple access fiber networks, *IEEE J. Selected Areas Commun.* **SAC-4,** Dec. 1986.
23. A. Albanese, Star network with collision avoidance circuits, *Bell Syst. Tech. J.* **62,** March 1983.
24. M.-S. Chen, N. R. Dono, and R. Ramaswami, A media-access protocol for packet-switched wavelength division multi-access metropolitan area networks, *IEEE J. Selected Areas Commun.* **SAC-8**(6), Aug. 1990.
25. A. Ganz and Z. Koren, WDM passive star protocols and performance analysis, 1991 IEEE INFOCOM Proc., Bal Harbor.
26. E. Arthurs, M. S. Goodman, H. Kobrinski, and M. P. Vecchi, HYPASS: An optoelectronic hybrid packet-switching system, *IEEE J. Selected Areas Commun.* **SAC-6,** Dec. 1988.
27. R. Chipalkatti, Z. Zhang, and A. S. Acampora, Protocols for optical star-coupled network using WDM: Performance and complexity study, *IEEE J. Selected Areas Commun.* **SAC-11**(4), May 1993.
28. T. E. Stern, Linear lightwave networks: How far can they go? 1990 IEEE GLOBECOM Conf. Rec., San Diego.

29. A. Bannister, L. Fratta, and M. Gerla, Topological design of the wavelength division optical network, 1990 IEEE INFOCOM Proc.
30. A. Bannister and M. Gerla, Design of the wavelength-division optical network, 1990 IEEE Int. Commun. Conf., Conf. Rec.
31. R. Gidron and A. Temple, TeraNet: A multihop multichannel ATM lightwave network, 1991 IEEE Int. Commun. Conf., Conf. Rec.
32. J.-F. Labourdette and A. S. Acampora, Logically rearrangeable multihop lightwave networks, *IEEE Trans. Commun.* **COM-39**(8), Aug. 1991.
33. K. Sivarajan and R. Ramaswami, Multihop lightwave networks based on de Bruijn graphs, 1991 IEEE INFOCOM Proc., Bal Harbor.
34. R. C. Alferness, Waveguide electro-optic switch arrays, *IEEE J. Selected Areas Commun.* **SAC-6**(7), Aug. 1988.
35. A. Papoalis, *Probability, Random Variables, and Stochastic Processes,* McGraw–Hill, New York, 1965.
36. J. M. Wozencraft and I. M. Jacobs, *Principles of Communication Engineering,* Wiley, New York, 1965.

8

Broadband Applications

8.1. Network Services Applications

The first seven chapter of this book were devoted to the concepts, system architectures, and technologies which underlie the field of broadband networks. In this final chapter, we shall qualitatively explore some of the multimedia applications that might benefit from, and drive the need for, wide-area, bandwidth-on-demand, broadband networks. These applications fall into two broad categories: network services and end-user applications. The latter can be further divided into three subcategories: people-to-people, people-to-machine, and machine-to-machine. Sometimes, an end-user application might span more than one of these subcategories.

Effective application of broadband networks may require a major shift in engineering emphasis from a "bandwidth is precious" to a "bandwidth is cheap" perspective. Whereas telecommunication applications have historically emphasized a need to conserve bandwidth (witness, for example, modulation formats for voiceband modems and digital radio which stress spectrum utilization efficiency, or audio/video codecs which seek to remove "redundancy" from the bit stream, thereby allowing speech or moving images to be communicated in a narrower bandwidth channel), broadband applications will seek to judiciously "squander" the enormous pool of inexpensive bandwidth provided by modern microelectronic and photonic technologies (discussed in Chapter 1) in an attempt to create a more "natural" communication environment. Naturalness is a key descriptor of broadband networks: can bandwidth be applied to increase the user friendliness and naturalness among people and machines communicating over remote distances?

Some representative network services which might be enabled by broadband networks include the support of lightweight protocols, improved reliability, simpler control software, and the creation of multiple virtual private networks. As mentioned in Chapter 1, communication protocols were created to permit information to be reliably transferred over remote distances via a telecommunication infrastructure that was prone to the introduction of distortion, much of which is the direct result of the scarcity of bandwidth. Design of a digital transmission link, for example, historically involved trade-offs among four basic parameters: data rate, bandwidth, signal-to-noise ratio available at the receiver, and bit error rate. A typical design problem involved optimizing with regard to one of these parameters, with the others held fixed. Since high data rates are usually sought (higher data rate usually translates into higher common carrier revenues) and both bandwidth and signal-to-noise ratio are often fixed, higher link bit error rates must often be accepted. The burden then falls on the communication protocol to detect the presence of errors and initiate retransmission attempts.

Traffic loading on multiuser networks employing such transmission links also tends to be high, again in an attempt to maximize network utilization efficiency and increase revenue. Here, the latency and dropped packet rate produced by the network may not always satisfy end-user applications, and again the protocols used most overcome these network deficiencies. Implementation of these protocols very often creates a bottleneck in the flow of information, causing the end-to-end throughput to be significantly lower than the rates at which the application would "naturally" operate.

In the broadband arena, much more bandwidth is available, and this bandwidth can be applied to permit transmission link operation at higher data rates and lower bit error rates. For example, powerful forward error correcting codes and higher-order modulation formats might be used. Also, the utilization efficiency of the network might be intentionally diminished: as bandwidth increases, the permissible traffic load would also increase, but by a smaller factor. Both latency and lost packet performance would thereby improve. With a less hostile communication environment, fewer layers of protocol protection will be needed, and the protocol bottleneck might be relaxed. Furthermore, the layers of protection which remain will be exercised less frequently, thereby improving throughput still further.

Abundant bandwidth can also be used to improve network reliability. In a narrowband network, the amount of bandwidth needed to support the execution of continually running, real-time sophisticated diagnostic routines might actually surpass the total network bandwidth, forcing the implementation of routines with compromised effectiveness. Furthermore, these routines are competing for network resources with revenue-bearing traffic, and again, reliability may be compromised further. In a broadband network, enhanced diagnostic routines would consume a far smaller percentage of network resources, and reliability could be significantly improved. These routines could rapidly exercise the network through its various states, generate dummy traffic loads to help detect regions of traffic congestion, provide real-time estimates of network latency, inject test sequences designed such that reception of an altered sequence unambiguously diagnoses a specific failure, and rapidly call in backup resources to enable hitless network self-healing.

It is even possible that broadband technology might reduce the enormous complexity of the software needed to operate a modern telecommunication network. Once again, the opportunity created by broadband technology is that of internationally underutilizing network transport resources (i.e., bandwidth) to permit simplification of the control algorithms. For example, as the size of a network grows, or as more users are attached, a proportionally greater pool of network transport resources are added, but the control complexity needed to maintain a constant utilization efficiency generally grows much faster. Perhaps, as networks grow and broadband technology is deployed, the emphasis should shift from preserving utilization efficiency to maintaining, at worst, only a proportionate increase in control software complexity. Then, as more equipment is added, a linearly increasing control burden could be shared among a linearly increasing supply of distributed control processors. Within such a distributed control environment, the burden-per-user port (and the software needed per user) would thereby remain fixed, with a penalty paid in overall utilization efficiency of the network transport facilities. This approach might indeed be attractive if distributed control algorithms could be developed such that the aggregate pool of transport resource always increases (although at a diminishing utilization efficiency) as more users are added, as opposed to actually decreasing as users are added beyond some critical point.

These ideas are further illustrated in Figure 8.1. Here, we show hypothetical plots depicting the growth in usable network resources (e.g., some maximum allowable applied load) as more user access stations (and more network facilities) are added. If each new access station brings some constant amount of new network facilities, then usable network resources would scale, at best, linearly with the number of access stations N. However, different control algorithms result in

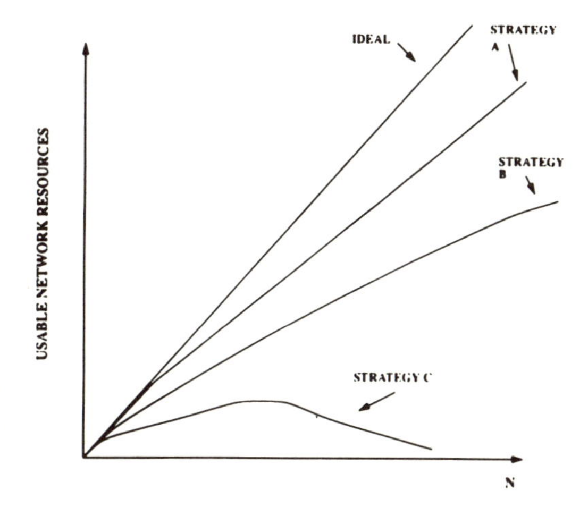

Figure 8.1. Representative plots of usable network resources versus number of access stations for different control strategies.

different facility utilization efficiencies, causing a departure from the ideal as shown. If bandwidth is "precious," a complex strategy producing a performance as represented by curve A might be needed; if bandwidth is "cheap," then a simpler strategy producing a performance as represented by curve B may suffice. A strategy which produces a performance as represented by curve C, however, may be deemed unacceptable because, here, the strategy is so inefficient as to actually cause a reduction in usable network resources as more facilities are added beyond some critical level.

We note in passing that the multihop lightwave network of Chapter 7 evokes some of these principles. With such a network, each newly added access station brings along with it some fixed number of new optical channels P. However (under the assumption of uniform traffic), the aggregate network capacity scales approximately as $N/\log N$, continually increasing as more stations are added but at a diminished utilization efficiency. Also, by exploiting dynamically changing traffic pattern nonuniformities, the utilization efficiency can be substantially improved, but, still, the utilization efficiency diminishes as the network grows larger.

The vast pool of resources provided by a broadband network could even be allocated to enable the creation of multiple virtual private networks sharing common physical facilities. For example, fixed bandwidth virtual paths could be created between each pair of end points of an ATM network which terminate at the multiple facilities of some large business user. That user could then manage traffic flow over those dedicated virtual paths to provide the requisite quality of service which satisfies that user's needs. The user could even implement a set of diagnostic routines to improve reliability and/or exercise backup facilities or alternate routing during periods of high congestion or equipment failure. Since the bandwidth of each such virtual path is dedicated to that user, the user could deploy its own virtual connection admission controller. Moreover, using the bandwidth-on-demand feature, the user could request the ATM networks' call processor to dynamically adjust the bandwidth of each virtual path. Virtual private network features such as these will generally lower the overall utilization efficiency of the ATM network since the bandwidth dedicated to the private network virtual paths is unavailable to serve other virtual paths carried on the same physical links, and statistical multiplexing of the virtual paths will be less effective, but, again, this may be perfectly acceptable in a "bandwidth is cheap" environment.

8.2. End-User Applications

If the ATM Adaptation Layer (AAL) were deployed on the user side of the user–network interface, then service to support end-user applications would be provisioned through a universal interface, an ATM port operating at a specified data speed by means of which the end-user could inject and receive ATM cells at a rate up to that permitted by the port speed. *Any* type of traffic (or combination of traffic types) could be presented through that interface, provided it is properly prepared by the AAL into ATM format.

The long-term goal of people-to-people broadband applications is to effectively use the network's bandwidth to drive the geographical separation among people toward zero, at least insofar as the movement of information is concerned. Multimedia connections would be used pervasively, allowing a combination of real-time voice and full-motion video, supplemented as needed by images and textual data. Presumably, the voice and video would involve only the communicators themselves, but the images and text could be drawn from a third-party archive. For example, the discussion between two conversants colocated in the same room might involve reference to a photograph appearing in a weekly newsmagazine. If the conversants were colocated and had in their possession a "hard copy" of the newsmagazine, then they could readily view the same photograph. To achieve the same "naturalness" over a remove distance, it might be necessary to request a copy of the photograph from a central archive for immediate viewing by the two now-separated conversants. Moreover, either party could "turn the page" and cause the new pictorial or text information to be simultaneously sent (in perfect synchronism, of course) to both conversants.

How much bandwidth is required to support "natural" transfer of images and full-motion video? An existing high-quality monochrome electronic publishing system supports image dimensions of up to 10.1 × 14.3 inches and provides a resolution of 600 dots per inch. Each image therefore contains approximately 50 million dots. If each analog dot is quantized to an 8-bit word, then the image contains about 400 million bits (uncompressed) and if the casual browser turns pages at the rate of 1 per second, then an access rate of 400 MB/sec is needed. Alternatively, a lower-resolution image might be transported in the browse mode, with a higher-resolution supplement delivered only when the user stops at a particular image. Such an approach might permit an access speed of 100 MB/sec if, in the browse mode, the resolution were lowered to 300 dots per inch and if it were acceptable for the user to wait 4 sec for delivery of a selected high-resolution image. Image compression might permit further reduction in required peak data rate, but caution is needed to ensure that (1) the cost of compression/expansion equipment remains modest; (2) the quality of the image not be compromised; and (3) the algorithmic speed of execution not compromise the naturalness of the service.

Assuming a peak rate of 100 MB/sec for browsing and that a user glances at ten images at a rate of one per second before selecting one for viewing at full resolution for, say, 2 min prior to resumption of browsing, we see that the resulting traffic profile is quite bursty. During the 13 sec required to browse through ten images and develop the full resolution for the one desired, a total of 1.3 Gbit of information is delivered; for the next 120 sec, no new information is delivered.

Thus, although the required peak rate is 100 MB/sec, the average rate consumed is only 1.3 GB/133 sec = 9.77 MB/sec. The *details* of the traffic profile (distribution of the number of browsed images, distribution of the dwell time on a chosen full-resolution image, etc.) would be needed to determine the number of "image connections" which could safely by multiplexed onto a fixed bandwidth virtual path while preserving the required quality of service (image delivery delay, visual quality degradation caused by cells lost to buffer overflow, etc.).

Transport of medical images may present a different set of requirements. Here, a color photograph may contain 1024 lines of resolution in each dimension, or about 1 million dots total, each quantized to a level of 24 bits for a total image information content (uncompressed) of about 25 Mbits. An X ray, on the other hand, may have a resolution of 4096 × 4096 with 16 bits of gray scale quantization, for a total information content of 268 Mbits. A set of X ray and color photographs may contain 10 to 50 images, and the physician's usage pattern (browse, study, retrieve prior image, etc.) will determine the traffic descriptors needed to provision and manage the network such that acceptable service quality is assured.

The requirements for continuous bit rate video traffic are dependent on the specified quality of the delivered signal. For standard uncompressed broadcast video, a speed of 120 MB/sec is needed, which can be reduced to the range of 3–6 MB/sec by means of video compression. For studio-quality video, the corresponding requirements are 210 and 10–30 MB/sec, respectively, while for broadcast HDTV, the requirements rise to 1500 and 20–30 MB/sec, respectively. Visualization imagery of the type that might be created by a supercomputer or image rendering workstation (1000 × 1000 lines, 60 frames/sec) may demand bit rates as high as 1.5–2.5 GB/sec, uncompressed, and 50–200 MB/sec compressed.

A good example of a natural, people-to-people broadband application involving the transport of both image and video might be video conferencing. Most video conferencing facilities involve conference tables, microphones, speakers or speakerphones, cameras, and video monitors. Typically, the cameras are voice-actuated, and the image of the person who is speaking in one such facility is projected to the monitors in the other facilities. "Side" discussions are not permitted, and the participants at one end of a video conference connection can see only the person who is speaking at the other end of the link. Broadband networks could substantially improve the quality and naturalness of a video conference. For example, rather than using discrete monitors, a lifelike "video wall" such as shown in Figure 8.2 might be used. The video wall at one end of the link would show the entire teleconferencing facility (including all of the participants) at the other end of the link. Images would be lifelike in size. Any participant could look at any other participant (at either end of the link), independently of who happens to be speaking. Documents could be transferred by means of high-speed scanners and printers. In fact, the illusion of geographical proximity might become so great that if the scanners and printers were wall-mounted, a person at one end of a link, in

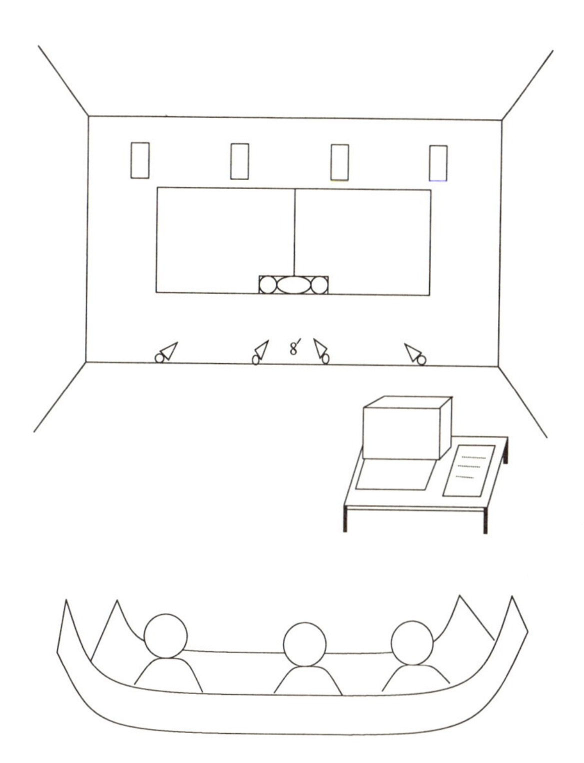

Figure 8.2. A "natural" videoconferencing facility employing a "video wall."

response to movement toward the video wall by a person holding a hard copy of a document at the other end of the link, would instinctively approach the wall to "accept" a hard copy of the document from the printer as it is electronically scanned from the other end! Information from a central archive could also be retrieved and distributed in hard-copy format. The video signal would be carried in high-resolution format, with a minimum of artifact-producing video compression; the use of video codecs which produce delay and "ghosting" in response to rapidly varying scenes would be totally avoided.

Interaction between people and machines in the broadband environment might typically involve electronic access to distributed data bases, image files, and video libraries by people sitting at workstations with high-resolution monitors. Among the more routine applications would be electronic access to a newspaper, journal, or magazine, and paging through this as conveniently and naturally as if hard copy were at hand. Among the more sophisticated applications would be medical imaging, whereby a physician sitting at a terminal could request access to a patient's files. These files might exist in geographically distributed data bases, and be recorded in multimodal formats (sonograms, magnetic resonance images, computerized axial tomography, etc.). Rather than querying these discrete sets of files, the physician's workstation might also be networked to a powerful supercomputer and rendering machine. The patient's files could then be transferred not to the physician but rather to the supercomputer which extracts and concurrently processes information from the multimodal files, providing the information to the rendering machine from which a three-dimensional image of an organ or bone could be created. It is this three-dimensional image which would then be delivered to the physician's workstation, and the physician could "rotate" the image by sending simple commands, in response to which the supercomputer and rendering machine would create and deliver a new image from a different perspective. The physician might even request a slow revolution of the image, slowing down or stopping the rotation as appropriate. Access bandwidth approaching the gigabit per second range might be needed (as described earlier) to permit rotation of such three-dimensional renderings.

Another example of three-dimensional medical rendering might involve radiation therapy planning, intended to optimize the radiation treatment of cancer patients by maximizing the irradiation of the tumor while minimizing damage to healthy tissue. Again, CAT and MRI scans would be supplied to a three-dimensional rendering machine. Using a joystick, the physician could select the beam shape, intensity, and direction from some multitude of radiation sources, and a supercomputer would find the resulting radiation pattern, which would then be superimposed on the three-dimensional organ rendering. When the physician has found the right radiation source placements, beam shapes, etc. the therapy plan can be finalized and put into effect.

Broadband networks might be used to support groupwork multimedia communication needs. Here, by drawing upon a suitable blend of voice, image, and video, and by allowing access to remote text and image data bases, the broadband network might enable the type of informal group meetings which typically occur among workers sharing the same physical premises, despite the fact that the workers are, in reality, dispersed among multiple geographical sites. These types of meetings typically occur frequently, and are often spontaneously arranged. They tend to be highly interactive, and the value derived from such meetings often does not require physical proximity. To be effective, a groupwork telecommunication service should support the conversation with an appropriate set of props, as needed, such as the ability to transfer documents and the ability to spontaneously write or draw on a readily available electronic pad with a paperlike "feel."

In the residential domain, broadband people-to-machine interaction might enable the delivery of entertainment video-on-demand. Rather than selecting from a limited menu of programs and starting times such as is provided by broadcast or cable TV, and also by various types of pay-per-view services, video-on-demand would ultimately permit a home viewer to select from a much larger menu and, for stored (nonlive) programs, the starting time could be arbitrarily chosen. Thus, a home viewer could select a personalized airing of a particular classic movie at a particular starting time, even if the neighbor down the street requested the same movie but beginning ten minutes earlier. For live programs, the viewer might control the selection of camera, viewing angle, and field of view. The user might even be able to select an arbitrary segment of full-motion video from a full 360° panorama, as illustrated in Figure 8.3. Here a cluster of video cameras is positioned at the center of the panorama. Each camera in the cluster is responsible for a particular arc-of-view. Then, using a joystick, the home user can select and combine partial segments from each of two contiguous cameras, thereby effectively being presented with the scene that would have been captured by a "virtual camera" pointed in *exactly* the direction that the user chooses to view. This might be accomplished by a joystick-driven digital signal processor which "stitches" together the scan lines produced by the appropriate two adjacent cameras.

The home user is likely to be less interested in being navigated through a vast information library by a set of menu-driven prompts, and more interested in choosing from a selected set of relevant information. Although home information delivery is currently broadcast in nature, it is beginning to migrate toward "person-

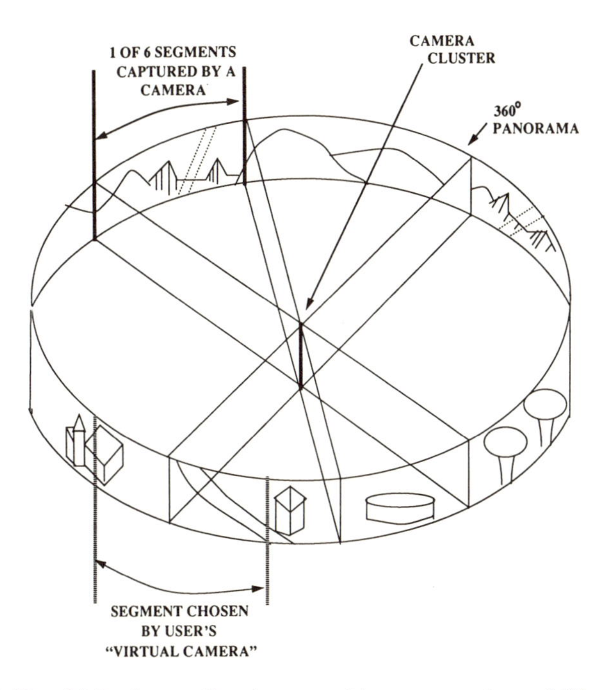

Figure 8.3. Use of "virtual camera" to choose an arbitrary segment from a 360° panorama.

alized" delivery requiring information filtering (user profiles) with simple, intuitive-to-use control interfaces.

The long-term goal of broadband network applications involving machine-to-machine interactions is to create a distributed computing environment with enough capacity and sufficiently low latency that geographically dispersed computers could cooperatively share work as effectively as if their components were housed in a common equipment rack, subject only to fundamental constraints such as the nonzero latency arising from the finite speed of light. Accomplishment of this objective involves bandwidth-on-demand virtual channels operating at data rates comparable to computer backplane speeds, uncongested network links and nodes which introduce little (if any) delay, and lightweight protocols that can fully exploit the high channel rates and low network delay. Although geographically dispersed machines could never cooperate as intimately as closely spaced machines because of the nonzero propagation delay, the goal is for this fundamental constraint to be the limiting factor, rather than some shortcoming of the broadband network. Collectively, the distributed processing afforded by these cooperating machines would come to create a "metacomputer" environment; thereby, in response to a user's request to execute an application, the network would select the distributed resources needed for the task from the pool of available resources, access the appropriate data bases, and deliver to the user only the requested results. A nearer-term example of machine-to-machine distributed processing might involve simulation of complex biomolecular dynamics. Here, the individual constituents of a complex molecule would each be assigned to a particular machine which tracks the dynamics of that constituent as it interacts with its neighbors. The equations of motion would be solved, taking into account the forces exerted by each constituent on every other constituent, and the steady-state molecular structure predicted. A chemist sitting at a workstation could apply heat and pressure, watching the results on a high-resolution monitor as a rendering machine produces three-dimensional molecular images.

We see that a vary array of new, bandwidth-intensive services and applications for the home and workplace will be enabled by wide-area broadband telecommunication networks. Multimedia connections will become dominant and will provide much-improved naturalness in communications among people and machines, although many issues related to protocols, systems, computer architectures, signal processing, and software will need to be addressed and resolved. (For example, when a distributed data base is simultaneously accessed and manipulated by a multitude of users, issues of concurrency arise: has the information requested from a remote data base been modified by another user before the request was received?) State-of-the-art advances in architectures and technologies for broadband networks, and technical feasibility demonstrations of advanced concepts, are now bringing issues such as these into crisp focus and much knowledge and understanding has been, and continues to be, developed.

Bandwidth will cease to be regarded as a precious network commodity and new custom applications will be developed by leading-edge telecommunication users, who will access a service-rich environment through universal, bandwidth-on-demand ports. Local Area Networks, Metropolitan Area Subnetworks, ATM, Broadband ISDN, and lightwave networks will provide pervasive broadband interconnections and access to remote resources, and will help to usher in an information age characterized by much-improved naturalness among people and machines worldwide.

References

1. *IEEE J. Selected Areas Commun.,* issue on B-ISDN Applications and Economics, **10**(9), Dec. 1992.
2. G. M. Nielson, Visualization in scientific and engineering computation, *IEEE Comput.* **24**(9), Sept. 1991.
3. L. D. Wittie, Computer networks and distributed systems, *IEEE Comput.* **24**(9), Sept. 1991.
4. M. N. Rarnsom and D. R. Spears, Applications of public gigabit networks, *IEEE Net. Mag.* **6**(2), March 1992.
5. J. Kohli, Medical imaging applications of emerging broadband networks, *IEEE Commun. Mag.* **27**(12), Dec. 1989.
6. W. Habermann, HDTV standards, *IEEE Commun. Mag.* **29**(8), Aug. 1991.
7. *IEEE J. Selected Areas Commun.,* issue on Speech and Image Coding, **10**(5), June 1992.
8. *IEEE J. Selected Areas Commun.,* issue on High Definition Television and Digital Video Communcations, **11**(1), Jan. 1993.

Index